T0183871

Lecture Notes in Computer Science 9667

Commenced Publication in 1973
Founding and Former Series Editors:
Gerhard Goos, Juris Hartmanis, and Jan van Leeuwen

More information about this series at http://www.springer.com/series/7407

Alexandra Bac · Jean-Luc Mari (Eds.)

Computational Topology in Image Context

6th International Workshop, CTIC 2016
Marseille, France, June 15–17, 2016
Proceedings

 Springer

Editors
Alexandra Bac
Aix-Marseille Université
Marseille
France

Jean-Luc Mari
Aix-Marseille Université
Marseille
France

ISSN 0302-9743   ISSN 1611-3349 (electronic)
Lecture Notes in Computer Science
ISBN 978-3-319-39440-4   ISBN 978-3-319-39441-1 (eBook)
DOI 10.1007/978-3-319-39441-1

Library of Congress Control Number: 2016939994

LNCS Sublibrary: SL1 – Theoretical Computer Science and General Issues

Printed on acid-free paper

This Springer imprint is published by Springer Nature
The registered company is Springer International Publishing AG Switzerland

Preface

The 6th International Workshop on Computational Topology in Image Context (CTIC 2016) took place in Marseille (France) from June 15 to 17, 2016. This conference addressed an incredibly large international audience considering the relatively small size of the community in computational topology: 35 papers were submitted originating from 15 different countries. Following a peer-reviewing process by two qualified reviewers, 24 papers were accepted and scheduled for either oral (19) or poster presentation (5). All of them appear in these proceedings.

The organization of this conference has been a rewarding experience for our research team G-Mod (LSIS laboratory) and for our research group on discrete geometry (G-Dis), part of the Research Federation in Computer Science and Interactions of Aix-Marseille (FRIIAM).

CTIC 2016 was the first edition to be endorsed by the International Association of Pattern Recognition (IAPR). It expresses an increasing interest of researchers in discrete mathematics and computer science for computational topology and its applications. This event was associated with the Technical Committee on discrete geometry IAPR-TC18. Moreover, CTIC 2016 was the second edition to be accepted for publication by Springer as a LNCS proceedings. The conference was also supported by our sponsoring institutions: Aix-Marseille Université, the LSIS laboratory, the FRIIAM Federation, the "Archimède" Excellence Laboratory (LabEx Archimède), the "Conseil Régional PACA", the "Conseil Départemental des Bouches-du-Rhône", and the City of Marseille. We also thank the engineering school "Polytech Marseille" at Aix-Marseille Université for hosting this event and providing all the necessary facilities.

The community dealing with computational topology grows a little bigger every year. CTIC was initially image-oriented when it was created in 2008 in Poitiers, France. But in 8 years, the topics moved slightly from nD images to more general topological objects, with application to genomics, cosmology, geology, or music analysis. Whenever it is possible to have a geometric representation of an abstract object or phenomenon, it is then possible to analyze its topology, with tools becoming more and more popular like persistent homology. The latter is actually an inescapable implement for extracting information in a structural way. This has led to a significant expansion of the number of papers dealing with persistence in the last years.

It has been a great honor for us to count on the participation of two international renowned researchers as invited speakers: Massimo Ferri (Professor of Geometry at the Engineering Faculty of the Bologna University, Department of Mathematics, Research Center for Mathematical Applications, Advanced Research Center for Electronic Systems "E. De Castro") and Pascal Lienhardt (Professor of Computer Science at the University of Poitiers, Computer Graphics team, XLIM-SIC, UMR CNRS 7252).

We would like to express our gratitude to the scientific committee members for their helpful comments, which enabled the authors to improve the quality of their contributions, and to Raphaël Maëstre for the design of the CTIC logo.

Finally, our warmest thanks go to the local Organizing Committee (Eric Remy, Aldo Gonzalez-Lorenzo, Ricardo Uribe Lobello) and to the conference secretary, Régine Martin, for their invaluable contribution to the organization of the event.

June 2016   Alexandra Bac
                                                                            Jean-Luc Mari

Organization

Program Committee

All organizers are members of the Université d'Aix-Marseille, France:

Alexandra Bac (Co-chair)
Aldo Gonzalez-Lorenzo
Jean-Luc Mari (Co-chair)
Eric Remy
Edouard Thiel
Ricardo Uribe Lobello

Scientific Committee

Alexandra Bac
Antonio Bandera
Reneta Barneva
Arindam Biswas
Isabelle Bloch
Srecko Brlek
Didier Coquin
Michel Couprie
Guillaume Damiand
Leila De Floriani
Isabelle Debled-Rennesson
Florent Dupont
Massimo Ferri
Fabien Feschet
Patrizio Frosini
Laurent Fuchs
Antonio Giraldo
Rocío González-Díaz
Aldo González-Lorenzo
María José Jiménez
Bertrand Kerautret
Reinhard Klette
Walter Kropatsch
Jacques-Olivier Lachaud
Pascal Lienhardt
Joakim Lindblad

Christophe Lohou
Jean-Luc Mari
Serge Miguet
Helena Molina-Abril
Marian Mrozek
Nicolas Normand
Darian Onchis
Nicolas Passat
Paweł Pilarczyk
Sanjoy Pratihar
Pedro Real
Eric Remy
Tristan Roussillon
Julio Rubio
Gabriella Sanniti di Baja
Henrik Schulz
Isabelle Sivignon
Natasa Sladoje
Michela Spagnuolo
Robin Strand
Edouard Thiel
Ricardo Uribe Lobello
Antoine Vacavant
Jose Antonio Vilches
Sophie Viseur

Contents

Invited Speakers

Progress in Persistence for Shape Analysis (Extended Abstract)

Massimo Ferri[✉]

Dip. di Matematica e ARCES, Univ. di Bologna, Bologna, Italy
massimo.ferri@unibo.it

Persistent topology mitigates the excessive freedom of topological equivalence by studying not just a topological space but a filtration of it. This makes it a very effective class of shape descriptors, with an impressive potential for applications in the image context, in particular when it comes to images of natural origin. Research in this field is lively and follows various threads. The talk will sample some recent results without any attempt to completeness.

1 Image Processing

Understanding the topology of an object out of a sampling — typically a digital picture — of it was at the origin of the Stanford flavour of persistence [10]. This raised very interesting issues of robustness with respect to noise (e.g. in [5,20]). An amazing segmentation algorithm in presence of noisy data is given in [22], where different segmentation options are suggested by the very nature of a persistence diagram (see Fig. 1).

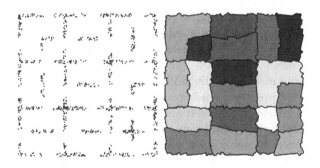

Fig. 1. A segmentation from noisy data [22]

2 Shape Analysis

Apart from the early applications to computer vision in the 90's (when persistence was still in its Size Theory era) there has always been research on 2D and 3D shapes using barcodes and persistence diagrams. More recent ones concern

© Springer International Publishing Switzerland 2016
A. Bac and J.-L. Mari (Eds.): CTIC 2016, LNCS 9667, pp. 3–6, 2016.
DOI: 10.1007/978-3-319-39441-1_1

- classification of hepatic lesions, where multidimensional filtering functions are used, and shown to be superior to the separate 1D ones [1],
- gait classification: four different filtrations in a sequence of silhouettes capture relations among the parts of a walking human body [23],
- analysis of hurricanes and galaxies: two different studies of natural spirals [2,3],
- analysis of brain artery trees: here persistent 1-cycles play a central role [6],
- retrieval of images of melanocytic lesions, with various colour-related filtering functions [16] (see Fig. 2).

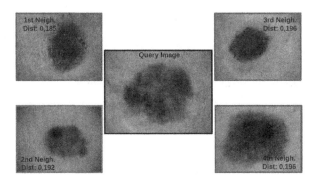

Fig. 2. Retrieval: A melanoma with a neighbourhood of nevi and melanomas [16]

3   Theoretical Progress

The simple remark that colour pictures are maps with range $\mathbb{R}^3$ suggests that the 1D indexing of "classical" persistence is not enough. Therefore the study of discontinuities of the persistent Betti numbers functions [11,13], of the monodromy phenomenon [12] and of the interleaving distance [24] in the multidimensional context are welcome.

Another, related measure of dissimilarity is the *natural pseudodistance* ("natural" with respect to persistence) between homeomorphic spaces endowed with filtering functions: It takes into account all possible homeomorphisms between the two spaces, and registers the infimum of the distorsions induced on the filtering functions. There are settings, in which it is convenient to limit the analysis to a subgroup of homeomorphisms [18]; this is an aspect of a new framework for shape analysis [17].

An interesting simplification comes from the use of *persistence landscapes* in statistical data analysis [9].

An extension of the theory, A_∞-persistence, makes the use of more sophisticated tools, like Massey products, possible [4].

Recent progress in the computation of the bottleneck distance between persistence diagrams [21] makes the representation through complex polynomials [14] less attractive, but the different impact of the various coefficients fosters its use in a "skimming" phase of image retrieval.

4 Not Only Images

Persistence has applications which go far beyond the image context: genetics [27], linguistics [28], neurosciences [26], music [7], robot navigation [8] take advantage of its power.

Acknowledgements. Work performed under the auspices of INdAM-GNSAGA.

References

1. Adcock, A., Rubin, D., Carlsson, G.: Classification of hepatic lesions using the matching metric. Comput. Vis. Image Underst. **121**, 36–42 (2014)
2. Banerjee, S.: Size functions in the study of the evolution of cyclones. Int. J. Meteorol. **36**(358), 39 (2011)
3. Banerjee, S.: Size functions in galaxy morphology classification. Int. J. Comput. Appl. **100**(3), 1–4 (2014)
4. Belchí, F., Murillo, A.: A_∞-persistence. Appl. Algebra Eng. Commun. Comput. **26**(1–2), 121–139 (2015)
5. Bendich, P., Edelsbrunner, H., Morozov, D., Patel, A., et al.: Homology and robustness of level and interlevel sets. Homology Homotopy Appl. **15**(1), 51–72 (2013)
6. Bendich, P., Marron, J., Miller, E., Pieloch, A., Skwerer, S.: Persistent homology analysis of brain artery trees. Ann. Appl. Stat. **10**(1), 198–218 (2016)
7. Bergomi, M.G.: Dynamical and topological tools for (modern) music analysis. Theses, Université Pierre et Marie Curie - Paris VI, December 2015. https://tel.archives-ouvertes.fr/tel-01293602
8. Bhattacharya, S., Ghrist, R., Kumar, V.: Persistent homology for path planning in uncertain environments. IEEE Trans. Robot. **31**(3), 578–590 (2015)
9. Bubenik, P., Dłotko, P.: A persistence landscapes toolbox for topological statistics. J. Symbolic Comput. (in press, 2016). http://dx.doi.org/10.1016/j.jsc.2016.03.009, arXiv:1207.6437 [math.AT]
10. Carlsson, G.: Topology and data. Bull. Am. Math. Soc. **46**(2), 255–308 (2009)
11. Cavazza, N., Ferri, M., Landi, C.: Estimating multidimensional persistent homology through a finite sampling. Int. J. Comput. Geom. Appl. **25**(03), 187–205 (2015)
12. Cerri, A., Ethier, M., Frosini, P.: A study of monodromy in the computation of multidimensional persistence. In: Gonzalez-Diaz, R., Jimenez, M.-J., Medrano, B. (eds.) DGCI 2013. LNCS, vol. 7749, pp. 192–202. Springer, Heidelberg (2013)
13. Cerri, A., Frosini, P.: Necessary conditions for discontinuities of multidimensional persistent Betti numbers. Math. Methods Appl. Sci. **38**(4), 617–629 (2015)
14. Di Fabio, B., Ferri, M.: Comparing persistence diagrams through complex vectors. In: Murino, V., Puppo, E. (eds.) ICIAP 2015. LNCS, vol. 9279, pp. 294–305. Springer, Heidelberg (2015)
15. Edelsbrunner, H., Morozov, D.: Persistent homology: Theory and practice. In: European Congress of Mathematics Kraków, 2–7 July 2012, pp. 31–50 (2014)
16. Ferri, M., Tomba, I., Visotti, A., Stanganelli, I.: A feasibility study for a persistent homology based k-nearest neighbor search algorithm in melanoma detection. arXiv preprint (2016)
17. Frosini, P.: Towards an observer-oriented theory of shape comparison. In: Eurographics Workshop on 3D Object Retrieval, pp. 1–4 (2016)

18. Frosini, P., Jabłoński, G.: Combining persistent homology and invariance groups for shape comparison. Discrete Comput. Geom. **55**(2), 373–409 (2016)

19. Frosini, P., Landi, C.: Size theory as a topological tool for computer vision. Pattern Recogn. Image Anal. **9**(4), 596–603 (1999)

20. Frosini, P., Landi, C.: Persistent Betti numbers for a noise tolerant shape-based approach to image retrieval. Pattern Recogn. Lett. **34**(8), 863–872 (2013)

21. Kerber, M., Morozov, D., Nigmetov, A.: Geometry helps to compare persistence diagrams. In: ALENEX 2016, pp. 103–112. SIAM (2016)

22. Kurlin, V.: A fast persistence-based segmentation of noisy 2D clouds with provable guarantees. Pattern Recognition Letters (2015)

23. Lamar-León, J., García-Reyes, E.B., Gonzalez-Diaz, R.: Human gait identification using persistent homology. In: Mejail, M., Gomez, L., Jacobo, J., Alvarez, L. (eds.) CIARP 2012. LNCS, vol. 7441, pp. 244–251. Springer, Heidelberg (2012)

24. Lesnick, M.: The theory of the interleaving distance on multidimensional persistence modules. Found. Comput. Math. **15**(3), 613–650 (2015)

25. Li, C., Ovsjanikov, M., Chazal, F.: Persistence-based structural recognition. In: Proceedings of the IEEE Conference on Computer Vision and Pattern Recognition, pp. 1995–2002 (2014)

26. Petri, G., Expert, P., Turkheimer, F., Carhart-Harris, R., Nutt, D., Hellyer, P., Vaccarino, F.: Homological scaffolds of brain functional networks. J. R. Soc. Interface **11**(101) (2014). 20140873. doi:10.1098/rsif.2014.0873

27. Platt, D.E., Basu, S., Zalloua, P.A., Parida, L.: Characterizing redescriptions using persistent homology to isolate genetic pathways contributing to pathogenesis. BMC Syst. Biol. **10**(1), 107 (2016)

28. Port, A., Gheorghita, I., Guth, D., Clark, J.M., Liang, C., Dasu, S., Marcolli, M.: Persistent topology of syntax. arXiv preprint (2015). http://arxiv.org/abs/1507.05134

Homology Computation During an Incremental Construction Process

Pascal Lienhardt$^{(\boxtimes)}$ and Samuel Peltier

Université de Poitiers, Laboratoire XLIM, UMR CNRS 7252, Poitiers, France
{pascal.lienhardt,samuel.peltier}@xlim.fr
http://www.xlim.fr/

Abstract. Controlling the construction of geometric objects is important for several Geometric Modeling applications. Homology (groups and generators) may be useful for this control. For such incremental construction processes, it is interesting to *incrementally* compute the homology, i.e. to deduce the homological information at step s of the construction from the homological information computed at step $s-1$. We here study the application of *effective homology* results [13] for such incremental computations.

Keywords: Homology · Simplicial and cellular combinatorial structures · Incremental computation

1 Introduction

Geometric Modeling deals with the representation and the construction of geometric objects. For instance, many representations and related construction operations have been conceived for CAD/CAM applications; since a modeled geometric object can be manufactured, it is necessary to control its construction, in order to detect any problem as soon as possible. According to the application, homology (i.e. homology groups and/or the correspondence between their generators and chains of cells of the object at each step of its construction) may be useful for controlling the construction.

Many works in Geometric Modeling deal with subdivided geometric objects, i.e. objects partitioned into cells of different dimensions, and many methods have been proposed for computing the homology of subdivided objects, e.g. based upon the Smith Normal Form of incidence matrices [1]. Such methods make it possible to compute the homology at each step of the construction of a geometric object; but the whole homological information has to be computed at each step, without taking advantage of the information known at the previous step. So, it seems interesting for such a process to *incrementally* compute the homology, i.e. to deduce the homology of the object at step s from the homology of the object at step $s-1$, according to the operation which is applied. Such incremental computation can be done by applying results of *effective homology* [13].

© Springer International Publishing Switzerland 2016
A. Bac and J.-L. Mari (Eds.): CTIC 2016, LNCS 9667, pp. 7–15, 2016.
DOI: 10.1007/978-3-319-39441-1_2

Although some aspects are close to the work described in [6], the approach here focuses on the application of construction operations and the related complexities. Since only the structure of an object has to be taken into account in order to compute its homology, it will be assumed that it is represented by a semi-simplicial set or a cellular combinatorial structure (incidence graph or combinatorial map). In the simplicial (resp. cellular) case, two basic operations (and the inverse operations) are studied[1]: cone and identification (resp. extension and identification), which make it possible to construct any semi-simplicial set (resp. any cellular combinatorial structure). At each step, some "homological information" is maintained in order to reduce the complexity of the homology computation. This "homological information" depends on the (subset of) combinatorial structure which is taken into account: for semi-simplicial sets and a subclass of cellular combinatorial structures, a "homological information" related to connected components is maintained; but it is necessary to maintain a "homological information" associated with cells and connected components for cellular combinatorial structures in general. More precision can be found in [3].

Notations. A chain complex (C, ∂) can be associated with a semi-simplicial set or a cellular combinatorial structure S, in such a way that there is a strong correspondence between them (for instance, a generator of the chain complex is associated with any simplex, and the boundary operator is deduced from the face operators). Chain complex $(H, 0)$ (or equivalently H) denotes the homology of S. The complexity of a chain complex, or a combinatorial structure, is related to the number of generators and the "cost" of the face or boundary operators (in terms of generators): for instance, the complexity of a n-dimensional semi-simplicial set (or its associated chain complex) is $\sum_{i=0}^{n} k_i c_i + \sum_{i=1}^{n} (i+1)k_i d_{i-1}$, where k_i is the number of i-simplices, c_i (resp. d_i) is the complexity of the representation of a i-simplex (resp. of a i-simplex in the boundary of a $(i+1)$-simplex).

2   Effective Homology Bases

This section is mainly based on the course notes of J. Rubio and F. Sergeraert [13] (see also [5]).

A *reduction* $\rho = ((C, \partial), (C^S, \partial^s), h, f, g)$ is a 5-tuple where (1) (C, ∂) and (C^S, ∂^s) are chain-complexes; (2) $f : (C, \partial) \rightarrow (C^S, \partial^s)$ and $g : (C^S, \partial^s) \rightarrow (C, \partial)$ are chain-complex morphisms; (3) $h : (C, \partial) \rightarrow (C, \partial)$ is a graded module morphism of degree $+1$. They satisfy[2]: (4) $gf = id_{C^S}$; (5) $fg + h\partial + \partial h = id_C$; (6) $hf = gh = hh = 0$. Reduction ρ will be sometimes symbolized by $(C, \partial) \overset{\rho}{\Longrightarrow} (C^S, \partial^s)$. The homologies of (C, ∂) and (C^S, ∂^s) are isomorphic, and the complexity of (C^S, ∂^s) is lower than that of (C, ∂); so, the homology of (C, ∂) is computed with a better complexity starting from (C^S, ∂^s). Several methods

---

[1] Obviously, other operations are useful for Geometric Modeling applications, and a similar study has to be done for each operation.

[2] fg denotes $g \circ f$.

are based on (composition of) elementary reductions[3] in order to simplify a chain complex before computing its homology [9,11].

A *homological equivalence* Υ is a pair of reductions $(C, \partial) \overset{\rho}{\Longleftarrow} (C^B, \partial^B) \overset{\rho^S}{\Longrightarrow} (C^S, \partial^s)$. This notion makes it possible to associate a "small" chain complex (C^S, ∂^s) with another chain complex (C, ∂), through a "bigger" one (C^B, ∂^B), even when no reduction $(C, \partial) \Longrightarrow (C^S, \partial^s)$ exists.

An *effective short exact sequence* $((C^0, \partial^0), (C^1, \partial^1), (C^2, \partial^2), i, j, r, s)$ is a diagram:

$$0 \overset{0}{\longrightarrow} (C^0, \partial^0) \underset{i}{\overset{r}{\rightleftarrows}} (C^1, \partial^1) \underset{j}{\overset{s}{\rightleftarrows}} (C^2, \partial^2) \overset{0}{\longrightarrow} 0$$

where (1) (C^0, ∂^0), (C^1, ∂^1), (C^2, ∂^2) are chain complexes; (2) i, j are chain complex morphisms; (3) r, s are graded module morphisms. They satisfy (4) $ir = id_{C^0}$; (5) $ri + js = id_{C^1}$; (6) $sj = id_{C^2}$. This notion is a key one for optimizing the homology computation at each step of an incremental construction process; indeed, if an effective short exact sequence can be associated with the applied operation, the following *SES theorem* can be applied (only the subparts of the theorem which are useful here are provided: cf. [13] page 71).

Theorem 1 (SES Theorem). *Let $((C^0, \partial^0), (C^1, \partial^1), (C^2, \partial^2), i, j, r, s)$ be a short exact sequence. Then, given two homological equivalences $\Upsilon^0 : (C^0, \partial^0) \Longleftarrow (C^{B0}, \partial^{B0}) \Longrightarrow (C^{S0}, \partial^{s0})$ and $\Upsilon^1 : (C^1, \partial^1) \Longleftarrow (C^{B1}, \partial^{B1}) \Longrightarrow (C^{S1}, \partial^{s1})$ (resp. $\Upsilon^2 : (C^2, \partial^2) \Longleftarrow (C^{B2}, \partial^{B2}) \Longrightarrow (C^{S2}, \partial^{s2})$), it is possible to deduce from i, j, r, s, Υ^0 and Υ^1 (resp. Υ^2) a homological equivalence $\Upsilon^2 : (C^2, \partial^2) \Longleftarrow (C^{B2}, \partial^{B2}) \Longrightarrow (C^{S2}, \partial^{s2})$ (resp. $\Upsilon^1 : (C^1, \partial^1) \Longleftarrow (C^{B1}, \partial^{B1}) \Longrightarrow (C^{S1}, \partial^{s1})$).*

If the applied operation "corresponds" to a short exact sequence, and if the operands of the operation are associated with homological equivalences, it is possible to deduce a homological equivalence, and thus a (homologically equivalent) "small" complex, for the result of the operation, providing a better complexity for the final homology computation[4]. Of course, it is also necessary to check the complexity of the application of the SES theorem in order to check that the whole computation is better than, say, a computation based upon the computation of the Smith Normal Form of the incidence matrices.

3   Application to Simplicial Structures

The SES theorem has been applied in order to reduce the cost of the homology computation [14]. For instance, Boltcheva *et al.* [5] applied it for the conception of the *Mayer-Vietoris (MV) algorithm*, which computes the homological

---

[3] An elementary reduction can be defined when a i-dimensional generator x appears in the boundary of a $(i + 1)$-generator y with a coefficient equal to 1 or -1. Note that reductions exist, which are not compositions of elementary reductions.

[4] For instance, when Υ^2 is deduced from i, j, r, s, Υ^0 and Υ^1, the generators of C^{S2} are that of C^{S0} and C^{S1}.

information of abstract simplicial complexes from the homological information of sub-complexes and their intersections. This algorithm has been applied for the *Manifold-Connected decomposition* of abstract simplicial complexes [6]. The basic idea is the following: let B and C be sub-complexes of A, such that $A = B \cup C$, and $\Upsilon_{B \cap C}$, Υ_B and Υ_C are homological equivalences associated with $B \cap C$, B and C; a short exact sequence $((B \cap C), (B \oplus C), A = (B \cup C), i, j, r, s)$ can be defined[5], and a homological equivalence associated with A can be computed by applying the SES theorem.

More basic operations are studied here: cone and identification (cf. [3] for more precisions). $Cone(A)$, the cone of A, consists in adding a new 0-simplex v to A, and, for each i-simplex σ, in adding a $(i + 1)$-simplex incident to σ and v. Obviously, a homological equivalence $\Upsilon : (Cone(A)) \Lleftarrow (Cone(A)) \Rrightarrow (X)$ can be defined, where X contains only one 0-dimensional generator (the homology of a cone is trivial), and Υ can be computed in linear time according to the size of A.

The basic identification operation $Ident(A, \sigma, \tau)$ consists, given two i-simplices σ and τ having the same boundary, in replacing them by a new simplex μ such that the boundary (resp. the star) of μ is the boundary of σ (resp. the union of the stars of σ and τ). This operation can be easily generalized in order to identify any two i-simplices, and more generally subsets of simplices.

Assume a homological equivalence $\Upsilon : (A) \Lleftarrow (A^B) \Rrightarrow (A^S)$ is associated with A. Given a set of simplices A^0 which have to be identified, a chain complex (A^0) can be computed, in which each i-dimensional generator corresponds to a basic identification of i-simplices (i.e. if k i-simplices are identified into one i-simplex, there are $(k - 1)$ corresponding i-generators in (A^0)). The boundary operator of (A^0) is deduced from the face operators of the simplices of A^0. So, a homological equivalence $\Upsilon^0 : (A^0) \Lleftarrow (A^0) \Rrightarrow (A^0)$ can be computed in linear time according to the size of A^0 (the size takes into account the simplices of A^0 and their face operators). A better homological equivalence $\Upsilon'^0 : (A^0) \Lleftarrow (A^0) \Rrightarrow (A'^0)$ may be deduced by applying a method for simplifying (A^0), i.e. which computes a reduction $(A^0) \Rrightarrow (A'^0)$.

Let $Ident(A, A^0)$ be the result of the identification of the simplices of A^0 in A. It is possible to compute a short exact sequence $Q = ((A^0), (A), (Ident(A, A^0)), i, j, r, s)$ in linear time according to the size of A. Thus, by applying the SES theorem, it is possible to compute a homological equivalence $\Upsilon^I : (Ident(A, A^0)) \Lleftarrow (I^B) \Rrightarrow (I^S)$ (cf. Fig. 1). Once again, a better homological equivalence $\Upsilon'^I : (Ident(A, A^0)) \Lleftarrow (I^B) \Rrightarrow (I'^S)$ may be deduced by applying a method for simplifying (I^S), i.e. which computes a reduction $(I^S) \Rrightarrow (I'^S)$. Then, the homology of $Ident(A, A^0)$ can be computed from (I'^S).

As said before, the homological equivalence Υ^0 can be computed in linear time according to the size of A^0, and the short exact sequence Q can be computed in linear time according to the size of A. The generators of (I^B) (resp. (I^S)) are the generators of (A^0) and of (A^B) (resp. (A'^0) and (A^S); this explains

---

[5] The chain complex associated with X is denoted (X).

why it is interesting to reduce (A^0) into (A'^0)). It is not possible to give a precise evaluation of the complexity related to the computation of the boundary operators of (I^B) and (I^S), and of the mappings of Υ^I, since this complexity depends on Υ, and thus depends also on the operations previously applied for constructing A. But the complexity can be evaluated for certain cases: for instance, when A is constructed by applying identifications, the whole computation is linear; moreover, if (I^B) (resp. I^S) is deduced by *modifying* (A^B) (resp. (A^S)), the computation is sub-linear; and the complexity of the whole construction (i.e. related to all computations for all steps) is linear according to the size of the initial object. In the general case, the complexity of the computation is clearly related to the size of the parts of the object which are identified, i.e. (A^0).

An informal argument for the interest of the approach is the fact that, for many construction processes, *local* operations are applied at each construction step; since the complexity is related to the modified parts, it is interesting to modify a previously known homological information rather than to compute it from scratch. Other precisions are given in Section 3 and Appendix 6.7 in [3]. In particular, note that the SES theorem applies for the inverse operation of the identification.

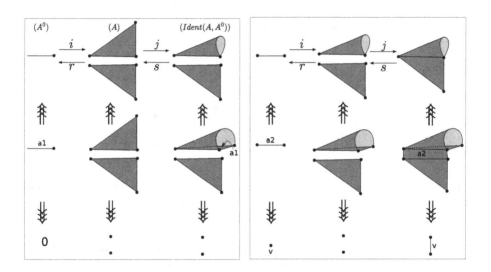

Fig. 1. Incremental computation of homological equivalences

At last, note that these results generalize in some sense the results described in [5,6], since the identification operation is a more basic operation than the gluing of connected components. In other words, gluing connected components can be achieved by identifying parts of their boundaries; but the identification operation can also be applied to subsets of simplices belonging to the same connected component.

4   Application to Cellular Structures

The previous results directly apply to other structures, as cubical [8] and simploidal [12] structures (as for simplices, the homology of any cube or simploid is trivial; moreover, classical results apply for the cartesian product, which corresponds in some way to the cone for simplicial objects). They can also be applied to cellular combinatorial structures as incidence graphs[6] [10] and combinatorial maps [7]. An incidence graph is represented in Fig. 2; results in [3] are obtained for combinatorial maps, which generalize incidence graphs in order to take multi-incidence into account[7], and these results directly apply to incidence graphs.

Any cellular combinatorial structure can be constructed by applying two basic operations: extension and identification. Given a connected n-dimensional structure, in which the dimension of all main cells is n, the extension consists in adding a new $(n + 1)$-cell incident to all n-cells. For instance, this operation applied to the incidence graph in Fig. 2 adds a volume incident to faces f_1 and f_2, the resulting "corresponding object" is a 3-ball. The identification operation consists in "merging" two cells having the same boundary; as for the simplicial case, this operation can be generalized in order to identify subsets of cells (this is constrained by the fact that two cells can be identified if they are isomorphic: for instance, it is not possible to identify a triangle and a square).

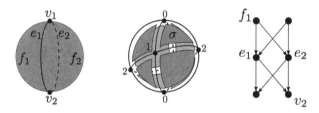

Fig. 2. A 2-sphere, subdivided into 2 faces, 2 edges and 2 vertices, together with its simplicial equivalent and the corresponding incidence graph.

In the general case, it is not possible to associate a "cellular" chain complex (A) with any cellular combinatorial structure A, i.e. a chain complex in which any generator corresponds to a cell, and such that the homology of (A) is the homology of A. It is thus not possible to directly apply the results obtained in the simplicial case to cellular combinatorial structures. Indeed, cellular combinatorial structures exist, in which cells cannot be associated with balls. For instance, applying the extension operation to an incidence graph which corresponds to

---

[6] These structures have been defined in different contexts; they are sometimes referred to as orders, Hasse diagrams, etc.

[7] It is not possible to unambiguously represent with incidence graphs "objects" in which a cell is incident several times to another one, but this is possible with combinatorial maps.

a subdivided torus or a Klein bottle does not produce a 3-ball. Moreover, any $(n + 1)$-cell can be defined as the result of the application of the extension operation to a n-dimensional cellular combinatorial structure; but we do not know how to decide in the general case whether a (simplicial or cellular) combinatorial structure corresponds to a n-sphere. So, we do not know how to decide in the general case whether a combinatorial cell corresponds to a ball.

But it is always possible to associate a *simplicial equivalent* $S(A)$ with any cellular combinatorial structure A; so, it is always possible to associate a "simplicial" chain complex $(S(A))$ with any cellular combinatorial structure A, i.e. a chain complex in which any generator corresponds to a simplex of $S(A)$, and such that the homology of $(S(A))$ is the homology of A. So, the results presented in Sect. 3 can be directly applied, by maintaining at each step of the construction of A its simplicial chain complex $(S(A))$. The problem here is the complexity of the approach, since it is possible that many simplices in $S(A)$ are associated with one cell in A. So, more efficient approaches have been investigated.

4.1   A Subclass of Cellular Combinatorial Structures

A subset of cellular combinatorial structures has been defined in [2] (cf. also [4]): each structure A of this subclass satisfies properties so that the cellular chain complex (A) is homologically equivalent to the simplicial chain complex $(S(A))$. So, the results presented in Sect. 3 can be directly applied to the structures of this subclass, associated with their corresponding cellular chain complexes, but it is necessary to check at each construction step whether the properties of the subclass are still satisfied (this control can be done easily without significative cost, due to the definition itself of the properties).

4.2   The General Case

The properties characterizing this subclass are sufficient but not necessary to ensure the equivalence between (A) and $(S(A))$. Moreover, it may be useful for some constructions to ignore these properties, even if only temporarily. It can thus be useful to handle A and $(S(A))$, even if it is less efficient than to handle A and (A). Even in this case, it is possible to optimize the computations. Indeed, the interior of each cell c in A corresponds to a subset of simplices in $S(A)$; a homological equivalence Υ^c can be associated with the interior of c when it is created, and it is possible to maintain Υ^c during the construction process in order to optimize the homology computation.

More precisely, assume a homological equivalence is associated with the interior of any cell and with any connected component. Assume c is created by an extension operation applied to a connected component C of A. This operation corresponds in $S(A)$ to a cone operation applied to $S(C)$, where the vertex v of the cone symbolizes cell c (in fact, the interior of c corresponds to the subset incident to v, so the structure of the interior of c is very close to the structure of $S(C)$, since c corresponds to a cone on $S(C)$). Each cell of C remains unchanged, so its homological equivalence is unchanged; the homological equivalence associated with the connected component after operation is trivially defined, since the resulting connected component is a cone; the homological equivalence associated with c can be easily deduced from the homological equivalence associated with C by applying the *perturbation lemmas* (cf. [13] pages 48–49). All computations can be performed in linear time according to the size of $S(C)$. Note that, at last, some reduction process may be applied to the small chain complex of Υ^c.

Let $Ident(A, A^0)$ be the result of the identification of cells of a subset A^0 in A. Let c be a cell of $Ident(A, A^0)$: either c does not result from the identification of cells, and its homological equivalence Υ^c is not modified; either it results from the identification of isomorphic cells, and Υ^c is simply a homological equivalence associated with one of these identified cells (all these homological equivalences are homologically equivalent, since the cells are isomorphic): so, nothing is really computed.

The homological equivalence associated with $S(Ident(A, A^0))$ is computed in the following way. A homological equivalence $\Upsilon^1_{A^0}$ is computed as the direct sum of homological equivalences corresponding to the identified cells; more precisely (and as for the simplicial case), if k isomorphic i-cells $\{c_j\}_{j \in [1,k]}$ are identified into one cell, there are in $\Upsilon^1_{A^0}$ $k-1$ copies of Υ^{c_j}, for some $j \in [1, k]$ (for instance, j may be chosen according to the complexity of Υ^{c_j}). It is now possible to deduce from $\Upsilon^1_{A^0}$ a homological equivalence $\Upsilon^2_{A^0}$ by "linking" the homological equivalences corresponding to cells accordingly to the boundary relations between cells in A; this can be done by applying the perturbation lemmas (cf. [13] pages 48–49), and the complexity is linear according to the size of $\Upsilon^1_{A^0}$ and n, where n is the highest dimension of a cell in A^0. Then, a reduction process can be applied to the small complex of $\Upsilon^2_{A^0}$, producing $\Upsilon^0 : C(A^0) \lll C^B(A^0) \Rrightarrow C^S(A^0)$.

Moreover, a short exact sequence $Q = (C(A^0), (S(A)), (S(Ident(A, A^0))), i, j, r, s)$ can be computed in linear time according to the size of $S(A)$ (as for the simplicial case, the complexity of the computation of the "interesting" information is sub-linear). The SES theorem can then be applied in order to deduce a homological equivalence Υ^I associated with $Ident(A, A^0)$; as for the simplicial case, a reduction process can be applied to the small complex of Υ^I, and a "better" homological equivalence Υ'^I can be deduced.

As for the simplicial case in Sect. 3, similar remarks about the complexity of the process can be done (the main difference with the simplicial case is the complexity of the computation of Υ^0, which depends also on the dimension of the identified cells). At last, note that a similar process can also be applied for the inverse of the identification operation.

Acknowledgments. Many thanks to Francis Sergeraert, Sylvie Alayrangues and Laurent Fuchs.

References

1. Agoston, M.K.: Algebraic Topology: A First Course. Pure and Applied Mathematics. Marcel Dekker Ed., New York (1976)
2. Alayrangues, S., Damiand, G., Lienhardt, P., Peltier, S.: A Boundary Operator for Computing the Homology of Cellular Structures. Research Report <hal-00683031-v2> (2011)
3. Alayrangues, S., Fuchs, L., Lienhardt, P., Peltier, S.: Incremental computation of the homology of generalized maps - an application of effective homology results. Research Report <hal-01142760-v2> (2015)
4. Basak, T.: Combinatorial cell complexes and Poincaré duality. Geom. Dedicata **147**(1), 357–387 (2010)
5. Boltcheva, D., Merino, S., Léon, J.-C., Hétroy, F.: Constructive Mayer-Vietoris Algorithm: Computing the Homology of Unions of Simplicial Complexes. INRIA Research Report RR-7471 (2010)
6. Boltcheva, D., Canino, D., Aceituno, S.M., Léon, J.-C., De Floriani, L., Hétroy, F.: An iterative algorithm for homology computation on simplical shapes. Comput. Aided Des. **43**(11), 1457–1467 (2011)
7. Damiand, G., Lienhardt, P.: Combinatorial Maps: Efficient Data Structures for Computer Graphics and Image Processing. A K Peters/CRC Press, Boca Raton (2014)
8. Dlotko, P., Kaczynski, T., Mrozek, M.: Computing the cubical cohomology ring. In: 3rd International Workshop on Computational Topology in Image Context, Chipiona, Spain, pp. 137–142 (2010)
9. Dlotko, P., Kaczynski, T., Mrozek, M., Wanner, T.: Coreduction homology algorithm for regular CW-complexes. Discrete Comput. Geom. **46**(2), 361–388 (2011)
10. Edelsbrunner, H.: Algorithms in Computational Geometry. Springer, New York (1987)
11. González-Díaz, R., Jiménez, M.J., Medrano, B., Real, P.: Chain homotopies for object topological representations. Discrete Appl. Math. **157**(3), 490–499 (2009)
12. Peltier, S., Fuchs, L., Lienhardt, P.: Simploidals sets - Definitions, operations and comparison with simplicial sets. Discrete App. Math. **157**, 542–557 (2009)
13. Rubio, J., Sergeraert, F.: Constructive homological algebra and applications. In: Genova Summer School on Mathematics - Algorithms - Proofs, Genova, Italy (2006)
14. The Kenzo program. https://www-fourier.ujf-grenoble.fr/~sergerar/Kenzo/

Main Contributions

Persistence-Based Pooling for Shape Pose Recognition

Thomas Bonis[1][✉], Maks Ovsjanikov[2], Steve Oudot[1], and Frédéric Chazal[1]

[1] DataShape Team, Inria Saclay, Palaiseau, France
Thomas.bonis@inria.fr
[2] Laboratoire d'Informatique de l'Ecole Polytechnique, Palaiseau, France

Abstract. In this paper, we propose a novel pooling approach for shape classification and recognition using the bag-of-words pipeline, based on topological persistence, a recent tool from Topological Data Analysis. Our technique extends the standard max-pooling, which summarizes the distribution of a visual feature with a single number, thereby losing any notion of spatiality. Instead, we propose to use topological persistence, and the derived persistence diagrams, to provide significantly more informative and spatially sensitive characterizations of the feature functions, which can lead to better recognition performance. Unfortunately, despite their conceptual appeal, persistence diagrams are difficult to handle, since they are not naturally represented as vectors in Euclidean space and even the standard metric, the bottleneck distance is not easy to compute. Furthermore, classical distances between diagrams, such as the bottleneck and Wasserstein distances, do not allow to build positive definite kernels that can be used for learning. To handle this issue, we provide a novel way to transform persistence diagrams into vectors, in which comparisons are trivial. Finally, we demonstrate the performance of our construction on the Non-Rigid 3D Human Models SHREC 2014 dataset, where we show that topological pooling can provide significant improvements over the standard pooling methods for the shape pose recognition within the bag-of-words pipeline.

Keywords: Shape recognition · Bag-of-words · Topological Data Analysis

1 Introduction

In the recent years, databases of 3-dimensional objects have been getting larger and larger. In order to automatically process these databases, many algorithms relying on retrieval have been proposed. However, for certain tasks, classification techniques can be more efficient. Efficient classification pipelines have been proposed for images and some elements of these techniques such as the bag-of-words methods [1] or feature learning using deep network architectures [2] have been used to perform retrieval and shape comparison. Traditionally, the bag-of-words method relies on extracting an unordered collection of descriptors from the

© Springer International Publishing Switzerland 2016
A. Bac and J.-L. Mari (Eds.): CTIC 2016, LNCS 9667, pp. 19–29, 2016.
DOI: 10.1007/978-3-319-39441-1_3

shapes we consider, which are then quantized into a set of vectors called "words". The information given by this quantization process is then summarized using a *pooling* scheme, which produces a vector usable by standard learning algorithms. Ideally, all the steps of this framework should be robust to transformations of the shape: translations, rotations, changes of scale, etc. Modern bag-of-words approaches for 3D-shapes usually rely on a pooling method called sum-pooling [1] which consists in taking the average of the value of each words across the shape.

Since its introduction for image processing in [3], the bag-of-words pipeline, which we present in Sect. 2, has been improved in various ways. Here, we focus on the pooling part of the framework. Apart from the traditional sum-pooling approach, a popular pooling method, called *max-pooling* introduced in [4], consists in taking the maximum of the value for each visual word. Several works have highlighted the improvement in accuracy obtained using this pooling scheme as well as its compatibility with the linear kernel for learning purposes, [4,5]. The strength of max pooling is due in part to its remarkable robustness properties. One of the main assumptions made in the bag-of-words approach is that the "word" values that compose the output of the encoding step, are, for a given class and a given word, i.i.d random variables. Refinements of the max-pooling scheme have been proposed under this assumption: for instance [6] proposed to consider the k highest values for each words to estimate the probability of at least k features being present in the object. However, the independence assumption of the word functions is unrealistic; for 3D shapes close vertices tend to have similar word functions, as illustrated in Fig. 1. Thus, in this example, the generalization proposed by [6] ends up capturing the same feature multiple times and providing multiple redundant values. On the other hand, pooling on different parts of an image [7] and 3D shape [8,9] has been proposed to take advantage of spatial information, an approach known as Spatial Pyramid Matching. This approach has drastically improved the performance of the bag-of-words procedures on multiple datasets, although it contradicts the identically distributed assumption, and lacks proper robustness guarantees.

In this work, we propose to see the word functions not as a unordered collection of random values but as a random function defined on the vertices of a graph (in our case, the mesh of the shape). Following this approach, we propose to use persistent homology to capture information regarding the global structure of the word functions which is not available for the traditional max-pooling approach.

Persistent homology was first introduced in the context of Topological Data Analysis under the name *size theory* [10]. It was later generalized to higher dimensions as *persistent homology* theory [11,12]. The 0-dimensional persistent homology of the superlevel-sets of a function encodes the prominence of the peaks of the function into a collection of points in the plane, called a *persistence diagram*. These diagrams enjoy strong robustness properties [13–15]. One option to compare persistence diagrams is to use a distance between diagrams such as the bottleneck distance and to use nearest-neighbor algorithms as it was

Fig. 1. Example of a word function obtained on two different shapes in the same pose and for the two different poses.

done by [16]. However, in this work, we aim at being able to use classification algorithms such as SVM or logistic regression that requires a Hilbert space structure, which is not the case of the space of persistence diagrams. One approach to tackle this issue is to make use of the "kernel trick" by using a positive-definite kernel in order to map the persistence diagrams into a Hilbert space. As recently shown by Reininghaus et al. [17], one cannot rely on natural distances such as the Wasserstein distance to build traditional distance-based kernels such as the Gaussian kernel. This led the authors to propose another kind of kernel. A major limitation of their approach, however, is that these types of kernel are non-linear and the complexity of the classification becomes linear with the size of the training set which causes scalability issues. Another approach to directly embed persistence diagrams into a Hilbert Space was proposed in [18]. However this embedding is highly memory-consuming as it maps a single diagram into a set of functions and is not appropriate for dealing with large datasets.

In this work, we propose to perform pooling by computing the persistence diagrams of each word function. We then map these persistence diagrams into $\mathbb{R}^d$ for some reasonable value of d -< 20- by considering the peaks with highest prominence. Since we provide a direct mapping of persistence diagrams into $\mathbb{R}^d$, we can use it for the pooling stage for the bag-of-words procedure and achieve good performance with respect to the classification phase. We call this pooling approach Topological Pooling. Since it relies on persistence diagrams, this method is stable with respect to most transformations the shape can undergo: translations, rotations, etc., as long as the descriptors used in input are also invariant to these transformations. Moreover, we show that this pooling approach is robust to perturbations of the descriptors. Finally we demonstrate the validity of our approach compared to both sum-pooling and max-pooling by performing pose recognition on the SHREC 2014 dataset.

2   The Bag of Words Pipeline

The bag-of-words pipeline consists of three main steps: feature extraction, coding and pooling. Here we describe each step briefly taking a functional point of view, and we also introduce the notations we will need to define our new pooling method. We will assume that the input to the pipeline is a set of M 3D-shapes G_i represented as triangle meshes with vertices V_i.

Feature extraction aims at deriving a meaningful representation of the shape: the feature function denoted as $\mathcal{F}_i : V_i \to \mathbb{R}^N$. It is usually done by computing local descriptors (such as HKS [19], SIHKS [20], WKS [21], Shape-net features [2], etc.) on each vertex of the mesh.

The purpose of *coding* is to decompose the values of the $\mathcal{F}_i$ by projecting them on a set of points $W = (w_k)_{k \in [[1,K]]} \in \mathbb{R}^N$ called a *codebook*. This allows to replace each feature function by a family of functions $(C_i : V_i \to \mathbb{R}^K)_{i \in [1,M]}$, called the *word functions*. In other words, for a coding procedure $Coding$ and codebook W, the C_i are defined through

$$C_i(V_i) = Coding(\mathcal{F}_i(V), W).$$

There exist various coding methods, such as Vector Quantization [22], Sparse Coding [4], Locally Constrained Linear Coding [23], Fisher Kernel [24] or Super-vector [25]. The codebook is usually computed using K-means but supervised codebook learning methods [5,23] generally achieve better accuracy. In the Sparse Coding approach, the one we use in this paper, W and C are computed on the training set following

$$\min_{(C_i)_{i \in [1,M]}, W} \sum_{i=1}^{M} \sum_{x \in V_i} \left(\|(\mathcal{F}_i(x)) - W C_i((\mathcal{F}_i(x)))\|_2^2 + \lambda \|C_i(\mathcal{F}_i(x))\|_1 \right),$$

with constraint $\|w_i\| \le 1$ and regularization parameter λ. During the testing phase, the optimization is only performed on C with the codebook already computed.

The *pooling* step aims at summarizing properties of the family $(C_i)_{i \in [1,M]}$ and representing them through a compact vector $(\mathcal{P}_i)_{i \in [1,M]}$ which can then be used in standard learning algorithms such as the SVM (Support Vector Machine). Usually, the pooling method depends on the coding scheme. For Vector Quantization, one traditionally uses sum-pooling:

$$\mathcal{P}_i = (SumPool(C_{i,1}), ..., SumPool(C_{i,K}))$$
$$= \left(\sum_{x \in V_i} (C_i(\mathcal{F}_i(x)))_1, ..., \sum_{x \in V_i} (C_i(\mathcal{F}_i(x)))_k \right).$$

Max-pooling was introduced along the Sparse Coding scheme by Yang *et al.* in [4]. With this pooling technique, we summarize a function by its maximum:

$$\mathcal{P}_i = (MaxPool(C_{i,1}), ..., MaxPool(C_{i,K}))$$
$$= (\max_{x \in V}(C_i(\mathcal{F}_i(x)))_1, ..., \max_{x \in V}(C_i(\mathcal{F}_i(x)))_K).$$

It is interesting to note that the max-pooling approach is more robust than the sum-pooling. Indeed, it is robust to usual transformations the shape can

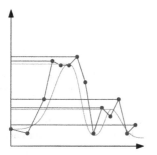 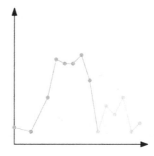

Fig. 2. A function f (red), a noisy approximation $\tilde{f}$ of f (blue) and their respective local maxima. Despite having a lot of local maxima, $\tilde{f}$ only has two "prominent peaks" (green and yellow). (Color figure online)

undergo: translations, rotations, changes of scales, etc. However, it is still quite limited as it summarizes a whole function by a single value. A natural idea is to not limit ourselves to the global maximum of the function but rather to capture all local maxima. On the other hand, in this naive form, the method results in a very unstable pooling vector since arbitrarily small perturbations of the word functions can create many local maxima, as shown in Fig. 2. Thus, a pooling approach consisting of taking the highest k local maxima is not stable. On the other hand, in the example shown in Fig. 2, we can see that, while there are a lot of local maxima for the noisy function, both functions show only two "prominent peaks". These notions of "peak" and "prominence" are properly defined in the 0-dimensional persistent homology framework which provides us with tools to derive a robust pooling method.

3   Introducing 0-dimensional Persistent Homology

0-persistent homology provides a formal definition of prominence and measures the prominence of each peak of a function f, with the promise that the most prominent ones are stable under small perturbations of f. We provide a brief overview of the computation of 0-dimensional persistent homology for the superlevel-sets of a function defined on a graph, and invite the reader to consult [11] for a more general introduction.

Let f be a function defined on the vertices of a finite graph $G = (V, E)$. In 0-dimensional persistent homology, one focuses on the evolution of the connectivity of the subgraphs F_α of G induced by the superlevel-sets of f: $F_\alpha = (\{v \in V \mid f(v) \geq \alpha\}, \{(u, v) \in E \mid \min(f(u), f(v)) \geq \alpha\})$, as α decreases from $+\infty$ to $-\infty$, as shown in Fig. 3. A vertex v is a *local maximum* if, for any edge (v, u) in E, we have $f(u) \leq f(v)$. A peak p corresponds to a local maximum $b_p = f(v_p)$ of f. We say that p is *born* at b_p, see Fig. 3(b). For a local maximum v_p, let $C(v_p, \alpha)$ be the connected component of v_p in F_α and let d_p be the largest value of α such that the maximum of f over $C(v_p, \alpha)$ is larger than b_p, we say that p *dies* at d_p. Intuitively, a peak dies when its connected component gets merged with the

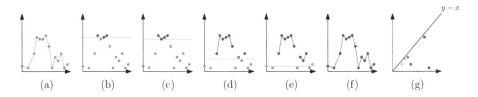

Fig. 3. Evolution of the connectedness of the superlevel-sets F_α of a function f in blue (a) as α (green) decreases from $+\infty$ to $-\infty$ (b–f). This evolution is then encoded in a persistence diagram (g).

one of another peak that has a higher maximum. Thus, there exists a vertex u_p which connects the two components such that $f(u_p) = d_p$. u_p is called a *saddle*, see Fig. 3(c). The "prominence" of p is then the difference $b_p - d_p$. The peak corresponding to the global maximum of f dies when α reaches the minimum value of f on G^1. Thus, a peak of f can be described by the couple (b_p, d_p). The set of such points (with multiplicity) in the plane is called a *persistence diagram*, denoted Δ_f, see Fig. 3(g).

Persistence diagrams are endowed with a natural metric called the *bottleneck distance*. The definition of this metric involves the notion of partial matching. A partial matching M between two diagrams Δ_1 and Δ_2 is a subset of $\Delta_1 \times \Delta_2$ such that each point of Δ_1 and Δ_2 appears at most once in M. The bottleneck cost $C(M)$ of a partial matching M between two diagrams Δ_1 and Δ_2 is the infimum of $\delta \geq 0$ that satisfy the following conditions:

- For any $(p_1, p_2) \in M$, $||p_1 - p_2||_\infty \leq \delta$, and
- For any other point (b, d) of Δ_1 or Δ_2, $b - d \leq 2\delta$.

The bottleneck distance between two diagrams D_1 and D_2, is then defined as:

$$d_B(\Delta_1, \Delta_2) = \inf_\delta \{\delta \mid \exists M, C(M) \leq \delta\}$$

Intuitively, the bottleneck distance can be seen as the cost of a minimum perfect matching between persistence diagrams (with possibility to match points to the diagonal $y = x$), where the cost is the length of the longest line, see Fig. 4. A remarkable property of persistence diagrams, proven by [13,15], is their robustness with respect to perturbations of f. Given two functions f and g defined on some graph G, we have:

$$d_B(\Delta_f, \Delta_g) \leq ||f - g||_\infty = \sup_{v \in V} |f(v) - g(v)| \tag{1}$$

In other words, if we compare the diagrams of a function f and of a noisy version of a function $\tilde{f}$ then each point $p \in D_{\tilde{f}}$ can either be matched to a point of D_f or it has a low prominence, see Fig. 4.

---

[1] This point is slightly different from the traditional persistent homology framework. Usually, the death value of the peak corresponding to the global maximum is set to $-\infty$.

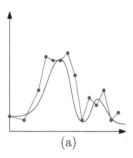
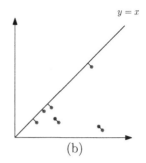

(a)   (b)

Fig. 4. (a): A real-valued function f (red) and a noisy approximation $\tilde{f}$ of f (blue).(b): Their respective persistence diagrams have close bottleneck distance. (Color figure online)

Computation As 0-dimensional persistence encodes the evolution of the connectivity of the superlevel-sets of a function, computing it can be done using a simple variant of a Union-find algorithm; in practice we use Algorithm 1 described by Chazal et al. [26], with parameter τ set to infinity. This algorithm has close to linear complexity in the number of vertices of the meshes; more precisely it has complexity $O(|V| \log(|V|) + |V|\alpha(|V|))$ where α is the inverse of the Ackermann function.

4   Using Persistence Diagrams for Pooling

As we previously mentioned at the end of Sect. 2, a simple idea to enhance the max-pooling approach is to consider the values of multiple local maxima. However, this can be highly unstable under small perturbations of the word functions. As we saw in Sect. 3, we can use persistence diagrams to deal with this issue. Given a persistence diagram Δ, we define the *prominence* p of a point $(b, d) \in \Delta$ by $p = b - d$; in other words, the prominence corresponds to the lifespan of a peak during the computation of the persistence diagram. Given a function f on a graph G, we define the infinite-dimensional Topological Pooling vector of f with $i - th$ coordinate given by

$$TopoPool(f)_i = p_i(\Delta_f),$$

where $p_i(\Delta_f)$ is the i-th highest prominence of the points of Δ_f if there is at least i points in Δ_f and 0 otherwise. Since the stability of persistence diagrams given in Eq. 1 implies the stability of the prominence of the points of Δ_f, such a construction yields some stability for our pooling scheme.

Proposition 1. *Let G be a graph and f and g two functions on a graph G with vertices V. Then, for any integer n, and any $0 < k < n$,*

$$|TopoPool(f)_k - TopoPool(g)_k| \leq 2 \sup_{x \in V} |f(x) - g(x)|$$

Fig. 5. The real SHREC 2014 dataset

Of course, in practice we cannot use an infinite-dimensional vector and we simply consider a truncation of this vector keeping n first coordinates, we denote such a truncated pooling vector "TopoPool-n". Using the notations of Sect. 2, given some $n > 0$, the pooling vectors $(\mathcal{P}_i)_{1 \leq i \leq M}$ we consider are

$$\mathcal{P}_i = (TopoPool - n((C_i(\mathcal{F}_i(x)))_1), ..., TopoPool - n((C_i(\mathcal{F}_i(x))))_K).$$

5 Experiments

In this section we evaluate the sum-pooling, the max-pooling and our topological pooling approaches on the SHREC 2014 dataset "Shape Retrieval of Non-Rigid 3D Human Models" [27], which we modify by applying a random rotation to each 3D shape. The dataset is composed of 400 meshes of 40 subjects taking 10 different poses (Fig. 5) and we wish to classify each of these meshes with respect to the pose taken by the subject. We consider both SIHKS features [20] and curvature-based features corresponding to the unary features from [28] and composed of 64 values corresponding to the curvatures, the Gaussian curvature, the mean curvature ... The coding step is performed using Sparse Coding [4] and the computation are performed performed using the SPAMS toolbox [29]. The learning part is done using a Support Vector Machine. We use 3 shapes per class for the training set, 2 for the validation set and 5 for the testing set. We compare the traditional sum-pooling with our TopoPool-n with different values for n -remark that $n = 1$ is equivalent to max-pooling- and under different codebook sizes. As a baseline, we also display the results obtained using a rigid Iterated Closest Point (ICP) [30] and a 1-nearest neighbour classification, which aims at iteratively minimizing the distance between two point clouds through rigid deformations. In our case it corresponds to finding the correct rotation to align the shapes as two shapes in a similar pose are close, however the approach can fail if it gets stuck in a local minimum and is not able to recover the correct rotation. We run the experiment a hundred times, selecting the training and testing sets at random. We display the mean accuracy over the multiple runs in Table 1.

The first noticeable fact about our experiments being the overall better results obtained by our Topological Pooling scheme compared to the max-pooling

Table 1. Mean accuracy obtained on the SHREC 2014 dataset.

Pooling / Codebook size	40	60	80	100	120	140	160	180	200
SIHKS features									
Sum-Pooling	0.53	0.56	0.60	0.60	0.58	0.62	0.61	0.60	0.60
TopoPool-1	0.46	0.55	0.53	0.54	0.58	0.59	0.63	0.64	0.64
TopoPool-5	0.69	0.71	0.69	0.70	0.73	0.70	0.74	0.73	0.72
TopoPool-10	0.70	0.71	0.71	0.69	0.72	0.71	0.73	0.74	0.72
TopoPool-15	0.72	0.73	0.71	0.70	0.74	0.71	0.74	**0.75**	0.71
TopoPool-20	0.72	0.73	0.70	0.72	0.73	0.72	0.73	0.75	0.73
Curvature features									
Sum-Pooling	0.80	0.80	0.84	0.85	0.88	0.88	0.87	0.88	0.89
TopoPool-1	0.39	0.56	0.56	0.57	0.64	0.69	0.69	0.73	0.76
TopoPool-5	0.63	0.79	0.80	0.80	0.82	0.85	0.86	0.87	0.86
TopoPool-10	0.74	0.85	0.85	0.86	0.86	0.87	0.89	0.89	0.88
TopoPool-15	0.78	0.85	0.87	0.87	0.88	0.89	0.89	**0.90**	0.90
TopoPool-20	0.79	0.88	0.88	0.88	0.88	0.89	0.90	0.90	0.89
ICP	0.55								

and to the sum-pooling for the SIHKS features. In the case of curvature features, Topological Pooling and sum-pooling gives similar accuracy results for large codebooks but in the case of smaller codebooks, Topological pooling gives much better results. It is interesting to notice that the gap between the different pooling scheme decreases as the size of the codebook increases. Indeed, the smaller the codebook, the richer each word function in terms of topology -and thus the richer each persistence diagrams will be-.

Regarding the running time of our experiment in the case of SIHKS features, online testing using the bag-of-words procedure with the largest codebook to a given shape takes around 40 seconds, where most of the time is devoted to computing the SIHKS. On the other hand, performing ICP between two shapes takes 6 seconds, thus the online testing time for a single shape with ICP is 6 times the cardinality of our training set seconds; in our case 5 minutes. On the other hand, with the ICP approach requires no offline training while the bag of words requires to compute the codebook, perform the whole bag-of-words pipeline on each training shape and compute the SVM which takes roughly 45 minutes. Overall we have to classify 350 shapes, the bag-of-words approach requires 4 hour and a half while the ICP approach requires more than a day.

6 Conclusion

In this paper, we proposed to use the canonical graph structure on shapes to capture neighborhood information between the different feature vectors. We then built discrete "word functions" on this graph instead of following the traditional approach of considering a collection of independent "word" vectors. We then proposed to consider new pooling features making use of this new information and generalizing the classical max-pooling approach by using the critical points of the "word functions". We proposed to use 0-dimensional persistent homology to ensure stability of a pooling output relying on these features. Finally, we designed a new pooling method relying on these new features and we experimentally showed that these features are efficient in a pooling context.

Acknowledgements. This work was supported by ANR project TopData ANR-13-BS01-0008. First author was supported by the French Délégation Générale de l'Armement (DGA). Second author was supported by Marie-Curie CIG-334283-HRGP, a CNRS chaire dexcellence, a chaire Jean Marjoulet from Ecole Polytechnique, and a Faculty Award from Google Inc.

References

1. Bronstein, A.M., Bronstein, M.M., Guibas, L.J., Ovsjanikov, M.: Shape google: Geometric words and expressions for invariant shape retrieval. ACM Trans. Graph. **30**, 1–20 (2011)
2. Masci, J., Boscaini, D., Bronstein, M.M., Vandergheynst, P.: Shapenet: Convolutional neural networks on non-euclidean manifolds. http://arxiv.org/abs/1501.06297
3. Fei-Fei, L., Pietro, P.: A bayesian hierarchical model for learning natural scene categories. In: Proceedings of the 2005 IEEE Computer Society Conference on Computer Vision and Pattern Recognition (CVPR 2005), CVPR 2005, vol. 2, pp. 524–531. IEEE Computer Society, Washington, DC (2005). http://dx.doi.org/10.1109/CVPR.2005.16
4. Yang, J., Yu, K., Gong, Y., Huang, T.: Linear spatial pyramid matching using sparse coding for image classification. In: IEEE Conference on Computer Vision and Pattern Recognition (CVPR) (2009)
5. Boureau, Y.-L., Bach, F., LeCun, Y., Ponce, J.: Learning mid-level features for recognition. In: Proceedings of CVPR (2010)
6. Liu, L., Wang, L., Liu, X.: In defense of soft-assignment coding. In: ICCV 2011, pp. 2486–2493 (2011)
7. Lazebnik, S., Schmid, C., Ponce, J.: Beyond bags of features: Spatial pyramid matching for recognizing natural scene categories. In: Proceedings of the 2006 IEEE Computer Society Conference on Computer Vision and Pattern Recognition, CVPR 2006, vol. 2, pp. 2169–2178 (2006)
8. López-Sastre, R.J., García-Fuertes, A., Redondo-Cabrera, C., Acevedo-Rodríguez, F.J., Maldonado-Bascón, S.: Evaluating 3D spatial pyramids for classifying 3D shapes. Comput. Graph. **37**, 473–483 (2013)
9. Li, C., Hamza, A.B.: Intrinsic spatial pyramid matching for deformable 3D shape retrieval. IJMIR **2**, 261–271 (2013)
10. Verri, A., Uras, C., Frosini, P., Ferri, M.: On the use of size functions for shape analysis. Biol. Cybern. **70**, 99–107 (1993)
11. Edelsbrunner, H., Harer, J.: Computational Topology - An Introduction. American Mathematical Society, New York (2010)

12. Zomorodian, A., Carlsson, G.: Computing persistent homology. Discrete Comput. Geom. **33**, 249–274 (2005)
13. Cohen-Steiner, D., Edelsbrunner, H., Harer, J.: Stability of persistence diagrams. In: Proceedings of 21st ACM Symposium Computer Geometry, pp. 263–271 (2005)
14. Chazal, F., Cohen-Steiner, D., Guibas, L.J., Glisse, M., Oudot, S.Y.: Proximity of persistence modules, their diagrams. In: Proceedings of 25th ACM Symposium Computer Geometry (2009)
15. Chazal, F., de Silva, V., Glisse, M., Oudot, S.: The structure and stability of persistence modules (2012). http://arxiv.org/abs/1207.3674
16. Li, C., Ovsjanikov, M., Chazal, F.: Persistence-based structural recognition. In: CVPR, pp. 2003–2010 (2014)
17. Reininghaus, J., Huber, S., Bauer, U., Kwitt, R.: A stable multi-scale kernel for topological machine learning. In: CVPR (2015)
18. Bubenik, P.: Statistical topology using persistence landscapes. JMLR **16**, 77–102 (2015)
19. Sun, J., Ovsjanikov, M., Guibas, L.: A concise, provably informative multi-scale signature based on heat diffusion. In: Proceedings of the Symposium on Geometry Processing, SGP 2009, pp. 1383–1392 (2009)
20. Bronstein, M.M., Kokkinos, I.: Scale-invariant heat kernel signatures for non-rigid shape recognition. In: Proceedings of CVPR (2010)
21. Bay, H., Ess, A., Tuytelaars, T., Van Gool, L.: Speeded-up robust features (SURF). Comput. Vis. Image Underst. **110**, 346–359 (2008)
22. Salton, G., McGill, M.J.: Introduction to Modern Information Retrieval. McGraw-Hill Inc., New York (1986)
23. Wang, J., Yang, J., Yu, K., Lv, F., Huang, T., Gong, Y.: Locality-constrained linear coding for image classification. In: IEEE Conference on Computer Vision and Pattern Recognition (CVPR) (2010)
24. Perronnin, F., Sánchez, J., Mensink, T.: Improving the fisher kernel for large-scale image classification. In: Daniilidis, K., Maragos, P., Paragios, N. (eds.) ECCV 2010, Part IV. LNCS, vol. 6314, pp. 143–156. Springer, Heidelberg (2010)
25. Zhou, X., Yu, K., Zhang, T., Huang, T.S.: Image classification using super-vector coding of local image descriptors. In: Daniilidis, K., Maragos, P., Paragios, N. (eds.) ECCV 2010, Part V. LNCS, vol. 6315, pp. 141–154. Springer, Heidelberg (2010)
26. Chazal, F., Guibas, L.J., Oudot, S.Y., Skraba, P.: Persistence-based clustering in riemannian manifolds. J. ACM **60**, 41 (2013)
27. Pickup, D., et al.: SHREC 2014 track: Shape retrieval of non-rigid 3D human models, EG 3DOR 2014 (2014)
28. Kalogerakis, E., Hertzmann, A., Singh, K.: Learning 3D mesh segmentation and labeling. ACM Trans. Graph. **29**, 102 (2010)
29. Mairal, J., Bach, F., Ponce, J., Sapiro, G.: Online learning for matrix factorization and sparse coding. J. Mach. Learn. Res. **11**, 19–60 (2010)
30. Besl, P.J., McKay, N.D.: A method for registration of 3-d shapes. IEEE Trans. Pattern Anal. Mach. Intell. **14**, 239–256 (1992)

Bijectivity Certification of 3D Digitized Rotations

Kacper Pluta[1,2(✉)], Pascal Romon[2], Yukiko Kenmochi[1], and Nicolas Passat[3]

[1] Université Paris-Est, LIGM, CNRS, ESIEE, Paris, France
kacper.pluta@univ-paris-est.fr, yukiko.kenmochi@esiee.fr
[2] Université Paris-Est, LAMA, UPEM, Paris, France
pascal.romon@u-pem.fr
[3] Université de Reims Champagne-Ardenne, CReSTIC, Reims, France
nicolas.passat@univ-reims.fr

Abstract. Euclidean rotations in $\mathbb{R}^n$ are bijective and isometric maps. Nevertheless, they lose these properties when digitized in $\mathbb{Z}^n$. For $n = 2$, the subset of bijective digitized rotations has been described explicitly by Nouvel and Rémila and more recently by Roussillon and Cœurjolly. In the case of 3D digitized rotations, the same characterization has remained an open problem. In this article, we propose an algorithm for certifying the bijectivity of 3D digitized rational rotations using the arithmetic properties of the Lipschitz quaternions.

1 Introduction

Rotations defined in $\mathbb{Z}^3$ are simple yet crucial operations in many image processing applications involving 3D data. One way of designing rotations on $\mathbb{Z}^3$ is to combine continuous rotations defined on $\mathbb{R}^3$ with a digitization operator that maps the result back into $\mathbb{Z}^3$. However, the digitized rotation, though uniformly close to its continuous sibling, often no longer satisfies the same properties. In particular, due to the alteration of distances between points—provoked by the digitization—the bijectivity is lost in general.

In this context, it is useful to understand which 3D digitized rotations are indeed bijective. "Simple" 3D digitized rotations, in particular those around one of the coordinate axes, possess the same properties as 2D digitized rotations. Therefore, an obvious subset of 3D bijective digitized rotations consists of the 2D bijective digitized rotations embedded in $\mathbb{Z}^3$. Nevertheless, the question of determining whether a non-simple 3D digitized rotation is bijective, remained open.

To our knowledge, few efforts were devoted to understand topological alterations of $\mathbb{Z}^3$ induced by digitized rotations. The contributions known to us were geared toward understanding these alterations in $\mathbb{Z}^2$: Andres and Jacob provided some necessary conditions under which 2D digitized rotations are bijective [5]; Andres proposed quasi-shear rotations which are bijective but possibly generate errors, particularly for angles around $\pi/2$ [1]; Nouvel and Rémila studied the discrete structure induced by digitized rotations that are not bijective but

© Springer International Publishing Switzerland 2016
A. Bac and J.-L. Mari (Eds.): CTIC 2016, LNCS 9667, pp. 30–41, 2016.
DOI: 10.1007/978-3-319-39441-1_4

generate no error [12,14]; moreover, they characterized the set of 2D bijective digitized rotations [13]. More recently, Roussillon and Cœurjolly used arithmetic properties of the Gaussian integers to give a different proof of the conditions for bijectivity of 2D digitized rotations [17]. On the other hand, more general 2D digitized rigid motions—rotations, translations and their compositions—were studied by Ngo et al. [9], with their impact on the topological properties of finite digital grids [10]. Moreover, Ngo et al. established some sufficient conditions for topology preservation under 2D digitized rigid motions [11]. Lately we provided a characterization of the set of 2D bijective digitized rigid motions [16].

In this article, our contribution is as follows. We consider an approach similar to that proposed by Roussillon and Cœurjolly to prove the conditions for bijectivity of 2D digitized rotations using arithmetic properties of Gaussian integers [17]—which are complex numbers whose real and imaginary parts are integers [4]. Indeed, the product of two complex numbers has a geometrical interpretation; more precisely, it acts as a rotation when the norm of the multiplier is one. In our work, we partially extend the results of Roussillon and Cœurjolly to 3D digitized rotations, employing Lipschitz quaternions, which play a similar role to Gaussian integers. However, due to the non-commutative nature of quaternions and their two-to-one relation with 3D rotations, the former approach has not succeeded yet to fully characterize the bijective digitized rotations. Nevertheless, we propose an algorithm which certifies whether a given digitized rotation, defined by a Lipschitz quaternion, is bijective. As a consequence, we cover all *the rational rotations*, i.e., those whose corresponding matrix representation contains only rational elements—since they correspond to rotations given by Lipschitz quaternions. From the point of view of the applications, excluding a rotation whose matrix has irrational elements is a minor issue, since computers mainly work with rational numbers. Moreover, using rational numbers ensures the exactness of the proposed certification algorithm.

This article is organized as follows. In Sect. 2, we recall the basic definitions of 3D rotations and Lipschitz quaternions. Section 3 provides our framework for studying the bijectivity of digitized rotations in $\mathbb{Z}^3$. In Sect. 4, we provide an algorithm certifying whether a given rational rotation is bijective or not when digitized in $\mathbb{Z}^3$. Finally, in Sect. 5, we conclude this article and provide some perspectives.

2   Digitized Rotations in Three Dimensions

A rotation in $\mathbb{R}^3$ is a bijective isometric map defined as

$$\left| \begin{aligned} \mathcal{U} : \mathbb{R}^3 &\to \mathbb{R}^3 \\ \mathbf{x} &\mapsto \mathbf{R}\mathbf{x} \end{aligned} \right. \tag{1}$$

where $\mathbf{R}$ is a 3D rotation matrix. Note that the matrix $\mathbf{R}$ can be obtained from a rotation angle and axis by Rodrigues' rotation formula [6,8,19] or from a quaternion [6,19].

2.1   Spatial Rotations and Quaternions

The proposed framework for bijectivity certification uses the formalism of quaternions. These are the elements of the set $\mathbb{H} = \{a + bi + cj + dk \mid a, b, c, d \in \mathbb{R}\}$ with the following properties:

$$i^2 = -1, \qquad j^2 = -1, \qquad k^2 = -1,$$
$$jk = -kj = i, \quad ki = -ik = j, \quad ij = -ji = k.$$

Similarly to the set of complex numbers, $\mathbb{H}$ possesses a division ring structure, albeit a non-commutative one. More precisely, for $p, q, r \in \mathbb{H}$:

- the conjugate of $q = a + bi + cj + dk$ is defined as $\bar{q} = a - bi - cj - dk$;
- the product of two quaternions, defined as

$$qp = (a_1 + b_1 i + c_1 j + d_1 k)(a_2 + b_2 i + c_2 j + d_2 k) =$$
$$a_1 a_2 - b_1 b_2 - c_1 c_2 - d_1 d_2 + (a_1 b_2 + b_1 a_2 + c_1 d_2 - d_1 c_2)i$$
$$+(a_1 c_2 - b_1 d_2 + c_1 a_2 + d_1 b_2)j + (a_1 d_2 + b_1 c_2 - c_1 b_2 + d_1 a_2)k,$$

    is not commutative, i.e. $qp \neq pq$, in general, although real numbers, i.e., quaternions such that $q = \bar{q}$ do commute with all others;
- the norm of q is defined as $|q| = \sqrt{q\bar{q}} = \sqrt{\bar{q}q} = \sqrt{a^2 + b^2 + c^2 + d^2}$;
- the inverse of q is defined as $q^{-1} = \frac{\bar{q}}{|q|^2}$, so that $qq^{-1} = q^{-1}q = 1$.

    Any point in $\mathbb{R}^3$ is represented by a pure imaginary quaternion: $\mathbf{x} = (x_1, x_2, x_3) \simeq x_1 i + x_2 j + x_3 k$. Then, any rotation $\mathcal{U}$ can be written as $\mathbf{x} \mapsto q\mathbf{x}q^{-1}$, where $\mathbf{x} \in \mathbb{R}^3$ [6,19]. The quaternion q is uniquely determined up to multiplication by a nonzero real number, and, if $|q| = 1$, up to a sign change: $q\mathbf{x}q^{-1} = (-q)\mathbf{x}(-q)^{-1}$; hence the correspondence between unit quaternions and rotation matrices is two-to-one. Note that for any unit norm quaternion $q = a + bi + cj + dk$, a rotation angle θ and an axis of rotation $\boldsymbol{\omega}$ are given as $\theta = 2\cos^{-1} a$, and $\boldsymbol{\omega} = \frac{(b,c,d)^t}{|(b,c,d)^t|}$, respectively. We refer the reader unfamiliar with quaternions to [2,6,19].

2.2   Digitized Rotations

According to Eq. (1), we generally have $\mathcal{U}(\mathbb{Z}^3) \not\subseteq \mathbb{Z}^3$. As a consequence, to define digitized rotations as maps from $\mathbb{Z}^3$ to $\mathbb{Z}^3$, we usually consider $\mathbb{Z}^3$ as a subset of $\mathbb{R}^3$, apply $\mathcal{U}$, and then combine the real results with a digitization operator

$$\left| \begin{array}{l} \mathcal{D} : \mathbb{R}^3 \quad\;\; \to \mathbb{Z}^3 \\ \quad (x, y, z) \mapsto \left(\lfloor x + \tfrac{1}{2} \rfloor, \lfloor y + \tfrac{1}{2} \rfloor, \lfloor z + \tfrac{1}{2} \rfloor \right) \end{array} \right.$$

where $\lfloor s \rfloor$ denotes the largest integer not greater than s. The digitized rotation is thus defined by $U = \mathcal{D} \circ \mathcal{U}_{|\mathbb{Z}^3}$. Due to the behavior of $\mathcal{D}$ that maps $\mathbb{R}^3$ onto $\mathbb{Z}^3$, digitized rotations are, most of the time, non-bijective. This leads us to define the notion of point status with respect to a given digitized rotation.

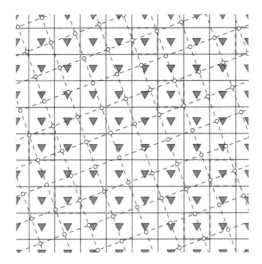

Fig. 1. Examples of three different point statuses: digitization cells corresponding to 0-, 1- and 2-points are in green, black and red, respectively. White dots indicate the positions of images of the points of the initial set $\mathbb{Z}^3$ by $\mathcal{U}$, embedded in $\mathbb{R}^3$, subdivided into digitization cells around the points of the final set $\mathbb{Z}^3$, represented by gray triangles. Note that, for readability purpose, $\mathcal{U}$ is a simple 3D digitized rotation such that $\theta = \frac{\pi}{9}, \boldsymbol{\omega} = (0, 0, 1)^t$. Therefore, as for 2D digitized rotations, only 0-, 1- and 2- point statuses are possible. Note that only one 2D slice of 3D space is presented.

Definition 1. *Let* $\mathbf{y} \in \mathbb{Z}^3$ *be an integer point. The set of preimages of* $\mathbf{y}$ *with respect to* U *is defined as* $M_U(\mathbf{y}) = \{\mathbf{x} \in \mathbb{Z}^3 \mid U(\mathbf{x}) = \mathbf{y}\}$, *and* $\mathbf{y}$ *is referred to as a* s-*point, where* $s = |M_U(\mathbf{y})|$ *is called the* status *of* $\mathbf{y}$.

Remark 1. In $\mathbb{Z}^3$, $|M_U(\mathbf{y})| \in \{0, 1, 2, 3, 4\}$ and one can prove that only points $\mathbf{p}, \mathbf{q} \in \mathbb{Z}^3$ such that $|\mathbf{p}-\mathbf{q}| < \sqrt{3}$ can be preimages of a 2-point; points $\mathbf{p}, \mathbf{q}, \mathbf{r} \in \mathbb{Z}^3$ forming an isosceles triangle of side lengths $1, 1$ and $\sqrt{2}$ can be preimages of a 3-point; points $\mathbf{p}, \mathbf{q}, \mathbf{r}, \mathbf{s} \in \mathbb{Z}^3$ forming a square of side length 1 can be preimages of a 4-point.

The non-injective and non-surjective behaviors of a digitized rotation result in the existence of s-points for $s \neq 1$. Figure 1 illustrates a simple 3D rotation which provokes 0- and 2- point statuse.

3 Bijectivity Certification

3.1 Set of Remainders

Let us compare the rotated digital grid $\mathcal{U}(\mathbb{Z}^3) = q\mathbb{Z}^3 q^{-1}$ with the grid $\mathbb{Z}^3$. The digitized rotation $U = \mathcal{D} \circ \mathcal{U}$ is bijective if and only if each digitization cell of $\mathbb{Z}^3$ contains one and only one rotated point of $q\mathbb{Z}^3 q^{-1}$; in other words,

$\forall \mathbf{y} \in \mathbb{Z}^3, |M_U(\mathbf{y})| = 1$. Let us denote by $\mathscr{C}(\mathbf{y})$ the digitization cell, i.e. the unit cube, centered at the point $\mathbf{y} = (y_1, y_2, y_3) \in \mathbb{Z}^3$:

$$\mathscr{C}(\mathbf{y}) = \left[y_1 - \frac{1}{2}, y_1 + \frac{1}{2} \right) \times \left[y_2 - \frac{1}{2}, y_2 + \frac{1}{2} \right) \times \left[y_3 - \frac{1}{2}, y_3 + \frac{1}{2} \right).$$

Instead of studying the whole source and target spaces, we study the set of remainders defined by the map

$$\left| \begin{array}{l} S_q : \mathbb{Z}^3 \times \mathbb{Z}^3 \to \mathbb{R}^3 \\ \quad (\mathbf{x}, \mathbf{y}) \quad \mapsto q\mathbf{x}q^{-1} - \mathbf{y}. \end{array} \right.$$

Then, the bijectivity of U can be expressed as

$$\forall \mathbf{y} \in \mathbb{Z}^3 \; \exists! \mathbf{x} \in \mathbb{Z}^3, S_q(\mathbf{x}, \mathbf{y}) \in \mathscr{C}(\mathbf{0}),$$

which is equivalent to the "double" surjectivity relation, used by Roussillon and Cœurjolly [17]:

$$\begin{cases} \forall \mathbf{y} \in \mathbb{Z}^3 \; \exists \mathbf{x} \in \mathbb{Z}^3, S_q(\mathbf{x}, \mathbf{y}) \in \mathscr{C}(\mathbf{0}) \\ \forall \mathbf{x} \in \mathbb{Z}^3 \; \exists \mathbf{y} \in \mathbb{Z}^3, S_q(\mathbf{x}, \mathbf{y}) \in q\mathscr{C}(\mathbf{0})q^{-1} \end{cases} \tag{2}$$

provided that both sets $S_q(\mathbb{Z}^3, \mathbb{Z}^3) \cap \mathscr{C}(\mathbf{0})$ and $S_q(\mathbb{Z}^3, \mathbb{Z}^3) \cap q\mathscr{C}(\mathbf{0})q^{-1}$ coincide; in other words, $S_q(\mathbb{Z}^3, \mathbb{Z}^3) \cap ((\mathscr{C}(\mathbf{0}) \cup q\mathscr{C}(\mathbf{0})q^{-1}) \setminus (\mathscr{C}(\mathbf{0}) \cap q\mathscr{C}(\mathbf{0})q^{-1})) = \varnothing$. Hereafter, we shall rely on Formula (2), and in the study of the bijectivity of digitized rotation U, we will focus on the values of S_q. More precisely, we will study the group $\mathcal{G}$ spanned by values of S_q:

$$\mathcal{G} = \mathbb{Z}q \begin{pmatrix} 1 \\ 0 \\ 0 \end{pmatrix} q^{-1} + \mathbb{Z}q \begin{pmatrix} 0 \\ 1 \\ 0 \end{pmatrix} q^{-1} + \mathbb{Z}q \begin{pmatrix} 0 \\ 0 \\ 1 \end{pmatrix} q^{-1} + \mathbb{Z} \begin{pmatrix} 1 \\ 0 \\ 0 \end{pmatrix} + \mathbb{Z} \begin{pmatrix} 0 \\ 1 \\ 0 \end{pmatrix} + \mathbb{Z} \begin{pmatrix} 0 \\ 0 \\ 1 \end{pmatrix}.$$

3.2   Dense Subgroups and Non-injectivity

The key to understanding the conditions that ensure the bijectivity of U is the structure of $\mathcal{G}$. For this reason, we start by looking at the image $\mathcal{G}$ of S_q, and discuss its density.

Proposition 2. *If one or more generators of $\mathcal{G}$ have an irrational term, then $\mathcal{G} \cap V$ is dense for some nontrivial subspace V. We say that $\mathcal{G}$ has a dense factor.*

On the contrary, we have the following result.

Proposition 3. *If all generators of $\mathcal{G}$ have only rational terms, then there exist vectors $\boldsymbol{\sigma}, \boldsymbol{\phi}, \boldsymbol{\psi} \in \mathcal{G}$ which are the minimal generators of $\mathcal{G}$.*

Proof. The generators of $\mathcal{G}$ are given by the rational matrix $\mathbf{B} = [\mathbf{R} \mid \mathbf{I}_3]$ where $\mathbf{I}_3$ stands for the 3×3 identity matrix. As $\mathbf{B}$ is a rational, full row rank matrix, it can be brought to its Hermite normal form $\mathbf{H} = [\mathbf{T} \mid \mathbf{0}_{3,3}]$, where $\mathbf{T}$ is a non-singular, lower triangular non-negative matrix and $\mathbf{0}_{3,3}$ stands for 3×3 zero matrix, such that each row of $\mathbf{T}$ has a unique maximum entry, which is located on the main diagonal[1] [18]. Note that the problem of computing the Hermite normal form $\mathbf{H}$ of the rational matrix $\mathbf{B}$ reduces to that of computing the Hermite normal form of an integer matrix: let s stand for the least common multiple of all the denominators of $\mathbf{B}$ which is given by $s = |q|^2$; compute the Hermite normal form $\mathbf{H}'$ for the integer matrix $s\mathbf{B}$; finally, the Hermite normal form $\mathbf{H}$ of $\mathbf{B}$ is obtained by $s^{-1}\mathbf{H}'$. The columns of $\mathbf{H}$ are the minimal generators of $\mathcal{G}$. Notice that the rank of $\mathbf{B}$ is equal to 3. Therefore, $\mathbf{H}$ gives a base $(\boldsymbol{\sigma}, \boldsymbol{\phi}, \boldsymbol{\psi})$, so that $\mathcal{G} = \mathbb{Z}\boldsymbol{\sigma} + \mathbb{Z}\boldsymbol{\phi} + \mathbb{Z}\boldsymbol{\psi}$. As $\mathbf{H}'$ gives an integer base, $s\mathcal{G}$ is an integer lattice. □

Lemma 4. *Whenever $\mathcal{G}$ is dense, the corresponding 3D digitized rotation is not bijective.*

Proof. Since $\mathcal{G}$ is dense, there exists $\mu = S_q(\mathbf{x}, \mathbf{y}) \in \mathcal{G} \cap \mathscr{C}(\mathbf{0})$, such that $\mu + \sigma = S_q(\mathbf{x} + i, \mathbf{y})$ also line in $\mathscr{C}(\mathbf{0})$. Then $\mathbf{x}$ and $\mathbf{x} + i$ are both preimages of $\mathbf{y}$ by U, which is therefore not bijective. □

When $\mathcal{G}$ is dense (see Fig. 2(a)), the reasoning of Nouvel and Rémila, originally used to discard 2D digitized irrational rotations as being bijective [13], shows that a corresponding 3D digitized rotation cannot be bijective as well. What differs from the 2D case is the possible existence of non-dense $\mathcal{G}$ with a dense factor (see Fig. 2(b)). In this context, we state the following conjecture.

Conjecture 1. *Whenever $\mathcal{G}$ has a dense factor, the corresponding digitized rotation is not bijective.*

Henceforth, we will assume that $\mathcal{G}$ is generated by rational vectors, and forms therefore a lattice (see Fig. 2(c)). In other words, corresponding rotations are considered as *rational*. The question now remains of comparing the (finitely many) points in $S_q(\mathbb{Z}^3, \mathbb{Z}^3) \cap \mathscr{C}(\mathbf{0})$ and $S_q(\mathbb{Z}^3, \mathbb{Z}^3) \cap q\mathscr{C}(\mathbf{0})q^{-1}$.

3.3 Lipschitz Quaternions and Bijectivity

To represent 2D rational rotations, Roussillon and Cœurjolly used Gaussian integers [17]. In $\mathbb{R}^3$, rational rotations are characterized as follows [3].

Proposition 5. *There is a two-to-one correspondence between the set of Lipschitz quaternions $\mathbb{L} = \{a + bi + cj + dk \mid a, b, c, d \in \mathbb{Z}\}$ such that the greatest common divisor of a, b, c, d is 1, and the set of rational rotations.*

Working in the framework of rational rotations allows us to turn to integers: $|q|^2\mathcal{G}$ is an integer lattice. As integer lattices are easier to work with from the computational point of view, we do scale $\mathcal{G}$ by $|q|^2$ in order to develop a certification algorithm.

---

[1] Note that the definition of Hermite normal form varies in the literature.

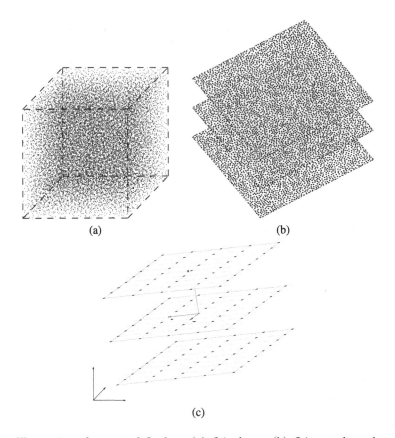

Fig. 2. Illustration of a part of $\mathcal{G}$ when: (a) $\mathcal{G}$ is dense; (b) $\mathcal{G}$ is not dense but has a dense factor – the set of points at each plane is dense while the planes are spaced by a rational distance; (c) $\mathcal{G}$ is a lattice. In the case of (a) and (b), only some random points are presented, for the sake of visibility. In (c), vectors $\boldsymbol{\sigma}, \boldsymbol{\phi}, \boldsymbol{\psi}$ are marked in red, blue and green, respectively (Color figure online).

Similarly to the former discussion, after scaling $\mathcal{G}$ by $|q|^2$, we consider the finite set of remainders, obtained by comparing the lattice $q\mathbb{Z}^3\bar{q}$ with the lattice $|q|^2\mathbb{Z}^3$, and applying the scaled version of the map S_q defined as

$$\left| \begin{array}{ll} \check{S}_q : \mathbb{Z}^3 \times \mathbb{Z}^3 \to \mathbb{Z}^3 \\ \quad (\mathbf{x}, \mathbf{y}) \quad \mapsto q\mathbf{x}\bar{q} - q\bar{q}\mathbf{y}. \end{array} \right. \tag{3}$$

Indeed, Formula (2) is rewritten as

$$\begin{cases} \forall \mathbf{y} \in \mathbb{Z}^3 \ \exists \mathbf{x} \in \mathbb{Z}^3, \check{S}_q(\mathbf{x}, \mathbf{y}) \in |q|^2\mathscr{C}(\mathbf{0}) \\ \forall \mathbf{x} \in \mathbb{Z}^3 \ \exists \mathbf{y} \in \mathbb{Z}^3, \check{S}_q(\mathbf{x}, \mathbf{y}) \in q\mathscr{C}(\mathbf{0})\bar{q}. \end{cases} \tag{4}$$

Note that the right hand sides of Formulae (3) and (4) are left multiples of q. As a consequence, we are allowed to divide them by q on the left, while keeping integer-valued functions. Let us define

$$\left| \begin{array}{l} S'_q : \mathbb{Z}^3 \times \mathbb{Z}^3 \to \mathbb{Z}^4 \\ \quad (\mathbf{x}, \mathbf{y}) \quad \mapsto \mathbf{x}\bar{q} - \bar{q}\mathbf{y}. \end{array} \right.$$

Then, the bijectivity of U is ensured when

$$\begin{cases} \forall \mathbf{y} \in \mathbb{Z}^3 \ \exists \mathbf{x} \in \mathbb{Z}^3, \ S'_q(\mathbf{x}, \mathbf{y}) \in \bar{q}\mathscr{C}(\mathbf{0}) \\ \forall \mathbf{x} \in \mathbb{Z}^3 \ \exists \mathbf{y} \in \mathbb{Z}^3, \ S'_q(\mathbf{x}, \mathbf{y}) \in \mathscr{C}(\mathbf{0})\bar{q}, \end{cases} \tag{5}$$

provided that both sets $S'_q(\mathbb{Z}^3, \mathbb{Z}^3) \cap \bar{q}\mathscr{C}(\mathbf{0})$ and $S'_q(\mathbb{Z}^3, \mathbb{Z}^3) \cap \mathscr{C}(\mathbf{0})\bar{q}$ coincide.

4   An Algorithm for Bijectivity Certification

In this section we present an algorithm which indicates whether a digitized rational rotation given by a Lipschitz quaternion is bijective or not. The strategy consists of checking whether there exists $\mathbf{w} \in ((\bar{q}\mathscr{C}(\mathbf{0}) \cup \mathscr{C}(\mathbf{0})\bar{q}) \backslash (\bar{q}\mathscr{C}(\mathbf{0}) \cap \mathscr{C}(\mathbf{0})\bar{q})) \cap \mathbb{Z}^4$ such that $\mathbf{w} = S'_q(\mathbf{x}, \mathbf{y})$. If this is the case, then the rotation given by q is not bijective, and conversely.

Because q is a Lipschitz quaternion, the values of S'_q span a sublattice $\check{\mathcal{G}} \subset \mathbb{Z}^4$. Therefore, given a Lipschitz quaternion $q = a + bi + cj + dk$, solving $S'_q(\mathbf{x}, \mathbf{y}) = \mathbf{w}$ with $\mathbf{x}, \mathbf{y} \in \mathbb{Z}^3$ for $\mathbf{w} \in \check{\mathcal{G}}$ leads to solving the following linear Diophantine system:

$$\mathbf{A}\mathbf{z} = \mathbf{w} \tag{6}$$

where $\mathbf{z}^t = (\mathbf{x}, \mathbf{y}) \in \mathbb{Z}^6$ and

$$\mathbf{A} = \begin{bmatrix} b & c & d & -b & -c & -d \\ a & -d & c & -a & -d & c \\ d & a & -b & d & -a & -b \\ -c & b & a & -c & b & -a \end{bmatrix}.$$

The minimal basis $(\check{\sigma}, \check{\phi}, \check{\psi})$ of $\check{\mathcal{G}}$ can be obtained from the columns of the Hermite normal form of the matrix $\mathbf{A}$. Since the rank of $\mathbf{A}$ is 3, we have $\check{\mathcal{G}} = \mathbb{Z}\check{\sigma} + \mathbb{Z}\check{\phi} + \mathbb{Z}\check{\psi}$.

Therefore, the problem amounts to: (i) finding the minimal basis $(\check{\sigma}, \check{\phi}, \check{\psi})$ of the group $\check{\mathcal{G}}$ by reducing the matrix $\mathbf{A}$ to its Hermite normal form; (ii) checking whether there exists a linear combination of these basis vectors $\mathbf{w} = u\check{\sigma} + v\check{\phi} + w\check{\psi}$, for $u, v, w \in \mathbb{Z}$ such that $\mathbf{w} \in (\bar{q}\mathscr{C}(\mathbf{0}) \cup \mathscr{C}(\mathbf{0})\bar{q}) \backslash (\bar{q}\mathscr{C}(\mathbf{0}) \cap \mathscr{C}(\mathbf{0})\bar{q})$.

To find points of $\check{\mathcal{G}}$ that violate Formula (5), we consider points $\mathbf{w} \in \mathbb{Z}^4 \cap \bar{q}\mathscr{C}(\mathbf{0})$ (or $\mathbf{w} \in \mathbb{Z}^4 \cap \mathscr{C}(\mathbf{0})\bar{q}$) such that $\mathbf{w} \notin \mathscr{C}(\mathbf{0})\bar{q}$ (or $\mathbf{w} \notin \bar{q}\mathscr{C}(\mathbf{0})$). Then, we verify whether $\mathbf{w}$ belongs to $\check{\mathcal{G}}$. The membership verification can be done in two steps. Step 1: we check if Eq. (6) has solutions, while verifying if the following holds:

$$aw_1 - bw_2 - cw_3 - dw_4 = 0,$$

where $\mathbf{w} = (w_1, w_2, w_3, w_4)$ and $q = a + bi + cj + dk$. Step 2: we check if Eq. (6) has integer solutions by solving it. This can be done by reducing the matrix

[$\mathbf{A} \mid \mathbf{w}$] to the Hermite normal form. Note that before iterating over points $\mathbf{w} \in \mathbb{Z}^4 \cap \bar{q}\mathscr{C}(\mathbf{0})$ (or $\mathbf{w} \in \mathbb{Z}^4 \cap \mathscr{C}(\mathbf{0})\bar{q}$), we can first reduce the matrix $\mathbf{A}$ to its Hermite normal form $\check{\mathbf{H}}$ and then reduce the augmented matrix [$\check{\mathbf{H}} \mid \mathbf{w}$], which is computationally less costly, as explained in the following discussion.

All the steps are summarized in Algorithm 1. Figure 3 presents sets of points $q\mathbf{w} \in q\mathscr{C}(\mathbf{0})\bar{q} \cup |q|^2\mathscr{C}(\mathbf{0})$ for some Lipschitz quaternions, which induce bijective digitized rational rotations, while Fig. 4 presents non-bijective cases. Finally, Table 1 lists some examples of Lipschitz quaternions that generate non-simple 3D bijective digitized rotations[2].

Algorithm 1. Checks if a given Lipschitz quaternion generates a 3D bijective digitized rotation.

Data: a Lipschitz quaternion $q = a + bi + cj + dk$ s.t. $gcd(a, b, c, d) = 1$.
Result: True if the digitized rotation given by q is bijective and false otherwise.
1 $\check{\mathbf{H}} \leftarrow$ HermiteNormalForm($\mathbf{A}$)
2 **foreach** $\mathbf{w} = (w_1, w_2, w_3, w_4) \in \mathbb{Z}^4 \cap \bar{q}\mathscr{C}(\mathbf{0})$ **do**
3 $\quad$ **if** $aw_1 - bw_2 - cw_3 - dw_4 = 0$ **and** $\{\mathbf{p} \mid \check{\mathbf{H}}\mathbf{p} = \mathbf{w}, \mathbf{p} \in \mathbb{Z}^3\} \neq \varnothing$ **then**
4 $\quad\quad$ **if** $\mathbf{w} \notin \mathscr{C}(\mathbf{0})\bar{q}$ **then**
5 $\quad\quad\quad$ **return false**

6 **return true**

The time complexity of Algorithm 1 is given as follows.

Step 1: reduction of the matrix $\mathbf{A}$ to the Hermite normal form can be done in a polynomial time [18]. For instance, one can apply the algorithm proposed by Micciancio and Warinschi [7] or its more recent, optimized version proposed by Pernet and Stein [15], whose running time complexity for full row rank matrices—with some slight modifications it can handle non-full row rank matrices—is $\mathcal{O}(mn^4 \log^2 N(\mathbf{A}))$, where n is the number of rows, m the number of columns and $N(\mathbf{A})$ stands for a bound on the entries of the matrix $\mathbf{A}$ [7]. Here $n = 4$ and $m = 6$. Thus, the time complexity of Step 1 is $\mathcal{O}(\log^2 N(\mathbf{A}))$.

Step 2: the number of points in $\mathbb{Z}^4 \cap \bar{q}\mathscr{C}(\mathbf{0})$ (resp. $\mathbb{Z}^4 \cap \mathscr{C}(\mathbf{0})\bar{q}$) is bounded by $|q|^3$. For each point, the time needed to reduce the matrix [$\check{\mathbf{H}} \mid \mathbf{w}$] to the Hermite normal form is $\mathcal{O}(n^4 \log^2 N([\check{\mathbf{H}} \mid \mathbf{w}]))$, where $n = 4$ and $N([\check{\mathbf{H}} \mid \mathbf{w}])$ is a bound on the entries of the matrix [$\check{\mathbf{H}} \mid \mathbf{w}$] [7]. Therefore, the time complexity of Step 2 is $\mathcal{O}(|q|^3 \log^2 N([\check{\mathbf{H}} \mid \mathbf{w}]))$. Note that determining whether $\mathbf{w} \notin \mathscr{C}(\mathbf{0})\bar{q}$ (or $\mathbf{w} \notin \bar{q}\mathscr{C}(\mathbf{0})$) can be done in a constant time while checking a set of inequalities.

Finally, we can conclude that the time complexity of Algorithm 1 is given by the complexity of Step 2, namely $\mathcal{O}(|q|^3 \log^2 N([\check{\mathbf{H}} \mid \mathbf{w}]))$.

---

[2] A complete list of Lipschitz quaternions in the range $[-10, 10]^4$, inducing bijective 3D digitized rotations can be downloaded from: http://dx.doi.org/10.5281/zenodo.50674

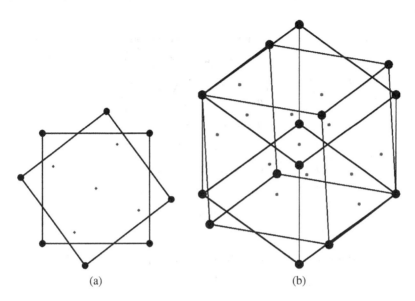

(a)   (b)

Fig. 3. Visualization of $q\mathbf{w} \in q\mathscr{C}(\mathbf{0})\bar{q} \cup |q|^2\mathscr{C}(\mathbf{0})$ together with $q\mathscr{C}(\mathbf{0})\bar{q}$ and $|q|^2\mathscr{C}(\mathbf{0})$, for (a) $q = 3 + k$ and (b) $q = 3 + 4i + k$, each of which induce bijective digitized rational rotation. Points $q\mathbf{w}$ are depicted as blue spheres (Color figure online).

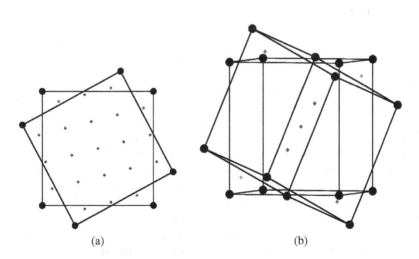

(a)   (b)

Fig. 4. Visualization of $q\mathbf{w} \in q\mathscr{C}(\mathbf{0})\bar{q} \cap |q|^2\mathscr{C}(\mathbf{0})$ – in blue, $q\mathbf{w} \in q\mathscr{C}(\mathbf{0})\bar{q} \setminus |q|^2\mathscr{C}(\mathbf{0})$ – in red, and $|q|^2\mathscr{C}(\mathbf{0}) \setminus q\mathscr{C}(\mathbf{0})\bar{q}$ – in green, for (a) $q = 4 + k$ and (b) $q = 2 - 3i - 2j - 5k$, each of which induces a non-bijective digitized rational rotations (Color figure online).

Table 1. Examples of Lipschitz quaternions which generate 3D bijective digitized rotations.

Lipschitz quaternion	Angle axis representation
$3 + 2i + j$	$\theta \approx 73.4°, \boldsymbol{\omega} = \left(\frac{2}{\sqrt{5}}, \frac{1}{\sqrt{5}}, 0\right)$
$5 + 4i + j$	$\theta \approx 79.02°, \boldsymbol{\omega} = \left(\frac{4}{\sqrt{17}}, \frac{1}{\sqrt{17}}, 0\right)$
$2 + i + j + k$	$\theta \approx 81.79°, \boldsymbol{\omega} = \left(\frac{1}{\sqrt{3}}, \frac{1}{\sqrt{3}}, \frac{1}{\sqrt{3}}\right)$
$4 + j + 3k$	$\theta \approx 76.66°, \boldsymbol{\omega} = \left(0, \frac{1}{\sqrt{10}}, \frac{3}{\sqrt{10}}\right)$
$3 + i + j + k$	$\theta \approx 60°, \boldsymbol{\omega} = \left(\frac{1}{\sqrt{3}}, \frac{1}{\sqrt{3}}, \frac{1}{\sqrt{3}}\right)$
$4 + i + j + k$	$\theta \approx 46.83°, \boldsymbol{\omega} = \left(\frac{1}{\sqrt{3}}, \frac{1}{\sqrt{3}}, \frac{1}{\sqrt{3}}\right)$
$5 + i + j + k$	$\theta \approx 38.21°, \boldsymbol{\omega} = \left(\frac{1}{\sqrt{3}}, \frac{1}{\sqrt{3}}, \frac{1}{\sqrt{3}}\right)$
$3 + 2i + 2j + 3k$	$\theta \approx 107.9°, \boldsymbol{\omega} = \left(\frac{2}{\sqrt{17}}, \frac{2}{\sqrt{17}}, \frac{3}{\sqrt{17}}\right)$
$-5 + 3i + 5j + 5k$	$\theta \approx 246.1°, \boldsymbol{\omega} = \left(\frac{3}{\sqrt{59}}, \frac{5}{\sqrt{59}}, \frac{5}{\sqrt{59}}\right)$
$5 - 4i + -5j + 5k$	$\theta \approx 116.8°, \boldsymbol{\omega} = \left(-2\sqrt{\frac{2}{33}}, -\frac{5}{\sqrt{66}}, \frac{5}{\sqrt{66}}\right)$
$10 - 10i + 10j + 9k$	$\theta \approx 118.4°, \boldsymbol{\omega} = \left(-\frac{10}{\sqrt{281}}, \frac{10}{\sqrt{281}}, \frac{9}{\sqrt{281}}\right)$
$-10 + 9i - 9j - 10k$	$\theta \approx 243.4°, \boldsymbol{\omega} = \left(\frac{9}{\sqrt{262}}, -\frac{9}{\sqrt{262}}, -5\sqrt{\frac{2}{131}}\right)$
$2 + 2i + j + 2k$	$\theta \approx 112.6°, \boldsymbol{\omega} = \left(\frac{2}{3}, \frac{1}{3}, \frac{2}{3}\right)$
$-2 - 2i - j + k$	$\theta \approx 258.5°, \boldsymbol{\omega} = \left(-\sqrt{\frac{2}{3}}, -\frac{1}{\sqrt{6}}, \frac{1}{\sqrt{6}}\right)$

5   Conclusion

In this article, we showed the existence of non-simple 3D bijective digitized rotations—ones for which a given rotation axis does not correspond to any of the coordinate axes.

The approach is similar to that used by Roussillon and Cœurjolly to prove the conditions for the bijectivity of 2D digitized rotations using Gaussian integers [17]. In our work, we used Lipschitz quaternions, which play a similar role to Gaussian integers. Due to the non-commutative nature of quaternions and their two-to-one relation with 3D rotations, the former approach has not succeeded yet to fully characterize the set of 3D bijective digitized rotations. Nevertheless, we proposed an algorithm that certifies whether a digitized rotation given by a Lipschitz quaternion q is bijective or not. The time complexity of proposed certification algorithm is $\mathcal{O}(|q|^3 \log^2 N([\check{\mathbf{H}} \mid \mathbf{w}]))$.

As a part of our future work, we would like to prove Conjecture 1 and find the general solution to Eq. (6), which allows us to characterize the set of 3D bijective digitized rotations. We may also consider images of finite sets (e.g. digital images or pieces of ambient space). The bijective digitized rotations found above will map bijectively any finite subset of $\mathbb{Z}^3$; but other (non-bijective) rotations may also be bijective when restricted to a given finite subset. Identifying those can be achieved by applying a similar algorithm to the one proposed by the authors in [16] for 2D rigid motions, though at a greater cost.

Acknowledgments. The authors express their thanks to Éric Andres of Université de Poitiers for his very helpful feedback and comments which allowed us to improve the article.

The research leading to these results has received funding from the Programme d'Investissements d'Avenir (LabEx Bézout, ANR-10-LABX-58).

References

1. Andres, E.: The quasi-shear rotation. In: Miguet, S., Ubéda, S., Montanvert, A. (eds.) DGCI 1996. LNCS, vol. 1176. Springer, Heidelberg (1996)
2. Conway, J., Smith, D.: On Quaternions and Octonions. Taylor & Francis, Ak Peters Series, Boca Raton (2003)
3. Cremona, J.: Letter to the editor. American Mathematical Monthly **94**(8), 757–758 (1987)
4. Hardy, G.H., Wright, E.M.: Introduction to the Theory of Numbers, vol. IV. Oxford University Press, Cambridge (1979)
5. Jacob, M.A., Andres, E.: On discrete rotations. In: DGCI. pp. 161–174 (1995)
6. Kanatani, K.: Understanding Geometric Algebra: Hamilton, Grassmann, and Clifford for Computer Vision and Graphics. CRC Press, Boca Raton (2015)
7. Micciancio, D., Warinschi, B.: A linear space algorithm for computing the Hermite Normal Form. In: ISSAC. pp. 231–236. ACM (2001)
8. Murray, R., Li, Z., Sastry, S.: A Mathematical Introduction to Robotic Manipulation. CRC Press, Boca Raton (1994)
9. Ngo, P., Kenmochi, Y., Passat, N., Talbot, H.: Combinatorial structure of rigid transformations in 2D digital images. Comput. Vis. Image Underst. **117**(4), 393–408 (2013)
10. Ngo, P., Kenmochi, Y., Passat, N., Talbot, H.: Topology-preserving conditions for 2D digital images under rigid transformations. J. Math. Imaging Vis. **49**(2), 418–433 (2014)
11. Ngo, P., Passat, N., Kenmochi, Y., Talbot, H.: Topology-preserving rigid transformation of 2D digital images. IEEE Trans. Image Process. **23**(2), 885–897 (2014)
12. Nouvel, B., Rémila, É.: On colorations induced by discrete rotations. In: Nyström, I., Sanniti di Baja, G., Svensson, S. (eds.) DGCI 2003. LNCS, vol. 2886, pp. 174–183. Springer, Heidelberg (2003)
13. Nouvel, B., Rémila, É.: Characterization of bijective discretized rotations. In: Klette, R., Žunić, J. (eds.) IWCIA 2004. LNCS, vol. 3322, pp. 248–259. Springer, Heidelberg (2004)
14. Nouvel, B., Rémila, E.: Configurations induced by discrete rotations: Periodicity and quasi-periodicity properties. Discrete Appl. Math. **147**(2–3), 325–343 (2005)
15. Pernet, C., Stein, W.: Fast computation of Hermite normal forms of random integer matrices. J. Number Theory **130**(7), 1675–1683 (2010)
16. Pluta, K., Romon, P., Kenmochi, Y., Passat, N.: Bijective rigid motions of the 2D Cartesian grid. In: Normand, N., Guédon, J., Autrusseau, F. (eds.) DGCI 2016. LNCS, vol. 9647, pp. 359–371. Springer, Heidelberg (2016). doi:10.1007/978-3-319-32360-2_28
17. Roussillon, T., Cœurjolly, D.: Characterization of bijective discretized rotations by Gaussian integers. Research report, LIRIS UMR CNRS 5205 (2016). https://hal.archives-ouvertes.fr/hal-01259826
18. Schrijver, A.: Theory of Linear and Integer Programming. Wiley, Chichester (1998)
19. Vince, J.: Quaternions for Computer Graphics. Springer, London (2011)

Morse Chain Complex from Forman Gradient in 3D with $\mathbb{Z}_2$ Coefficients

Lidija Čomić[(⊠)]

Faculty of Technical Sciences, University of Novi Sad, Novi Sad, Serbia
comic@uns.ac.rs

Abstract. A Forman gradient V on a cell complex Γ enables efficient computation of the homology of Γ: the Morse chain complex defined by critical cells of V and their connection through gradient V-paths is equivalent to the homology of chain complex defined by cells of Γ and the immediate boundary relation between them.

We propose an algorithm that computes the boundary operator of the Morse chain complex associated with Forman gradient V defined on a regular cell 3-complex Γ. The algorithm computes the boundary operator with coefficients in $\mathbb{Z}_2$, and encodes it in the form of the boundary matrix. Our algorithm is incremental: as it progresses through a filtration of Γ induced by V, it computes the boundary operator for each critical cell reached in the filtration order.

Keywords: Morse chain complex · Forman gradient

1 Introduction

Available scientific data sets are of increasing quantity and quality, thus generating the need for efficient computational methods for the topological analysis of shapes represented as such complexes and of functions defined on them.

Forman theory has been established as a versatile and widely applied tool in many research fields, such as computational topology, computer graphics, scientific visualization, molecular shape analysis, and geometric modeling [1,2,7–9,11,13]. To be able to exploit its theoretical results, starting from a scalar field f given on the vertices of a regular cell complex Γ, a Forman gradient V is defined on Γ. Many algorithms that construct such gradient have been proposed, and the connection between critical points of scalar field f and critical cells of the associated Forman gradient V has been established in 2D [9] and 3D [13].

The Morse chain complex $\mathcal{M}$ of a Forman gradient V on a cell complex Γ enables the calculation of its (persistent) homology [4,12]. The chain groups of $\mathcal{M}$ are defined by the critical k-cells of V and the boundary operator is defined through gradient V-paths connecting them.

We propose here an iterative algorithm, which computes the boundary operator $\partial_{\mathcal{M}}$ and boundary matrices B_k with coefficients in $\mathbb{Z}_2$ from a Forman gradient V on a regular 3-complex Γ in $\mathbb{R}^3$. The algorithm updates the Morse chain complex at each step of the Forman gradient traversal. Thus, it produces the Morse

© Springer International Publishing Switzerland 2016
A. Bac and J.-L. Mari (Eds.): CTIC 2016, LNCS 9667, pp. 42–52, 2016.
DOI: 10.1007/978-3-319-39441-1_5

chain complex not only for the given cell complex Γ, but also for each subcomplex F_i in a (non -unique) filtration $\emptyset = F_0 \subseteq F_1 \subseteq ... \subseteq F_M = \Gamma$ induced by V.

In Sect. 2 we give some basic notions on Forman theory. In Sect. 3 we describe our algorithm for computing the Morse chain complex $\mathcal{M}$. In Sect. 4 we summarize the paper with a brief discussion.

2   Background Notions

We review some basic notions on Forman gradient and on the associated Morse chain complex. We focus on regular cell 3-complexes. Recall that a cell d-complex in $\mathbb{R}^m$ is a finite set Γ of cells in $\mathbb{R}^m$ such that (i) the cells in Γ are pairwise disjoint, and (ii) for each cell $a \in \Gamma$, the boundary of a is a disjoint union of cells in Γ. The maximum of dimensions of cells in Γ is d. A complex is constructed inductively by starting from a discrete set of points and attaching discs of nondecreasing dimension along their boundaries. Each attaching map is continuous, homeomorphic on the interior of discs, and maps the boundary of the disc to a union of lower-dimensional discs. A complex is regular if each attaching map is a homeomorphism. The immediate boundary of a k-cell a in Γ is composed of $(k-1)$-cells incident to a (called faces of a). The set of k-cells in Γ is denoted as Γ_k. The total number of cells in Γ is denoted as n.

2.1   Forman Gradient

A *discrete vector field* V on a regular cell complex Γ is a collection of pairs (a, b), such that

- a is a k-cell, and b is a $(k+1)$-cell of Γ,
- a is a face of b (denoted as $a < b$), and
- each cell in Γ is in at most one pair of V.

Thus, V can be seen as a mapping $V : \Gamma \to \Gamma \cup \{\emptyset\}$. If $(a, b) \in V$, then $V(a) = b$, and (from the third condition of the previous definition) $V(b) = \emptyset$.

A *V-path* is a sequence $a_1, b_1, a_2, b_2, ..., a_{r+1}$ of k-cells a_i and $(k+1)$-cells b_j, $i = 1, .., r+1$, $j = 1, .., r$, such that $(a_i, b_i) \in V$, $b_i > a_{i+1}$, and $a_i \neq a_{i+1}$. V-path $a_1, b_1, ..., a_{r+1}$, $r > 0$, is *closed* if $a_{r+1} = a_1$. Sequence a_1 is a *stationary V-path*.

A discrete vector field V is called a *Forman (discrete) gradient* if and only if there are no closed V-paths in V. A *critical cell* of V of index k is a k-cell c which does not appear in any pair of V. In other words, a cell c is critical if $V(c) = \emptyset$, and $c \notin ImV$. We denote as C the set of critical cells, and as C_k the set of critical k-cells.

In Fig. 1, two Forman gradients V_1 and V_2 are illustrated. The pairing between a k-cell a and a $(k+1)$-cell b is indicated by an arrow starting at a and pointing towards b. Both gradients V_1 and V_2 have two critical vertices (labeled 1 and 2) and one critical edge (labeled c). Gradient V_2 has also one critical face (labeled D) and one critical 3-cell (labeled v).

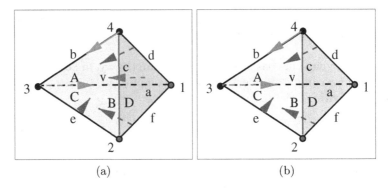

Fig. 1. Forman gradient (a) V_1 and (b) V_2 defined on a complex Γ with one 3-cell, four triangles, six edges and four vertices (a solid tetrahedron). Arrows indicate the pairing between cells (green for vertices and edges, blue for edges and faces, red for faces and 3-cells). (Color figure online)

2.2   Morse Chain Complex

The homology of cell complex Γ with $\mathbb{Z}_2$ coefficients can be computed as the homology of the chain complex with chain groups defined by the k-cells in Γ, and the boundary operator defined for each k-cell a in Γ as the set of all $(k-1)$-cells in its immediate boundary. The homology of Γ is equivalent to the homology of the Morse chain complex $\mathcal{M}$ induced by a Forman gradient V on Γ [5]. The chain groups of $\mathcal{M}$ are defined by the critical cells of V, and the boundary operator $\partial_{\mathcal{M}}$ is defined by the parity of gradient V-paths connecting them: a critical $(k-1)$-cell c is in the boundary $\partial_{\mathcal{M}}(d)$ of critical k-cell d in the Morse chain complex $\mathcal{M}$ if there is an odd number of V-paths connecting some $(k-1)$-cell incident to d in Γ to c, i.e., for $d \in C_k$

$$\partial_{\mathcal{M}}(d) = \sum_{\substack{c \in C_{k-1} \\ \text{there is an odd number of } V\text{-paths} \\ \text{starting at a cell } e \in \Gamma_{k-1}, e<d, \text{ and ending at } c}} c$$

Complex $\mathcal{M}$ has fewer cells than complex Γ, implying that homology computation on $\mathcal{M}$ requires less time than homology computation on Γ, if the boundary operator $\partial_{\mathcal{M}}$ can be computed efficiently. In the next section, we propose an iterative algorithm that computes this boundary operator and the boundary matrices B_k of the Morse chain complex $\mathcal{M}$, not only for Γ, but also for subcomplexes F_i of Γ in a filtration induced by the topological order defined by Forman gradient V.

3   Extraction Algorithm

The input of the algorithm is a regular cell 3-complex Γ in $\mathbb{R}^3$, endowed with a Forman gradient V. The gradient V induces a filtration $\emptyset = F_0 \subset F_1 \subset .. \subset$

$F_M = \Gamma$ of Γ, which is computed in the preprocessing step of the algorithm. In the main loop of the algorithm, the boundary operator ∂_M on the Morse chain complex M is computed iteratively, while traversing a filtration induced by V. In the post processing step, boundary matrices B_k, $1 \leq k \leq 3$, are constructed from boundary operator ∂_M.

Recall that a cell complex is regular if all the attaching maps are homeomorphisms, i.e., if there are no identifications on the boundaries of attached cells. We are interested in computing the boundary operator ∂_M of the associated Morse chain complex M and the boundary matrices B_k, $1 \leq k \leq 3$, with coefficients in $\mathbb{Z}_2$. Thus, there is no need to consider the orientation of cells: the incidence coefficient between incident cells of consecutive dimension in Γ is equal to 1. Most complexes used in shape modeling, computer graphics, or image processing, such as cubical and simplicial complexes, are regular.

3.1   Filtration

A Forman gradient V on a cell complex Γ can be encoded in a directed acyclic graph (DAG) $G = (N, A)$. Each node in N corresponds either to a critical cell of V, or to a pair of cells in V, i.e., $N = \{\{c\} : c \in C\} \cup \{\{a, b\} : (a, b) \in V\}$. There is an arc in A connecting node $m_1 \in N$ to node $m_2 \in N$ if a cell in node m_2 is in the boundary of a cell in node m_1.

The DAG G encodes a partial order on the set N of nodes, which can be extended to a (non-unique) total order, called *topological order* of the DAG [3]. When the nodes in N are sorted in ascending topological order as $m_1 \leq m_2 \leq \ldots \leq m_M$, then no cell in Γ comes before any cell in its boundary.

Each subsequence $m_1, \ldots m_i$ corresponds to a subcomplex F_i of Γ. The topological order induces the filtration $\emptyset = F_0 \subseteq F_1 \subseteq \ldots \subseteq F_M = \Gamma$ of Γ, where each F_i, $1 \leq i \leq M$, is obtained from F_{i-1} by adding to it the cells in m_i. Thus

- $F_i = F_{i-1} \cup \{c\}$, where c is a critical cell of V, or
- $F_i = F_{i-1} \cup \{a, b\}$ where $(a, b) \in V$.

For the Forman gradient V_1 illustrated in Fig. 1(a), one possible topological order is e.g.

$$\{\{1\}, \{2\}, \{3, a\}, \{4, b\}, \{c\}, \{d, A\}, \{e, B\}, \{f, C\}, \{D, v\}\}.$$

For Forman gradient V_2 in Fig. 1(b), one possible topological order is

$$\{\{1\}, \{2\}, \{3, a\}, \{4, b\}, \{c\}, \{d, A\}, \{e, B\}, \{f, C\}, \{D\}, \{v\}\}.$$

The corresponding filtrations of complex Γ are illustrated in Fig. 2.

3.2   Boundary Operator

For each critical edge c, $\partial_M(c)$ is either empty, or it consists of two distinct critical vertices. As the gradient lines connecting critical vertices and edges never split, they can be extracted by tracing the Forman gradient V starting from the

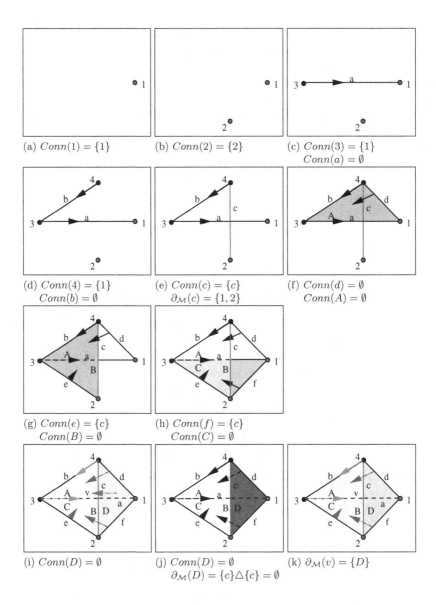

Fig. 2. (a)–(h) The common subcomplexes in the filtrations induced by the topological order of Forman gradients V_1 and V_2 illustrated in Fig. 1, and the updates of sets $Conn$ and $\partial_{\mathcal{M}}$ performed by the extraction algorithm. (i) The final complex obtained by adding the paired face D and 3-cell v to the complex illustrated in (h) and the last step of the extraction algorithm for Forman gradient V_1 illustrated in Fig. 1(a). (j) and (k) The final complex obtained by adding critical face D and critical 3-cell v to the complex in (h) and the last steps of the extraction algorithm for Forman gradient V_2 illustrated in Fig. 1(b)

endpoints of each critical edge c until critical vertices are reached. If the two reached critical vertices are distinct, then there is a unique gradient path from c to each of them, and they both belong to $\partial_{\mathcal{M}}(c)$. Otherwise, if the same critical vertex is reached from both endpoints of c, then it is reached through two distinct gradient paths from c, and $\partial_{\mathcal{M}}(c)$ is empty.

Dually, gradient lines connecting critical 3-cells and faces never merge, and can be extracted by backtracking V starting from critical faces until critical 3-cells are reached. Each critical face d belongs to $\partial_{\mathcal{M}}(v)$ of two distinct critical 3-cells, or it does not belong to $\partial_{\mathcal{M}}(v)$ for any critical 3-cell v, depending on whether the two reached critical 3-cells are distinct or the same, respectively. Thus, boundary operator for critical edges and critical 3-cells (and boundary matrices B_1 and B_3) can be computed directly from V in a straightforward fashion. For completeness, we will include their computation in the algorithm through the same technique used for the computation of boundary operator for faces (and boundary matrix B_2).

The interesting and challenging part of the algorithm is the extraction of boundary operator for critical faces in the Morse chain complex, and we describe this part of the algorithm in greater detail. The algorithm is iterative. It traverses the cells of the complex in ascending order determined by the Forman gradient V and the induced filtration F and updates two sets ($Conn$ and $\partial_{\mathcal{M}}$) associated with relevant edges and faces.

The edges that contribute to boundary operator for faces are critical edges and edges that are paired with a face, while the edges paired with a vertex do not contribute to it. The faces that contribute to boundary operator for critical faces are critical faces of V, and those that are paired with an edge. The latter will be processed at the same step of the algorithm as the edge they are paired with. The algorithm stores for each current reached edge a the set $Conn(a)$ of all critical edges c that can be reached from a following the Forman gradient V, through an odd number of gradient V-paths.

If a is a critical edge, then the only critical edge that can be reached from a is a itself, through a unique stationary path of length 0 ($Conn(a) = \{a\}$). This unique path from a to a consists of a only.

If edge a is paired with some face b in V, then each V-path starting at a is of the form $a, b, e, ...$, where e is an edge incident to face b in Γ. The only critical edges that can be reached from a are those that can be reached from some of the edges e. In other words, the set of all gradient V-paths that start at edge a and connect edge a to some critical edge can be obtained by adding edge a and face b at the beginning of each gradient V-path that starts at some edge e incident to face b in Γ and ends at some critical edge. Such edges e are those that are not paired with a vertex in V: they are either critical edges, or edges that are paired with some face in V. The information contained in the edges e in the boundary of b is propagated to edge a. If a critical edge c cannot be reached through a V-path from some edge e incident to face b in Γ, then it cannot be reached from edge a, and it does not belong to $Conn(a)$. If c can be reached through a V-path from some edge e incident to b in Γ, then the total number of V-paths from a to c that pass through e is equal to the total number of V-paths from e to c. This is due to the fact that there is exactly one V-path from a to e: it is the path a, b, e. The total number of paths from a to c (mod 2), i.e., the parity of the number of such paths, is equal to the sum (mod 2) of the number of paths from some edge e incident to face b in Γ to c. The sum is taken over all edges e. Thus, the set $Conn(a)$ of all critical edges that can be reached from edge a through an odd number of V-paths can be obtained as the symmetric difference of sets $Conn(e)$ over all edges e incident to face b in Γ.

When a critical face d is reached by the algorithm, critical edges c in the sets $Conn(e)$ associated with the edges e incident to critical face d in Γ are used to compute the boundary operator $\partial_{\mathcal{M}}(d)$. With the similar reasoning as above, we conclude that $\partial_{\mathcal{M}}(d)$ can be obtained as the symmetric difference of sets $Conn(e)$ over all edges e incident to d in Γ.

We give a more formal pseudo-code-like description of the algorithm, and we illustrate its steps in Fig. 2. At step i, i.e., when complex F_i is reached in the filtration F, the following actions are performed depending on $D_i = F_i - F_{i-1}$:

$\mathbf{D_i = \{c\}, c \in C_0}$

– set $Conn(c) = \{c\}$

For example, after the addition of critical vertex 1 to empty complex F_0, the set $Conn(1)$ of critical vertices that are connected to critical vertex 1 through an odd number of gradient V-paths contains only vertex 1 (see Fig. 2(a)), and similarly for critical vertex 2 (Fig. 2(b)).

$\mathbf{D_i = \{a, b\}, (a, b) \in V, a \in \Gamma_0, b \in \Gamma_1}$

– set $Conn(a) = Conn(a_1)$, where a_1 is the other endpoint of edge b
– set $Conn(b) = \emptyset$

For example, when vertex 3 and edge a (that are paired in V) are reached and added to the complex, the set $Conn(3)$ of critical vertices connected to vertex 3 contains critical vertex 1 (see Fig. 2(c)). Similarly, when vertex 4 and edge b are added, the set $Conn(4)$ contains critical vertex 1 (Fig. 2(d)). The sets $Conn(a)$ and $Conn(b)$ are empty.

$\mathbf{D_i = \{c\}, c \in C_1}$

– set $Conn(c) = \{c\}$, $\partial_{\mathcal{M}}(c) = \emptyset$
– for each of the two vertices v incident to c in Γ do $\partial_{\mathcal{M}}(c) = \partial_{\mathcal{M}}(c) \triangle Conn(v)$

For example, the two vertices incident to critical edge c in Γ are 2 and 4. Since $Conn(2) = \{2\}$, $Conn(4) = \{1\}$ and $1 \neq 2$, the boundary $\partial_{\mathcal{M}}(c)$ of c contains critical vertices 1 and 2. The set $Conn(c)$ of critical edges that are connected to c contains only edge c (see Fig. 2(e)).

$\mathbf{D_i = \{a, b\}, (a, b) \in V, a \in \Gamma_1, b \in \Gamma_2}$

– set $Conn(a) = \emptyset$
– for each edge e incident to face b in Γ do $Conn(a) = Conn(a) \triangle Conn(e)$
– set $Conn(b) = \emptyset$

For example, there are no critical edges in the set $Conn(d)$ of edge d paired with face A in V, because the other two edges a and b incident to face A in Γ are paired with a vertex in V: no critical edge can be reached from edge d through V (see Fig. 2(f)). The set $Conn(e)$ for edge e paired with face B in V contains critical edge c, because c is incident to face B in Γ, and the remaining edge b incident to A is paired with a vertex (Fig. 2(g)). The set $Conn(f)$ for edge f paired with face C contains critical edge c, because the other two edges incident to face C in Γ are a and e. Edge a is paired with a vertex $(Conn(a) = \emptyset)$ and $Conn(e) = \{c\}$ (Fig. 2(h)).

$$\mathbf{D_i} = \{\mathbf{a, b}\}, (\mathbf{a, b}) \in \mathbf{V}, \mathbf{a} \in \mathbf{\Gamma_2}, \mathbf{b} \in \mathbf{\Gamma_3}$$

– set $Conn(a) = \emptyset$
– for each face f incident to 3-cell b in Γ do $Conn(a) = Conn(a) \triangle Conn(f)$

For example, $Conn(D) = \emptyset$, because each of the remaining faces A, B and C incident to 3-cell v in Γ is paired with an edge, and hence no critical face is connected to any of them through V_1 (see Fig. 2(i)).

$$\mathbf{D_i} = \{\mathbf{d}\}, \mathbf{d} \in \mathbf{C_2}$$

– set $Conn(d) = \{d\}$
– set $\partial_{\mathcal{M}}(d) = \emptyset$
– for each edge e incident to d in Γ do $\partial_{\mathcal{M}}(d) = \partial_{\mathcal{M}}(d) \triangle Conn(e)$

For example, $\partial_{\mathcal{M}}(D) = \emptyset$, because there are two gradient paths starting at an edge incident to D in Γ and ending at c: one starts at edge f, and the other at edge c (see Fig. 2(j)).

$$\mathbf{D_i} = \{\mathbf{v}\}, \mathbf{c} \in \mathbf{C_3}$$

– set $\partial_{\mathcal{M}}(v) = \emptyset$
– for each face f incident to v in Γ do $\partial_{\mathcal{M}}(v) = \partial_{\mathcal{M}}(v) \triangle Conn(f)$

For example, $\partial_{\mathcal{M}}(v) = \{D\}$, since there is one gradient path from a face incident to 3-cell v in Γ to critical face D: it is the stationary path D (see Fig. 2(k)).

3.3   Boundary Matrices

There is a 1-1 correspondence between rows in B_k and critical $(k-1)$-cells of V, and between columns of B_k and critical k-cells of V. Boundary matrices are computed from the boundary operator in a straightforward manner.

For the 2-complex Γ_1 and the Forman gradient illustrated in Fig. 2(i), the computed boundary matrices are

$$B_1 = \begin{bmatrix} 1 \\ 1 \end{bmatrix} \text{ and } B_2 = \begin{bmatrix} 0 \end{bmatrix}.$$

The rows of matrix B_1 correspond to critical vertices 1 and 2, respectively, and the column corresponds to critical edge c.

The row of matrix B_2 corresponds to critical edge c and the column corresponds to critical face D.

For the 3-complex Γ and Forman gradient V_1 illustrated in Fig. 1(a) and in Fig. 2(j), the (only nontrivial) boundary matrix is

$$B_1 = \begin{bmatrix} 1 \\ 1 \end{bmatrix}.$$

For the 3-complex Γ and Forman gradient V_2 in Figs. 1(b) and 2(k), the boundary matrices are

$$B_1 = \begin{bmatrix} 1 \\ 1 \end{bmatrix} \quad B_2 = \begin{bmatrix} 0 \end{bmatrix} \quad \text{and} \quad B_3 = \begin{bmatrix} 1 \end{bmatrix}.$$

The row of matrix B_3 corresponds to critical face D and the column corresponds to critical 3-cell v.

3.4   Analysis

The preprocessing step of the proposed algorithm finds a topological order in the DAG $G = (N, A)$ induced by Forman gradient V on the complex Γ. The number $|N|$ of nodes in N is in $O(n)$, where n is the total number of cells in Γ. The number $|A|$ of arcs in A in a DAG is at most $|N| \cdot |N - 1|/2$. Thus, $|A|$ is in $O(n^2)$. Kahn's algorithm finds a topological order of the nodes in N in time $O(|N| + |A|) = O(n^2)$ [3].

If Γ is a cubical complex, then each k-cell has a constant number of $(k - 1)$-cells in its immediate boundary (six for 3-cells, four for edges and two for edges). Thus, the number $|A|$ of arcs in A is in $O(n)$, and the preprocessing step takes $O(n)$ time in the worst case.

Proposition 1. *The proposed algorithm correctly extracts the boundary operator $\partial_{\mathcal{M}}$ and boundary matrices B_k from the Forman gradient V on a regular 3D cell complex Γ.*

Proof. We need to show that the extracted boundary operator $\partial_{\mathcal{M}}$ is correct and does not depend on the filtration order. The algorithm maintains the following invariant: if the sets $Conn$ and $\partial_{\mathcal{M}}$ are correct for complex F_{i-1}, then the application of the corresponding step of the algorithm produces the correct sets $Conn$ and $\partial_{\mathcal{M}}$ for complex F_i, $1 \leq i \leq n$. This follows from the discussion in Sect. 3.2, and the fact that the initial complex F_0 is empty. Thus, for each sub complex F_i, the algorithm computes the correct sets $Conn$ and $\partial_{\mathcal{M}}$. The last complex in every filtration induced by some topological order is Γ, implying that the output of the algorithm is correct and independent of the filtration order.

Proposition 2. *The time cost of the extraction algorithm is in $O(nhc)$, where h is the maximum cardinality of the set of cells forming the immediate boundary of cells in Γ and c is the total number of critical cells of V.*

Proof. The algorithm iterates over the filtration induced by V, and the number of complexes F_i in the filtration is in $O(n)$. The time cost for each step of the algorithm depends on the set D_i, and can be broken in two parts. The first consists of initialization of sets $Conn$ and/or $\partial_{\mathcal{M}}$, which can be done in constant time. The second part is due to loop through $O(h)$ $(k-1)$-cells incident to the processed k-cell (critical k-cell c if $D_i = \{c\}$, $c \in C_k$, or higher-dimensional cell b if $D_i = \{(a,b)\}$, $(a,b) \in V$), and the computation of symmetric difference of $O(h)$ sets each containing $O(c)$ elements.

If Γ is a cubical 3-complex, then h is constant $(h = 6)$, and the extraction algorithm runs in time $O(nc) = O(n^2)$ (the total number of critical cells of V may be linear in the total number of cells in Γ).

The alternative algorithms for the extraction of Morse chain complex with $\mathbb{Z}_2$ coefficients have been proposed in [6,13]. Both algorithms first construct a Forman gradient on a (cubical) cell 3-complex Γ, and then the Morse chain complex induced by it.

For each critical k-cell d of V, the algorithm in [13] follows all gradient paths that start at d using a breadth first search, and counts those that connect d to another critical $(k-1)$-cell c. First, the $(k-1)$-cells incident to d in Γ that are paired with some k-cell in V are enqueued. Then, for each $(k-1)$ cell a in the queue that is paired with a k-cell b in V, each non-critical $(k-1)$-cell e that is incident to b in Γ and that is paired with a k-cell in V is enqueued and subsequently processed by the algorithm. The gradient paths connecting edges and faces may (branch and) merge, causing the possible multiple traversal of cells: when processing a critical k-cell d, each $(k-1)$-cell e that can be reached from d through a V-path may be enqueued and processed multiple times.

The algorithm in [6] improves on the previous one by not allowing this multiple traversal. It first extracts all $(k-1)$-cells that can be reached from a critical k-cell d by traversing Forman gradient V and deleting the visited cells, thus preventing the multiple traversal of cells. Then, from each critical $(k-1)$ cell c that can be reached from d, V-paths connecting d and c are backtracked and their number is counted (mod 2). The reported computational complexity of algorithms in [13] and [6] is in $O(n^2)$ and $O(cn)$, respectively.

Both algorithms in [13] and [6] compute the boundary operator $\partial_{\mathcal{M}}$ and boundary matrices for the given cubical 3-complex Γ with the Forman gradient V. Unlike ours, these algorithms do not adapt straightforwardly to the computation of the same boundary information for all intermediate complexes in a filtration induced by V.

4   Conclusions

We have presented an iterative algorithm that extracts the boundary operator $\partial_{\mathcal{M}}$ and boundary matrices B_k, $k = 1, 2, 3$ for homology computation over $\mathbb{Z}_2$ of the Morse chain complex $\mathcal{M}$ of a regular 3D cell complex Γ endowed with a Forman gradient V. The algorithm progresses through a filtration of Γ induced by V, and computes this data not only of the Morse chain complex of Γ but also of each of the subcomplexes in the filtration of Γ.

Our present work includes the extension of the algorithm presented here to computation of boundary operator and boundary matrices with coefficients in $\mathbb{Z}$ for cell complexes in arbitrary dimension. We are also developing a specialization of the extraction algorithm to cubical complexes. The structure of cubical complexes allows for implicit encoding of its cells, which can be accessed through their combinatorial coordinates [10]. We will utilize this encoding for efficient implementation of the extraction algorithm. We plan to investigate the computation of persistent homology of Γ using the extracted boundary matrices.

References

1. Cazals, F., Chazal, F., Lewiner, T.: Molecular shape analysis based upon the Morse-Smale complex and the Connolly function. In: Proceedings of the Nineteenth Annual Symposium on Computational Geometry, pp. 351–360 (2003)
2. Čomić, L., Mesmoudi, M.M., De Floriani, L.: Smale-like decomposition and Forman theory for discrete scalar fields. In: Debled-Rennesson, I., Domenjoud, E., Kerautret, B., Even, P. (eds.) DGCI 2011. LNCS, vol. 6607, pp. 477–488. Springer, Heidelberg (2011)
3. Cormen, T., Leiserson, C., Rivest, R.: Introduction to Algorithms. MIT Press, Cambridge (1990)
4. Edelsbrunner, H., Letscher, D., Zomorodian, A.: Topological persistence and simplification. Discrete Comput. Geom. **28**(4), 511–533 (2002)
5. Forman, R.: Morse theory for cell complexes. Adv. Math. **134**, 90–145 (1998)
6. Günther, D., Reininghaus, J., Wagner, H., Hotz, I.: Efficient computation of 3D Morse-Smale complexes and persistent homology using discrete Morse theory. Vis. Comput. **28**(10), 959–969 (2012)
7. Gyulassy, A., Bremer, P.-T., Hamann, B., Pascucci, V.: A practical approach to Morse-Smale complex computation: scalability and generality. IEEE Trans. Vis. Comput. Graph. **14**(6), 1619–1626 (2008)
8. Gyulassy, A., Natarajan, V., Pascucci, V., Hamann, B.: Efficient computation of Morse-Smale complexes for three-dimensional scalar functions. IEEE Trans. Vis. Comput. Graph. **13**(6), 1440–1447 (2007)
9. King, H., Knudson, K., Mramor, N.: Generating discrete Morse functions from point data. Exp. Math. **14**(4), 435–444 (2005)
10. Kovalevsky, V.A.: Geometry of Locally Finite Spaces (Computer Agreeable Topology and Algorithms for Computer Imagery). Editing House Dr. Bärbel Kovalevski, Berlin (2008)
11. Lewiner, T., Lopes, H., Tavares, G.: Applications of Forman's discrete Morse theory to topology visualization and mesh compression. Trans. Vis. Comput. Graph. **10**(5), 499–508 (2004)
12. Munkres, J.R.: Elements of Algebraic Topology. Addison-Wesley, Redwood City (1984)
13. Robins, V., Wood, P.J., Sheppard, A.P.: Theory and algorithms for constructing discrete Morse complexes from grayscale digital images. IEEE Trans. Pattern Anal. Mach. Intell. **33**(8), 1646–1658 (2011)

Parallel Homology Computation of Meshes

Guillaume Damiand[1](✉) and Rocio Gonzalez-Diaz[2]

[1] Univ Lyon, CNRS, LIRIS, UMR5205, F-69622 Lyon, France
guillaume.damiand@liris.cnrs.fr
[2] Dpto. de Matemática Aplicada I, Universidad de Sevilla, 41012 Sevilla, Spain
rogodiuses@gmail.com

Abstract. In this paper, we propose a method to compute, in parallel, the homology groups of closed meshes (i.e., orientable 2D manifolds without boundary) represented by combinatorial maps. Our experiments illustrate the interest of our approach which is really fast on big meshes and which obtains good speed-up when increasing the number of threads.

Keywords: Homology groups computation · 2D combinatorial maps · Parallel algorithm

1 Introduction

With the rapid increase of the amount of data produced in recent decades, availability of efficient tools to analyze these data is of great importance. Homology computation is a basic tool that helps to identify connected components, holes and voids in the given data. Nevertheless, the design of efficient algorithms for computing homology on large data is still a challenging task nowadays.

In [11], two parallel algorithms to compute homology of large simplicial complexes on multicore machines and GPUs were presented. The given complex is decomposed into different partitions. A map from simplices to their boundaries and coboundaries is constructed. This step takes up the highest percentage of the total execution time. Then, parallel algebraic reductions of cells [8] that reduce the size of the chain complex while maintaining its homology are performed. The modification of boundaries and coboundaries is also time-consuming. Finally, reduced chain complexes are merged together and algebraic reductions are then performed sequentially to compute the homology of the input complex.

In this paper, we use 2D combinatorial maps to represent meshes (i.e., orientable 2D manifolds) avoiding the time-consuming step of constructing and modifying boundaries and coboundaries of cells. Besides, instead of using algebraic structures (i.e., chain complexes) our data structures are always combinatorial maps. The principle of our method is to merge, in parallel, the faces of the mesh while the topology is preserved. To achieve this goal, faces of the object are dispatched in clusters. Edges shared by two faces in the same cluster can be processed safely in parallel. The edges shared by two faces in different clusters are then processed sequentially. In order to test quickly (in almost constant

© Springer International Publishing Switzerland 2016
A. Bac and J.-L. Mari (Eds.): CTIC 2016, LNCS 9667, pp. 53–64, 2016.
DOI: 10.1007/978-3-319-39441-1_6

time) if two adjacent faces can be merged, union find trees are used. At the end of the process, a minimal 2D combinatorial map gives a direct representation of homology generators of the given mesh.

Section 2 recalls the related material regarding combinatorial maps and homology computation on these maps. Section 3 proposes a parallel algorithm to compute a minimal combinatorial map from which we can directly obtain homology groups and generators. Some experimental and computational results are presented in Sect. 4. Finally, we summarize the paper with a brief discussion in Sect. 5.

2 Preliminary Notions

2.1 2D Combinatorial Maps

A 2D combinatorial map [6,9], called 2-map, is a model of representation of a mesh, which is composed by *i-cells*: *vertices* or 0-cells associated with points, *edges* or 1-cells which link two vertices, and *faces* or 2-cells which are bounded by cycle of edges. Two cells are *incident* if one cell belongs to the boundary of the other one; while two *i*-cells c_1 and c_2 are *adjacent* if it exists one $(i-1)$-cell incident to both c_1 and c_2. Note that, in the meshes considered here, it is possible to mix different types of faces. For example, we can find triangles and quadrangles in a same mesh. The mesh shown in Fig. 1a, has 5 faces, 14 edges and 12 vertices. The 2-map shown in Fig. 1b describing such mesh has 20 darts.

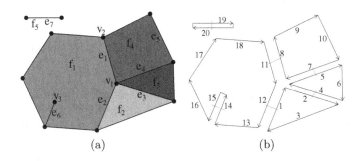

(a)   (b)

Fig. 1. (a) Example of a mesh. (b) The corresponding 2-map.

An edge e is *dangling* if it is adjacent to only one edge (like edge e_6 in Fig. 1a). In this case, one vertex incident to e is incident to no other edge than e. An edge e' is *isolated* if it has no adjacent edge (like edge e_7 in Fig. 1a). Note that an isolated edge is a special case of a dangling edge. In this case, the two vertices incident to e' are incident to no other edge than e'. An edge e'' incident to two different faces is called *inner* (like edge e_1 in Fig. 1a). Such an edge is necessarily not dangling, nor isolated. Reciprocally, a dangling or isolated edge is necessarily incident to only one face.

In a 2-map, the different elements of a mesh are encoded by *darts* and links between these darts. A dart is an orientation of an edge. If an edge separates two faces, it is described by two darts in the 2-map which represent the two possible orientations of the edge. A dart belongs exactly to one vertex, one edge and one face, and thus each cell of the mesh is described by a set of darts in the 2-map. For example, in Fig. 1b, we can see the 2-map which represents the mesh given in Fig. 1a. This 2-map has 20 darts drawn in Fig. 1b as oriented numbered segments. Vertex v_1 of the mesh is described by the set of darts $\{2, 5, 8, 12\}$; edge e_1 by $\{8, 11\}$, and face f_1 by $\{11, 12, 13, 14, 15, 16, 17, 18\}$.

Note that a 2-map is equivalent to the well-known half-edge data structure [10,12]. The main interest of combinatorial maps is that they can be defined in any dimension which allows us to envisage to extend this work in higher dimension.

A 2-map represents the topological part of the mesh, i.e. its subdivision into cells and all the incidence and adjacency relations between these cells. It is possible to associate information to cells thanks to the notion of *attribute*. We speak about i-attributes for information associated to i-cells (for example colors to faces, or mutexes to vertices).

Several operations are defined on 2-maps in order to build, traverse and update meshes. Among these operations, *edge removal* and *edge contraction* will be used in this paper in order to compute the homology of a mesh. The removal of an inner edge (like edge e_1 in Fig. 1a) consists in removing the edge from the mesh by merging its two incident faces. In Fig. 1a, removing edge e_1 will merge faces f_1 and f_4 in only one face with 9 edges. Removing a dangling edge (like edge e_6 in Fig. 1a) will also remove the vertex incident only to the dangling edge (this avoids isolated vertex). In Fig. 1a, removing edge e_6 will remove also vertex v_3. Lastly removing an isolated edge will also remove its two incident vertices (like edge e_7 in Fig. 1a) and its incident face (because this face was described by only this isolated edge and its two incident vertices). Edge contraction consists in contracting an edge by merging its two incident vertices (when they exist). In Fig. 1a, contracting edge e_1 will merge vertices v_1 and v_2 in only one vertex, then face f_1 becomes a pentagon with a dangling edge and f_4 a triangle.

2.2   Homology Computation

Homology can be thought as a method for defining k-dimensional holes (connected components, tunnels, voids) in a given mesh. We can think in a cycle as a closed submanifold and a boundary as the boundary of a submanifold. Then, a homology class (which represents a hole) is an equivalence class of cycles modulo boundaries. Homology groups (i.e. the groups of homology classes) are defined from an algebraic structure called chain complex composed by a set of groups $\{C_k\}$, where each C_k is the group of k-chains generated by all the k-cells, and a set of homomorphisms $\{\partial_k : C_k \to C_{k-1}\}$, called boundary operator, describing the boundaries of k-chains. This way, a k-dimensional homology generator (called Hk generator) is a k-cycle which is not the boundary of any $(k + 1)$-chain. Thus, in principle, the manipulation of the boundary operator is needed to compute homology.

An algorithm allowing to compute the minimal 2-map describing an initial mesh without boundary while preserving the homology is described in [7]. Its principles consist first in removing all the inner edges: this merges all the faces of the mesh in only one face. The second step removes all the dangling edges: a dangling edge is removed when it is found, and the possible path of dangling edges is followed in order to ensure that all the dangling edges are removed. For computing $H0$ generators, we keep in a stack one vertex for each isolated edge before to remove it from the 2-map. The last step of the algorithm consists in contracting all non-loop edges. This allows us to obtain the minimal representation of the initial mesh with 1 vertex for each connected component which is not a sphere, plus some loops incident to these vertices depending on the amount of H1 generators of the corresponding mesh. Observe that we do not need to compute or store the boundary operator for computing the homology of the mesh. The resulting minimal 2-map gives a direct representation of homology generators. $H0$ generators are directly obtained by the set of vertices in the stack plus the set of vertices in the 2-map; $H1$ generators are directly obtained by the set of edges in the 2-map which are all loops and $H2$ generators are directly obtained by the set of faces in the initial 2-map.

3   Parallel Algorithm

The goal of this section is to propose a parallel version of the previous algorithm allowing to compute the minimal 2-map corresponding to a given mesh. When writing a parallel algorithm, the main issue is the concurrent access of the shared memory. In order to allow efficient parallel access, we use here a *distributed version* of 2-maps.

A distributed 2-map M is a 2-map where darts are distributed in several *clusters*, each face of the 2-map belonging to exactly one cluster. To describe that a face belongs to a cluster, we associate all darts of this face with this cluster. In this way, clusters are sets of darts which form a partition of the set of all the darts, and which can also be seen as a partition of the set of all the faces of the 2-map. An edge of M is said *critical* if it belongs to different clusters[1], otherwise it is said *non-critical*.

The main interest of a distributed 2-map is to allow to process in parallel the different clusters since they are independent. However, it allows only to process in parallel all the non-critical edges. This property is used in Algorithm 1, the algorithm which is the parallel simplification of a given 2-map. This algorithm has four steps:

- step *1a* and step *1b* to remove the inner edges; *1a* in parallel for non-critical edges (by cluster); *1b* sequentially for the whole set of critical edges of the 2-map;
- step *2* to remove dangling edges in parallel (for the global 2-map);
- step *3* to contract non loop edges sequentially (for the global 2-map).

---

[1] Since an edge e in a 2-map without boundary is a set of two darts, e belongs to different clusters if its two darts belong to two different clusters.

Algorithm 1. Parallel simplification of a closed mesh in its minimal representation.

Input: A distributed 2-map M representing a closed mesh.
Output: The minimal 2-map corresponding to M, computed in parallel.

1a **parallel for each** *non-critical edge e of* M **(by cluster) do**
  $\quad$ **if** *e is an inner edge* **then** remove e;

1b **foreach** *critical edge e of* M **(sequentially for the global 2-map) do**
  $\quad$ **if** *e is an inner edge* **then** remove e;

2 **parallel for each** *edge e of* M **(for the global 2-map) do**
  $\quad$ **while** *e is dangling* **do**
  $\quad\quad$ **if** *e is isolated* **then** push one vertex of e in S; Remove e;
  $\quad\quad$ **else**
  $\quad\quad\quad$ $e' \leftarrow$ one edge adjacent to e;
  $\quad\quad\quad$ lock the vertex v incident to e and e';
  $\quad\quad\quad$ remove e; $\quad e \leftarrow e'$;
  $\quad\quad\quad$ unlock v;

3 **foreach** *edge e of* M **(sequentially for the global 2-map) do**
  $\quad$ **if** *e is not a loop* **then**
  $\quad\quad$ contract e;

3.1   Inner Edge Removals

The step which does the inner edge removals is split in two parts (for non-critical
edges and critical edges). Indeed, it is not possible to process in parallel the critical
edge as illustrated in Fig. 2. In this example, a mesh representing a torus is simplified (cf. Fig. 2a). This 2-map has two different clusters (darts in the first cluster
are drawn in red and darts in the second cluster in blue). After the removal of the
three inner edges $e1$, $e2$ and $e4$, the 2-map shown in Fig. 2b is obtained. If edges $e3$

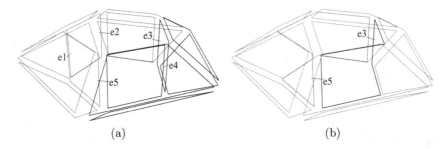

(a)   (b)

Fig. 2. Example of 2-map simplification illustrating why it is not possible to process
critical edges in parallel for inner edge removal step. (a) Initial 2-map representing a
torus. (b) 2-map obtained after the removal of the three inner edges of the initial 2-map
$e1$, $e2$ and $e4$ (arrows on darts are not drawn in this figure). (Color figure online)

and $e5$ (which are criticals) are considered in parallel, they are both inner edges and thus they will be both removed. This would result in a wrong 2-map because the face resulting of the union of the 5 initial faces is not homeomorphic to a disk (here this is an annulus). Considering critical edges sequentially solves this problem: if $e3$ is considered, it is an inner edge and it is thus removed. Then when $e5$ will be considered, it will be no more an inner edge (because it will be incident twice to the same face) and thus it will be kept.

Note that this problem can not occur for non-critical edges (i.e. edges between two faces belonging to the same cluster). Indeed, in this case the two faces can be merged by only one thread, the one which processes this cluster. This justifies why we can process non-critical edges in parallel in step *1a* of Algorithm 1.

For these two first steps, we avoid all the possible concurrent accesses because:

– darts of different clusters are distributed in different containers in the 2-map. The **parallel for each** is done by associating one thread to each cluster, which allows to remove these darts in parallel during step *1a*;
– critical edges (having two darts belonging to two different clusters) are processed sequentially.

3.2   Dangling Edge Removals

The same solution (process first non-critical edges in parallel then critical edges sequentially) can not be used for dangling edges, as illustrated in Fig. 3. This example shows the result of the two first steps of our simplification algorithm (steps *1a* and *1b*) on the mesh representing a torus (given in Fig. 2a). We have a 2-map having only one face (otherwise we would have some inner edges), some edges belong to H1 generators, i.e. to a cycle (in purple in Fig. 3b) but other edges do not (in black in Fig. 3b). The black edges need to be removed since they do not belong to H1 generators. This is done by the dangling edge removal step. In this example, all the edges 45, 78, gh, no and de are dangling and will be removed. However edges 69 and $3a$ are not yet dangling: they will become dangling after the removal of other dangling edges.

Two problems can arise if we want to process dangling edges in parallel cluster by cluster:

1. if two threads process edges in parallel (a first thread for red edges, a second for blue edges), we can consider simultaneously edges 45 and 69 (if the thread that processes red edges has already removed edge 78). If the two edges are removed simultaneously, the two threads will follow the path of dangling edges and they both will try to remove the same edge $3a$;
2. to solve this problem, we can restrict each thread to process only the edges in its cluster. In this case, if edge 45 is considered by the blue thread before edge 69 by the red thread, the blue thread will not remove edge $3a$ because it is not yet dangling.

Our solution, given in Algorithm 1 step *2* is to process all edges in parallel without taking into account the clusters. When a thread founds a dangling edge e,

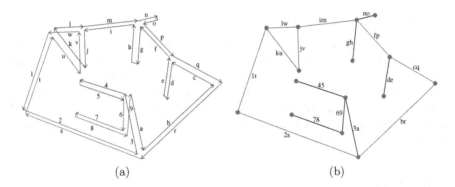

Fig. 3. Example of dangling edge removal. (a) 2-map obtained from the initial 2-map given in Fig. 2a at the end of steps *1a* and *1b*. All inner edges were removed. Darts are labeled by numbers and letters. (b) Cellular representation representing by a graph all the vertices and edges of (a). Each edge of the graph is labeled by the two labels of its two darts in (a). Purple edges belong to H1 cycles while black edges not.

it removes e and follows the possible path of dangling edges starting from e. In order to avoid two threads to follow the same path, it is enough to add a mutex on each vertex of the 2-map, and to lock the mutex of the vertex which is not only incident to the dangling edge before removing the edge.

Now let us reconsider the previous example (given in 3) where two threads are processing edges 45 and 69 simultaneously. Only one thread will be able to lock the vertex $\{4, 6, a\}$; let us suppose it is the blue thread. This thread will remove edge 45, and then it will stop to follow the path of edges since the next edge in the path is not yet dangling. Then the second thread will now be able to take the mutex and will remove edge 69, but then the next edge $3a$ is dangling and it will be processed by this thread.

These mutexes solve the problem of concurrent access while guaranteeing that each edge that does not belong to an H1 generator will be removed. Indeed, these edges either belong to only one path of dangling edges (like edge 69 in the example) and these edges will be removed when the extremity of the path will be considered; or these edges belong to several paths (like edge $3a$ in the example) and these edges will be considered exactly once by the last thread that will take the mutexes of the vertices incident to the edges.

Lastly, darts are not directly deleted from the 2-map during this step because the two darts of the current edge can belong to two different clusters, and two threads can want to delete simultaneously two darts belonging to the same cluster. Instead of adding mutexes on the containers of darts, we do not delete the darts when dangling edges are removed but we only mark the darts to delete thanks to a specific Boolean mark. Then marked darts are deleted after having finished to process all dangling edges. These deletions can be done in parallel by iterating simultaneously through all the different containers.

3.3 Non Loop Edge Contraction

The last step of Algorithm 1 is the contraction of all the non-loop edges (an edge is a loop if it is incident twice to the same vertex). As we will see in our experiments in the next section, since this step is very fast (because the number of edges to contract is very small regarding the number of edges to remove), we decide in this work not to parallelize this step. Thus we use the classical algorithm which consists in iterating through each edge e, test if e is not a loop and contract it when it is the case.

4 Experiments

We have implemented our two algorithms, sequential and parallel version, by using the CGAL implementation of combinatorial maps [2] and the additional layer, called linear cell complex, which additionally represents the geometry [3]. All our experiments were run on an Intel®i7-4790 CPU, 4 cores @ 3.60 GHz with 32 Go RAM. We have tested our algorithms by comparing the sequential and the parallel version, and compared also our sequential version with RedHom [1][2].

For our tests, we used the six meshes shown in Fig. 4, having between 703.512 and 10.000.000 faces. All these meshes have only one connected component, except *Blade* which has 295 connected components because it contains many small isolated closed meshes inside the blade. For the parallel algorithm, clusters are built during the load of the off files. We divide the bounding box of the mesh in $8 \times 8 \times 8$ cubes, and fix the cluster of a face of the mesh depending on the cube which contains the minimal point of the face (these clusters are represented by different colors in Fig. 4).

Firstly we have compared our sequential algorithm and the sequential one implemented in *RedHom*. As we can see in Fig. 5a, our method is much faster than *RedHom*: 22 times faster in average, the minimum gain is $3, 3$ times for DrumDancer, and the maximum gain is 97 times for Blade. We observe that *RedHom* requires more time for meshes with big genus and/or big number of connected components while the complexity of our method is only related to the number of cells and not to the genus nor to the number of connected components.

Secondly we have compared our two algorithms: the sequential and the parallel one. For the parallel version, we used eight threads because we ran the tests in a four core CPU, but with hyper-threading enabled. As we can observe in Fig. 5b, the parallel version is in average $2, 4$ times faster than the sequential one, the minimum gain is $1, 7$ times for Blade and the maximum gain is 3 times for Neptune. We observe that the gain is better for bigger time of the sequential algorithm, while this gain becomes small for very small time. However, this result is interesting because it is more important to speed up the bigger times than the smaller ones.

---

[2] We did not compare our solution with [11] because we were not able to compile their parallel version.

	#0-cells	#1-cells	#2-cells	#H0	#H1	#H2
(a) Blade	882.954	2.648.082	1.765.388	295	330	295
(b) DrumDancer	1.335.436	4.006.302	2.670.868	1	0	1
(c) Neptune	2.003.932	6.011.808	4.007.872	1	6	1
(d) HappyBuddha	543.652	1.631.574	1.087.716	1	208	1
(e) Iphigenia	351.750	1.055.268	703.512	1	8	1
(f) ThaiStatue	4.999.996	15.000.000	10.000.000	1	6	1

Fig. 4. The six meshes used in our experiments. Colors show the different clusters. Black curves show the H1 generators computed by our algorithm. The table gives the number of i-cells, #i-cells, and the number of Hi generators, #Hi, for $i = 0, 1, 2$. (Color figure online)

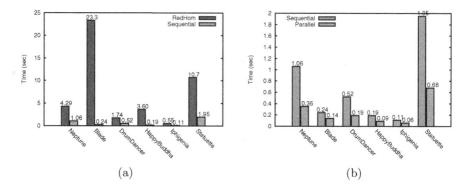

(a)   (b)

Fig. 5. (a) Comparison of the times for the homology computation between RedHom and our sequential version. (b) Comparison of the times for the homology computation between our sequential and parallel versions (8 threads for the parallel version).

This result was confirmed by our next experiment where we have used our parallel algorithm by progressively increasing the number of threads from 1 to 8. The results shown in Fig. 6 confirm that the gain is more important when the time is bigger.

In order to study more precisely our algorithm, we have computed the number of edges contracted and removed during the different parts of our algorithms, both for the sequential version and the parallel version. Results are given in Table 1. Firstly we can see that the biggest number of edges is for the inner edge removal step (S1 and P1a), and the second biggest number is for the dangling edge removal step (S2 and P2). This justifies the interest of parallelizing these two steps.

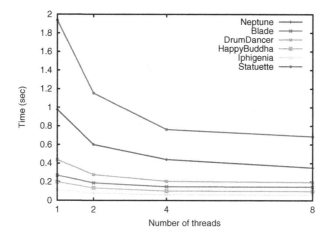

Fig. 6. Times for the homology computation depending on the number of threads: 1, 2, 4 and 8.

Secondly we can see that the number of critical edges processed by the parallel algorithm is very small as well as the time spend by this step (P1b). This shows that our solution to process these edges sequentially does not penalize the efficiency of our method. Lastly, we can see that the number of contracted edges is small as well as the time spend by this step (S3 and P3) which shows that the interest of parallelizing this step is less important (even if we can probably have a small gain here).

Table 1. Number of edges removed and contracted during the different steps of our algorithms, comparing the sequential and the parallel ones (8 threads for the parallel version). The different steps are: for the sequential algorithm: S1 is the inner edge removal, S2 is the dangling edge removal, S3 is the edge contraction; for the parallel algorithm: P1a is the inner edge removal of non critical edges, P1b is the inner edge removal of critical edges, P2 is the dangling edge removal and P3 is the edge contraction. All the rows give the number of removed or contracted edges, except the last line which gives the time (in seconds). The last two lines are the means for the six meshes.

	Sequential			Parallel			
	S1	S2	S3	P1a	P1b	P2	P3
Blade	1,765,093	857,663	24,996	1,764,486	607	858,825	23,834
DrumDancer	2,670,867	1,335,435	0	2,669,774	1,093	1,335,435	0
Neptune	4,007,871	1,991,051	12,880	4,006,454	1,417	1,998,477	5,454
HappyBuddha	1,087,715	531,494	12,157	1,085,360	2,355	529,280	14,371
Iphigenia	703,511	350,189	1,560	702,319	1,192	350,693	1,056
Statuette	9,999,999	4,994,414	5,581	9,991,484	8,515	4,995,786	4,209
Mean	3,372,509	1,676,708	9,529	3,369,980	2,530	1,678,083	8,154
Time	0.363s	0.278s	0.077s	0.113s	0.009s	0.093s	0.078s

5 Conclusion

In this paper, we propose a parallel algorithm for computing the homology of orientable 2D manifolds without boundary represented by a 2-map. Our experiments illustrate that the implemented version of the algorithm is computationally convenient on big meshes and with good speed-up when increasing the number of threads.

We plan to extend our work to non-orientable manifolds once the package implementing generalized maps will be integrated in CGAL. Another possible extension is the parallel homology computation on manifolds with boundary. Probably we will need to add to the algorithm a special case for border edges. Finally, extension in nD could be given based on the theoretical results for removal and contraction operations in any dimension given in [4, 5].

Acknowledgments. This research was partially supported by Spanish project MTM2015-67072-P and by the French National Agency (ANR), project SoLStiCe ANR-13-BS02-0002-01. We also thank the anonymous reviewers for their valuable comments.

References

1. Redhom. http://redhom.ii.uj.edu.pl/
2. Damiand, G.: Combinatorial maps. In: CGAL User and Reference Manual. 3.9th edn (2011). http://www.cgal.org/Pkg/CombinatorialMaps
3. Damiand, G.: Linear cell complex. In: CGAL User and Reference Manual. 4.0edn (2012). http://www.cgal.org/Pkg/LinearCellComplex
4. Damiand, G., Gonzalez-Diaz, R., Peltier, S.: Removal operations in nD generalized maps for efficient homology computation. In: Ferri, M., Frosini, P., Landi, C., Cerri, A., Di Fabio, B. (eds.) CTIC 2012. LNCS, vol. 7309, pp. 20–29. Springer, Heidelberg (2012)
5. Damiand, G., Gonzalez-Diaz, R., Peltier, S.: Removal and contraction operations in nD generalized maps for efficient homology computation. CoRR, abs/1403.3683 (2014)
6. Damiand, G., Lienhardt, P.: Combinatorial Maps: Efficient Data Structures for Computer Graphics and Image Processing. A K Peters/CRC Press, Boca Raton (2014)
7. Damiand, G., Peltier, S., Fuchs, L.: Computing homology for surfaces with generalized maps: application to 3D images. In: Bebis, G., Boyle, R., Parvin, B., Koracin, D., Remagnino, P., Nefian, A., Meenakshisundaram, G., Pascucci, V., Zara, J., Molineros, J., Theisel, H., Malzbender, T. (eds.) ISVC 2006. LNCS, vol. 4292, pp. 235–244. Springer, Heidelberg (2006)
8. Kaczyński, T., Mrozek, M., Ślusarek, M.: Homology computation by reduction of chain complexes. Comput. Math. Appl. **35**(4), 59–70 (1998)
9. Lienhardt, P.: N-Dimensional generalized combinatorial maps and cellular quasi-manifolds. Int. J. Comput. Geom. Appl. **4**(3), 275–324 (1994)
10. Mäntylä, M.: An Introduction to Solid Modeling. Computer Science Press, College Park (1988)
11. Murty, N.A., Natarajan, V., Vadhiyar, S.: Efficient homology computations on multicore and manycore systems. In: 2013 20th International Conference on High Performance Computing (HiPC), pp. 333–342, December 2013
12. Weiler, K.: Edge-based data structures for solid modeling in curved-surface environments. Comput. Graph. Appl. **5**(1), 21–40 (1985)

Computing the Overlaps of Two Maps

Jean-Christophe Janodet[1(✉)] and Colin de la Higuera[2]

[1] IBISC Lab, University of Evry, 23 Bd de France, 91037 Evry, France
janodet@ibisc.univ-evry.fr
[2] LINA Lab, UMR 6241, University of Nantes, 2 R de la Houssinière,
44322 Nantes, France
cdlh@univ-nantes.fr

Abstract. Two combinatorial maps M_1 and M_2 overlap if they share a sub-map, called an overlapping pattern, which can be extended without conflicting neither with M_1 nor with M_2. Isomorphism and subisomorphism are two particular cases of map overlaps which have been studied in the literature. In this paper, we show that finding the largest connected overlap between two combinatorial maps is tractable in polynomial time. On the other hand, without the connectivity constraint, the problem is $\mathcal{NP}$-hard. To obtain the positive results we exploit the properties of a product map.

Keywords: 2D semi-open combinatorial maps · Overlaps · Overlapping patterns · Product map

1   Introduction

2D-combinatorial maps are algebraic structures which allow to describe and work with plane graphs, that is, embeddings of planar graphs, with applications in Image Processing for instance [1]. Using such structures has allowed to establish several algorithmic properties. *E.g.*, it is possible to decide whether two drawings of planar graphs, or two maps, are isomorphic or not in quadratic time [2–4]. Moreover, deciding whether a *pattern* (*i.e.*, a drawing made of a connected subset of faces) appears in a map is also tractable in quadratic time; this property is interesting since determining whether a connected graph is a sub-graph of a planar graph is known to be an $\mathcal{NP}$-complete problem [4–6]. So focusing on a particular drawing of a planar graph (among possibly an exponential number of possibilities) is very helpful from an algorithmic point of view. It has also been shown that searching for a *disconnected*-pattern, built from several disconnected patterns, is an $\mathcal{NP}$-hard problem [4].

A related problem consists in finding large common patterns in two maps. In order to get a common pattern, one must eliminate subsets of faces from both maps and obtain the same pattern up to an isomorphism. *E.g.*, in Fig. 1, map (*c*)

C. de la Higuera—The author Wishes to acknowledge the support of University of Kyoto.

A. Bac and J.-L. Mari (Eds.): CTIC 2016, LNCS 9667, pp. 65–76, 2016.
DOI: 10.1007/978-3-319-39441-1_7

is the maximum common pattern of maps (a) and (b); the eliminated faces are shown with dotted lines. It has been proved in [7] that computing such large common patterns is an $\mathcal{NP}$-hard problem, even when patterns are connected.

In this paper, we constrain the definition of common patterns: we now require the pattern to be *extendable* to both maps when adding *independent* groups of faces. Independence means that if any new face is added in a connected way to the common pattern, then the result ceases to be a pattern of one of the two maps; moreover, if any face is added to the pattern in order to get one of the maps, then no face can be added at the same place in the pattern to get the second map. See Fig. 1(d) for an example. Every common pattern that has this property results of an overlap between both maps, where pairs of faces of both maps were merged together: such an overlap defines an *overlapping pattern*.

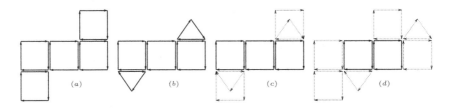

Fig. 1. Maps (a) and (b) have map (c) as maximal common pattern, and map (d) as maximal overlapping pattern. Dotted lines are construction features, and do not belong to the maps.

Notice that the overlapping patterns can be smaller than the maximal common patterns. On the other hand, while the latter are not tractable in polynomial time [7], we show in this paper that computing any connected overlap is tractable in linear time, and enumerating all of the connected overlap is a quadratic-time problem. It follows that finding the largest connected overlap can also be done in polynomial time and space. In contrast, we prove that finding large possibly disconnected overlaps is $\mathcal{NP}$-hard.

Finally, in terms of applications, every maximal overlapping pattern O yields a distance defined by: $d(M_1, M_2) = \mathrm{size}(M_1) + \mathrm{size}(M_2) - 2.\mathrm{size}(O)$. If one can find any maximal overlap in polynomial time, distance d is efficiently computable, and may then be used as an efficient rough approximation to tighter $\mathcal{NP}$-hard graph edit distances.

In Sect. 2, we recall the definitions of full and semi-open maps. The overlaps are introduced in Sect. 3, and the overlapping patterns in Sect. 4. The polynomial problems, related to the existence and the enumeration of the connected overlaps, are investigated in Sect. 5. The correctness of both algorithms is sketched with no detail, due to the lack of space. The case of disconnected overlaps is discussed in Sect. 6. Finally, Sect. 7 concludes the paper.

2   Combinatorial Maps

Definition 1. *Let D be a finite set of* darts. *A 2D full combinatorial map is a triple $M = (D, \alpha, \beta)$ such that (1) $\alpha : D \to D$ is a 1-to-1 mapping (i.e., a permutation over D), and (2) $\beta : D \to D$ is a 1-to-1 mapping such that for all $d \in D$, $\beta(\beta(d)) = d$ (i.e., an involution over D). Two darts d and d' such that $d' = \alpha(d)$ or $d' = \beta(d)$ are respectively said α-sewn or β-sewn.*

Figure 2 shows an example of a full map. Notice that, given a dart d, the *face* which is incident to d is obtained by iterating α. E.g., in Fig. 2, the face incident to dart 12 is described with set $\{12, 13, 14, 15\}$ and we have $\alpha(12) = 13$, $\alpha(13) = 14$, $\alpha(14) = 15$ and $\alpha(15) = 12$. Similarly, the *edges* and the *vertices* of a full map are respectively introduced as the orbits of permutations α and $\beta \circ \alpha$.

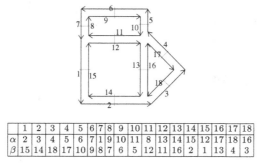

	1	2	3	4	5	6	7	8	9	10	11	12	13	14	15	16	17	18
α	2	3	4	5	6	7	1	9	10	11	8	13	14	15	12	17	18	16
β	15	14	18	17	10	9	8	7	6	5	12	11	16	2	1	13	4	3

Fig. 2. An example of full map. The darts are represented by numbered black segments. Two α-sewn darts are drawn consecutively, and two β-sewn darts are drawn concurrently and in reverse orientation, with little gray segment between them.

All the faces of a full map are defined, which is irrelevant if this map is expected to overlap with others; some faces must be invisible, so function β must be partially defined. This leads us to introduce *semi-open maps*, simply called *maps* throughout the rest of this paper [8]. The idea is to implicitly add a new element ε to the set of darts, and allow any dart to be β-linked with ε whenever such a dart has no adjacent face. Figure 3 shows an example.

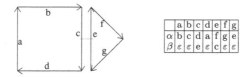

	a	b	c	d	e	f	g
α	b	c	d	a	f	g	e
β	ε	ε	e	ε	c	ε	ε

Fig. 3. An example of (semi-open) map. Darts a, b, d, f and g are not β-sewn.

Definition 2. *Let D be a finite set of darts and $\varepsilon \notin D$ a fresh implicit dart. A semi-open map, or simply map, is a triple $M = (D, \alpha, \beta)$ such that*

- $\alpha : D \cup \{\varepsilon\} \to D \cup \{\varepsilon\}$ *is a 1-to-1 mapping with $\alpha(\varepsilon) = \varepsilon$;*
- $\beta : D \cup \{\varepsilon\} \to D \cup \{\varepsilon\}$ *is a* partial involution *[9, Definition 4], that is, a mapping such that (1) $\beta(\varepsilon) = \varepsilon$ and (2) for all $d \in D$, if $\beta(d) \neq \varepsilon$, then $\beta(\beta(d)) = d$.*

The complexity of the problems that we address on the maps is often related to connectivity. For instance, searching for a connected pattern (subset of contiguous faces) in a map is a quadratic problem, whereas searching for a disconnected pattern (subset of independent patterns) is $\mathcal{NP}$-complete [9].

Definition 3. *Let $M = (D, \alpha, \beta)$ be a semi-open map. Any subset $U \subseteq D$ is* connected *in map M if for all $d, d' \in U$, there exists a sequence $d_0, d_1, \ldots d_n \in U$ such that (1) $d_0 = d$ and (2) $d_n = d'$ and (3) for all $0 \leq k < n$, we have $d_{k+1} = \gamma_k(d_k)$ with $\gamma_k = \alpha$ or $\gamma_k = \beta$. We say that map M is* connected *if set D itself is connected in map M.*

3   The Overlaps of Two Maps

An overlap is a maximal one-to-one *matching*. Let $M_1 = \langle D_1, \alpha_1, \beta_1 \rangle$ and $M_2 = \langle D_2, \alpha_2, \beta_2 \rangle$ be two fixed semi-open maps.

Definition 4. *A (one-to-one)* matching *is a function $h : U_1 \to U_2$ such that*

1. $U_1 \subseteq D_1$ *and $U_2 \subseteq D_2$,*
2. h *is bijective,*
3. *for all $d_1 \in U_1$ and $d_2 = h(d_1)$,*
   - $\alpha_1(d_1) \in U_1$ *if and only if $\alpha_2(d_2) \in U_2$, and $h(\alpha_1(d_1)) = \alpha_2(d_2)$;*
   - $\beta_1(d_1) \in U_1$ *if and only if $\beta_2(d_2) \in U_2$, and $h(\beta_1(d_1)) = \beta_2(d_2)$.*

An example is given in Fig. 4. Notice that $\varepsilon \notin U_1$ and $\varepsilon \notin U_2$ (since $\varepsilon \notin D_1$ and $\varepsilon \notin D_2$); nevertheless, for convenience reasons, we shall implicitly suppose that $h(\varepsilon) = \varepsilon$.

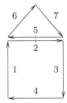
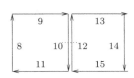

Fig. 4. Consider the maps M_1 and M_2 above. Mapping $h = \{1 \mapsto 8, 2 \mapsto 9, 6 \mapsto 13\}$ is a matching, whereas mapping $h' = \{1 \mapsto 8, 2 \mapsto 9, 6 \mapsto 10\}$ is not; indeed, we have $\alpha_2(9) = 10$, but $\alpha_1(2) \neq 6$. Any matching must preserve the seams of *both* maps.

Definition 5. *Let* $h : U_1 \to U_2$ *be a one-to-one matching as in Definition 4. We say that h is an* overlap *if:*

1. *for all* $d_1 \in U_1$, *we have* $\alpha_1(d_1) \in U_1$, *and*
2. *for all* $d_1 \in U_1$ *and* $d_2 = h(d_1)$, *if* $\beta_1(d_1) \neq \varepsilon$ *and* $\beta_2(d_2) \neq \varepsilon$, *then* $\beta_1(d_1) \in U_1$ *and* $\beta_2(d_2) \in U_2$.

We say that h is a connected overlap *if subset* U_1 *is connected in map* M_1. *We say that h is a* disconnected overlap *otherwise.*

The first condition above implies that if a dart d_1 of M_1 matches a dart d_2 of M_2, then the whole face incident to d_1 must match the whole face incident to d_2. The second condition means that if darts d_1 and d_2 match together and both have adjacent faces, then opposite β-sewn darts, and thus adjacent faces must also match together. In consequence, each pieces of both maps M_1 and M_2 must be as large as possible, that is, the matching must be "maximal". See Figs. 5 and 6 for examples.

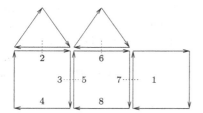 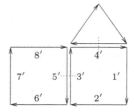

Fig. 5. An example of connected overlap $h = \{1 \mapsto 1', 2 \mapsto 2', \ldots 8 \mapsto 8'\}$. Notice that no overlap can be built with darts 5 and 4' matched together.

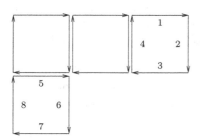 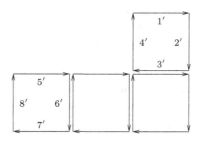

Fig. 6. An example of disconnected overlap $h = \{1 \mapsto 1', 2 \mapsto 2', \ldots 8 \mapsto 8'\}$.

4   Properties of the Overlapping Patterns

Given two maps M_1 and M_2 and an overlap $h : U_1 \to U_2$, the elimination of all the darts, but those from sets U_1 and U_2, in maps M_1 and M_2 respectively, defines two isomorphic sub-maps of M_1 and M_2. In other words, an overlap defines an *overlapping common pattern*:

Definition 6. *Map $P = \langle D, \alpha, \beta \rangle$ is a pattern of map $M = \langle D', \alpha', \beta' \rangle$ if there exists a one-to-one matching $\varphi : D \to V$ (with $V \subseteq D'$). Moreover, any map P is a common pattern of maps M_1 and M_2 if P is a pattern of both M_1 and M_2.*

As direct consequences of Definitions 4 and 5, the overlapping patterns have the following properties:

– Every overlapping pattern of M_1 and M_2 is a pattern of both M_1 and M_2;
– Given every pair of darts (d_1, d_2), there exists at most one connected overlap, and thus at most one connected overlapping pattern, in which d_1 and d_2 are matched together;
– An overlapping pattern is *maximal* in the following sense: if we add to this pattern something and the result is a semi-open map, then this map is no longer a sub-map of M_1 or of M_2.

Concerning computational issues, we have provided in [3] an algorithm in $\mathcal{O}(|D_1| \times |D_2|)$ time to decide whether two (possibly non-connected) maps were isomorphic or not. An extension of this algorithm can be used to prove that a connected map is a pattern of another map. In both cases, due to the technical necessity for the matching h to commute with bijections α and β, there are no more than $|D_1|$ possible matching functions, and the algorithms actually enumerate all of them in $\mathcal{O}(|D_1| \times |D_2|)$ time. Concerning the second problem, the fact that the first map is connected is crucial, since this problem is $\mathcal{NP}$-complete for disconnected patterns [7].

With respect to finding large common connected patterns, it has unfortunately been shown in [7] that this problem is $\mathcal{NP}$-hard too. But looking for large overlapping patterns is simpler: whereas in general the pattern can be placed anywhere in the maps, we now insist on the pattern to somehow be placed on the border of both maps, corresponding to the part where they overlap.

In order to illustrate and discuss this point, we turn to strings. In terms of strings, common patterns would be common sub-strings of two given strings: given $u, v \in \Sigma^*$, a pattern is a string $w \in \Sigma^+$ such that $u = lwr$ and $v = l'wr'$ for some $l, l', r, r' \in \Sigma^*$.

On the other hand, an overlap defines a string w as above, such that $u = lwr$ and $v = l'wr'$ but with the added condition that $l = \epsilon$ or $l' = \epsilon$ on one hand, $r = \epsilon$ or $r' = \epsilon$ on the other. This means that exactly 4 cases are possible:

1. $u = lw$ and $v = wr'$, w is a suffix of u and a prefix of v,
2. $u = wr'$ and $v = l'w$, w is a prefix of u and a suffix of v,
3. $u = lwr$ and $v = w$, v is a sub-string of u,
4. $u = w$ and $v = l'wr'$, u is a sub-string of v.

So the problem is simpler.

5   Finding Connected Overlaps Efficiently

In this section, we show that it is possible to efficiently find the largest connected overlap of two maps.

5.1 A Linear Time and Space Algorithm to Check Whether a Connected Overlap Exists

Let $M_1 = \langle D, \alpha_1, \beta_1 \rangle$ and $M_2 = \langle D_2, \alpha_2, \beta_2 \rangle)$ be two semi-open maps. The first problem we tackle consists in determining whether two darts $d_1 \in D_1$ and $d_2 \in D_2$ can match together in an overlap. This is the purpose of Algorithm 1.

This procedure performs a parallel traversal of both maps M_1 and M_2, starting from the darts d_1 and d_2, which are grouped together into a couple (d_1, d_2). The procedure uses the α- and β-functions of both maps to discover new couples of darts from the couples that have been discovered so far. It more precisely builds a candidate overlap h such that $h[d_1] = d_2$. So initially, $h[d_1]$ is set to d_2, whereas $h[d]$ are set to nil for all other darts d.

Each time a couple (a, a') is discovered from another couple (d, d') by using the α-functions, we check whether both darts a and a' have ever been met. If either a or a' have already been visited through the traversal, we carefully check basic conditions which ensure us to get a valid matching h at the end of the algorithm. Otherwise, $h[a]$ is set to a' and the couple (a, a') is further used to discover new darts. With respect to β-functions, the same principle holds, but more cases of failure can occur, as the darts d or d' may have no adjacent face.

Note that Algorithm 1 returns a single Boolean value. Nevertheless, one can easily modify the procedure and get the connected overlap h as a certificate, in case of success. The following theorem claims the correctness of this algorithm. We shall sketch a proof just after Theorem 3.

Theorem 1. *Algorithm 1 is correct, that is:*

- *If CHECKCONNECTEDOVERLAP(M_1, M_2, d_1, d_2) returns* **true**, *then the array h that is built by the procedure encodes a connected overlap $h : U_1 \rightarrow U_2$ such that $h(d_1) = d_2$.*
- *If CHECKCONNECTEDOVERLAP(M_1, M_2, d_1, d_2) returns* **false**, *then no overlap $h : U_1 \rightarrow U_2$ exists such that $h(d_1) = d_2$.*

With respect to the complexity of the algorithm, we have:

Theorem 2. *Algorithm 1 runs in $\mathcal{O}(\min(|D_1|, |D_2|))$ time and $\mathcal{O}(\max(|D_1|, |D_2|))$ space.*

Proof. Suppose that $|D_1| \leq |D_2|$ without loss of generality. Then the while loop of Algorithm 1 is iterated at most $|D_1|$ times. Indeed, (1) at each iteration, exactly one dart $d \in D_1$ is removed from the stack S, within a couple (d, x) for some $x \in D_2$, and (2) each dart $d \in D_1$ enters S at most once, within a couple (d, x) for any $x \in D_2$: d enters S only if $h[d] = nil$, and before entering S, $h[d]$ is set to dart x. So the algorithm runs in $\mathcal{O}(\min(|D_1|, |D_2|))$ time. As for space issues, notice that the arrays h and g respectively have $|D_1|$ and $|D_2|$ entries, while stack S never contains more than $\min(|D_1|, |D_2|)$ couples of darts. $\square$

Algorithm 1. CHECKCONNECTEDOVERLAP(M_1, M_2, d_1, d_2)

Input: Two semi-open maps $M_1 = \langle D_1, \alpha_1, \beta_1 \rangle$ and $M_2 = \langle D_2, \alpha_2, \beta_2 \rangle$, and an initial couple of darts $(d_1, d_2) \in D_1 \times D_2$

Output: true if a connected overlap h exists such that $h(d_1) = d_2$, false otherwise

Variables: Two arrays $h : D_1 \rightarrow D_2$ and $g : D_2 \rightarrow D_1$ (where $g = h^{-1}$) both initialized with nil, and a stack S which is initially empty

1 $h[d_1] \leftarrow d_2$; $g[d_2] \leftarrow d_1$; push (d_1, d_2) in S ;

2 **while** S is not empty **do**

3   pop a couple of darts (d, d') from S ;

4      $a \leftarrow \alpha_1(d)$; $a' \leftarrow \alpha_2(d')$;

5   **if** $h[a] = nil$ **and** $g[a'] = nil$ **then**

6          $h[a] \leftarrow a'$; $g[a'] \leftarrow a$; push (a, a') in S ;

7   **else if** $h[a] \neq a'$ **or** $g[a'] \neq a$ **then**

8   return false;

9      $b \leftarrow \beta_1(d)$; $b' \leftarrow \beta_2(d')$;

10   **if** $b \neq \varepsilon$ **and** $b' \neq \varepsilon$ **then**

11   **if** $h[b] = nil$ **and** $g[b'] = nil$ **then**

12              $h[b] \leftarrow b'$; $g[b'] \leftarrow b$; push (b, b') in S ;

13   **else if** $h[b] \neq b'$ **or** $g[b'] \neq b$ **then**

14   return false;

15   **else if** $b = \varepsilon$ **and** $b' \neq \varepsilon$ **and** $g[b'] \neq nil$ **then**

16   return false ;

17   **else if** $b \neq \varepsilon$ **and** $b' = \varepsilon$ **and** $h[b] \neq nil$ **then**

18   return false;

19 return true; // Array h may be returned as a certificate

Notice that Algorithm 1 exploits the same key idea as the algorithms developed in [3] to solve the map isomorphism and sub-isomorphism problems. We nevertheless improve them as the failure cases are detected during the traversal of the maps, and no further verification stage is needed after the traversal.

5.2 A Quadratic Time and Space Algorithm to Get All the Connected Overlaps

Let $M_1 = \langle D, \alpha_1, \beta_1 \rangle$ and $M_2 = \langle D_2, \alpha_2, \beta_2 \rangle$ be two semi-open maps. In Sect. 5.1, we have given a procedure that checks whether two darts $d_1 \in D_1$ and $d_2 \in D_2$ can match together in a connected overlap. To achieve this goal, Algorithm 1 performs a parallel traversal of both maps M_1 and M_2, starting from couple (d_1, d_2), and using the α- and β-functions of both maps to investigate new couples of darts from the couples that have been visited so far. Clearly, this procedure may visit any couple of $D_1 \times D_2$. So we use this set to define a new map, denoted $M_1 \otimes M_2$, and call the *product map* of maps M_1 and M_2:

Definition 7. *The* product *of two maps M_1 and M_2 is $M_1 \otimes M_2 = \langle D_1 \times D_2, \alpha, \beta \rangle$ where, for all $(d_1, d_2) \in D_1 \times D_2$:*

$$\alpha(d_1, d_2) = (\alpha_1(d_1), \alpha_2(d_2)), \text{ and}$$

$$\beta(d_1, d_2) = \begin{cases} (\beta_1(d_1), \beta_2(d_2)) & \text{if } \beta_1(d_1) \neq \varepsilon \text{ and } \beta_2(d_2) \neq \varepsilon, \\ \varepsilon & \text{otherwise.} \end{cases}$$

For instance, consider the maps of Fig. 7, respectively made of 7 and 6 darts. The product map $M_1 \otimes M_2$, which has 42 darts, contains 6 connected components. Clearly, two kinds of connected components appear. The four small components will be called *real*, and the two large ones, *imaginary*:

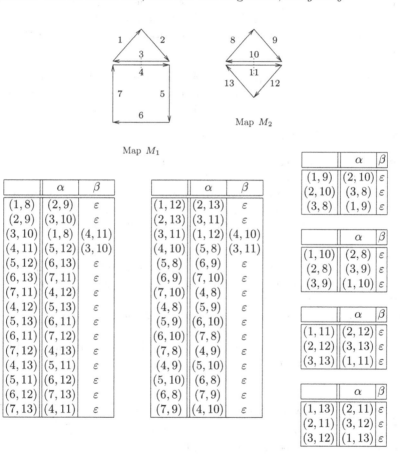

Map M_1

Map M_2

	α	β
$(1, 8)$	$(2, 9)$	ε
$(2, 9)$	$(3, 10)$	ε
$(3, 10)$	$(1, 8)$	$(4, 11)$
$(4, 11)$	$(5, 12)$	$(3, 10)$
$(5, 12)$	$(6, 13)$	ε
$(6, 13)$	$(7, 11)$	ε
$(7, 11)$	$(4, 12)$	ε
$(4, 12)$	$(5, 13)$	ε
$(5, 13)$	$(6, 11)$	ε
$(6, 11)$	$(7, 12)$	ε
$(7, 12)$	$(4, 13)$	ε
$(4, 13)$	$(5, 11)$	ε
$(5, 11)$	$(6, 12)$	ε
$(6, 12)$	$(7, 13)$	ε
$(7, 13)$	$(4, 11)$	ε

	α	β
$(1, 12)$	$(2, 13)$	ε
$(2, 13)$	$(3, 11)$	ε
$(3, 11)$	$(1, 12)$	$(4, 10)$
$(4, 10)$	$(5, 8)$	$(3, 11)$
$(5, 8)$	$(6, 9)$	ε
$(6, 9)$	$(7, 10)$	ε
$(7, 10)$	$(4, 8)$	ε
$(4, 8)$	$(5, 9)$	ε
$(5, 9)$	$(6, 10)$	ε
$(6, 10)$	$(7, 8)$	ε
$(7, 8)$	$(4, 9)$	ε
$(4, 9)$	$(5, 10)$	ε
$(5, 10)$	$(6, 8)$	ε
$(6, 8)$	$(7, 9)$	ε
$(7, 9)$	$(4, 10)$	ε

	α	β
$(1, 9)$	$(2, 10)$	ε
$(2, 10)$	$(3, 8)$	ε
$(3, 8)$	$(1, 9)$	ε

	α	β
$(1, 10)$	$(2, 8)$	ε
$(2, 8)$	$(3, 9)$	ε
$(3, 9)$	$(1, 10)$	ε

	α	β
$(1, 11)$	$(2, 12)$	ε
$(2, 12)$	$(3, 13)$	ε
$(3, 13)$	$(1, 11)$	ε

	α	β
$(1, 13)$	$(2, 11)$	ε
$(2, 11)$	$(3, 12)$	ε
$(3, 12)$	$(1, 13)$	ε

Fig. 7. Given the maps M_1 and M_2, we build the product map $M_1 \otimes M_2$ and display its connected components. The two big components are imaginary, and the four small ones are real. Couple $(1, 8)$ belongs to an imaginary component, so following Theorem 3, no overlap exists such that darts 1 and 8 match. Conversely, couple $(1, 9)$ is in a real component and mapping $\{1 \mapsto 9, 2 \mapsto 10, 3 \mapsto 8\}$ is a connected overlap.

Definition 8. *A connected component* $C = \langle D, \alpha, \beta \rangle$ *of product map* $M_1 \otimes M_2$ *is said* real *if for all* $(d_1, d_2), (d'_1, d'_2) \in D$,

1. $d_1 = d'_1$ *iff* $d_2 = d'_2$, *and*
2. $\beta_1(d_1) = d'_1$ *iff* $\beta_2(d_2) = d'_2$.

A connected component which is not real is said imaginary.

Remark 1. The reader may wonder why no condition addressing the α-functions is given in Definition 8. Actually, a consequence of Condition 1 is that for all $(d_1, d_2), (d'_1, d'_2) \in D$, we have $\alpha_1(d_1) = d'_1$ iff $\alpha_2(d_2) = d'_2$. Indeed, let $(d_1, d_2) \in D$ and suppose that $(\alpha_1(d_1), d'_2) \in D$; as component C is connected, we have $\alpha(d_1, d_2) = (\alpha_1(d_1), \alpha_2(d_2)) \in D$; so using Condition 1, we deduce that $d'_2 = \alpha_2(d_2)$. Such a proof cannot be given for Condition 2, due to the fact that d_1 or d_2 can β-free.

Connected overlaps of maps M_1 and M_2 on the one hand, and real connected components of product map $M_1 \otimes M_2$ on the other hand, are strongly related. Indeed, we get the following result, which proceeds from the definitions:

Theorem 3. *Let* M_1 *and* M_2 *be two semi-open maps.*

- *For each connected overlap* $h : U_1 \to U_2$, *there exists a real connected component* C *of product map* $M_1 \otimes M_2$ *whose set of darts is* $\{(d, h(d)) : d \in U_1\}$;
- *Conversely, for every real connected component* $C = \langle D, \alpha, \beta \rangle$ *of product map* $M_1 \otimes M_2$, *set* D *is the graph of a connected overlap of* M_1 *and* M_2, *that is to say, if we fix* $h(d_1) = d_2$ *for all* $(d_1, d_2) \in D$, *then* h *is a connected overlap.*

Note that we get Theorem 1 as a consequence of Theorem 3. Indeed, Algorithm 1 actually traverses the connected component where initial couple of darts (d_1, d_2) stands; if the traversal returns `false`, then we get an evidence that the component is imaginary, so no connected overlap h exists such that $h(d_1) = d_2$. Otherwise, we can show that the component is real and the array h is a connected overlap.

As a consequence of Theorem 3, we also deduce an efficient algorithm to enumerate all the connected overlaps (see Algorithm 2).

Algorithm 2. GETALLCONNECTEDOVERLAPS(M_1, M_2)

Input: Two semi-open maps $M_1 = \langle D, \alpha_1, \beta_1 \rangle$ and $M_2 = \langle D_2, \alpha_2, \beta_2 \rangle)$
Output: All the connected overlaps of M_1 and M_2

1 Compute product map $M_1 \otimes M_2$; // see `Definition 7`
2 Select all the real connected components; // see `Definition 8`
3 Return all the connected overlaps; // see `Theorem 3`

Theorem 4. *Algorithm 2 is correct and runs in* $\mathcal{O}(|D_1| \cdot |D_2|)$ *time and space.*

Proof. The correctness follows from Theorem 3. As for the complexity, the product map $M_1 \otimes M_2$ has $|D_1| \cdot |D_2|$ darts. Obviously, the computation of functions α and β, and the computation of the connected components, and the selection of the real connected components, are in linear time and space with respect to $|D_1| \cdot |D_2|$. $\qquad\square$

5.3   Finding the Largest Overlap of Two Maps

It is straightforward to use the results from the previous algorithm to return the largest overlap, provided this one is connected.

Furthermore a direct procedure exists to find the largest possibly disconnected overlap: consider a graph whose nodes are the connected overlaps between maps M_1 and M_2, each with a weight indicating how many darts they concern, and an edge indicates that two overlaps are compatible (do not contain common darts). Finding a clique with maximum sum of weights gives the largest possibly disconnected overlap. This procedure clearly runs in exponential time, and the following section shows that we cannot hope better.

6   Finding Large Disconnected Overlaps Is Intractable

We consider the following problem:

Name LARGE_DISCONNECTED_OVERLAP;
Instance An integer N, and two semi-open maps M_1 and M_2;
Problem Does there exist a disconnected overlap $h : U_1 \to U_2$ s.t. $|U_1| \geq N$?

We get the following result:

Theorem 5. *Problem* LARGE_DISCONNECTED_OVERLAP *is $\mathcal{NP}$-complete.*

Proof. Basically, Problem LARGE_DISCONNECTED_OVERLAP is in class $\mathcal{NP}$ since any certificate h can easily be verified. Now consider following problem:

Name DISCONNECTED_PATTERN;
Instance Two semi-open maps M_1 and M_2;
Problem Is map M_1 a disconnected pattern of map M_2?

One can prove, by reduction from SEPARABLE_PLANAR_3SAT [10, Lemma 1], that this problem is $\mathcal{NP}$-complete. We do not give the details here: the proof is essentially the same as that provided in [7, Sect. 4][1]. We can finally reduce DISCONNECTED_PATTERN to LARGE_DISCONNECTED_OVERLAP: we simply need to fix $N = |D_1|$. Indeed, map M_1 is a pattern of map M_2 iff an overlap $h : U_1 \to U_2$ exists with $|U_1| \geq |D_1|$ (that is, $U_1 = D_1$). $\qquad\square$

---

[1] Actually, Problem DISCONNECTED_PATTERN, which is defined for semi-open maps, is an instance of Problem INDUCED_SUBMAP_ISOMORPHISM, which is defined for nG-maps, but proved $\mathcal{NP}$-complete by using 2G-maps, and gadgets that can easily be redefined in terms of semi-open maps. Thus rewriting such a proof is of no relevance.

7   Conclusion and Future Works

Computing the overlaps has two advantages over computing the common patterns of two maps: on one hand, the optimisation problems are tractable (Sect. 5), and on the other, the overlaps are maximal objects (Sect. 4).

The overlaps allow us also to consider super-maps, where a super-map of M_1 and M_2 is a map of which both M_1 and M_2 are patterns. Then the smallest common super-map is obtained by adding to the largest overlap the faces which belong to M_1 and M_2 but are not matched.

Super-maps offer interesting possibilities as smallest common super-maps would be constructable in polynomial time whereas their duals, the largest common sub-maps, are not.

Finally, notice that the techniques introduced in this paper can, with no difficulty, be extended to *open* maps[2], nD-maps and n-Gmaps [1].

References

1. Damiand, G., Lienhardt, P.: Combinatorial Maps: Efficient Data Structures for Computer Graphics and Image Processing. A K Peters/CRC Press (2014)
2. Cori, R.: Un code pour les graphes planaires et ses applications. Ph.D. thesis, Université Paris 7 (1973)
3. Damiand, G., de la Higuera, C., Janodet, J.-C., Samuel, É., Solnon, C.: A polynomial algorithm for submap isomorphism. In: Torsello, A., Escolano, F., Brun, L. (eds.) GbRPR 2009. LNCS, vol. 5534, pp. 102–112. Springer, Heidelberg (2009)
4. de la Higuera, C., Janodet, J.C., Samuel, E., Damiand, G., Solnon, C.: Polynomial algorithms for open plane graph and subgraph isomorphisms. Theor. Comput. Sci. **498**, 76–99 (2013)
5. Dorn, F.: Planar subgraph isomorphism revisited. In: Proceedings of 27th International Symposium on Theoretical Aspects of Computer Science (STACS 2010), vol. 5 of LIPIcs., Schloss Dagstuhl - Leibniz-Zentrum fuer Informatik, pp. 263–274(2010)
6. Eppstein, D.: Subgraph isomorphism in planar graphs and related problems. J. Graph Algorithms Appl. **3**(3), 1–27 (1999)
7. Solnon, C., Damiand, G., de la Higuera, C., Janodet, J.C.: On the complexity of submap isomorphism and maximum common submap problems. Pattern Recogn. **48**(2), 302–316 (2015)
8. Poudret, M., Arnould, A., Bertrand, Y., Lienhardt, P.: Cartes combinatoires ouvertes. Research Notes 2007–1, Laboratoire SIC E.A. 4103, F-86962 Futuroscope Cedex - France (2007)
9. Damiand, G., Solnon, C., de la Higuera, C., Janodet, J.C., Samuel, E.: Polynomial algorithms for subisomorphism of nD open combinatorial maps. Comput. Vis. Image Underst. **115**(7), 996–1010 (2011)
10. Lichtenstein, D.: Planar formulæ and their uses. SIAM J. Comput. **11**(2), 329–343 (1982)

---

[2] for which the α-functions may be *partial permutations* [8].

Topological Descriptors for 3D Surface Analysis

Matthias Zeppelzauer[1]([⊠]), Bartosz Zieliński[2]([⊠]), Mateusz Juda[2],
and Markus Seidl[1]

[1] Media Computing Group, Institute of Creative Media Technologies,
St. Poelten University of Applied Sciences, Matthias-Corvinus Strasse 15,
3100 St. Poelten, Austria
{m.zeppelzauer,markus.seidl}@fhstp.ac.at
[2] The Institute of Computer Science and Computer Mathematics,
Faculty of Mathematics and Computer Science, Jagiellonian University,
ul. Łojasiewicza 6, 30-348 Kraków, Poland
{bartosz.zielinski,mateusz.juda}@uj.edu.pl

Abstract. We investigate topological descriptors for 3D surface analysis, i.e. the classification of surfaces according to their geometric fine structure. On a dataset of high-resolution 3D surface reconstructions we compute persistence diagrams for a 2D cubical filtration. In the next step we investigate different topological descriptors and measure their ability to discriminate structurally different 3D surface patches. We evaluate their sensitivity to different parameters and compare the performance of the resulting topological descriptors to alternative (non-topological) descriptors. We present a comprehensive evaluation that shows that topological descriptors are (i) robust, (ii) yield state-of-the-art performance for the task of 3D surface analysis and (iii) improve classification performance when combined with non-topological descriptors.

Keywords: 3D surface classification · Surface topology analysis · Surface representation · Persistence diagram · Persistence images

1 Introduction

With the increasing availability of high-resolution 3D scans, topological surface description is becoming increasingly important. In recent years, methods for sparse and dense 3D scene reconstruction have progressed strongly due to availability of inexpensive, off-the-shelf hardware (e.g. Microsoft Kinect) and the development of robust reconstruction algorithms (e.g. structure from motion techniques, SfM) [5,28]. Since 3D scanning has become an affordable process the amount of available 3D data has increased significantly. At the same time, the reconstruction accuracy has increased strongly, which enables 3D reconstructions with sub-millimeter resolution [27]. The high resolution enables the accurate description of a 3D surface's geometric micro-structure, which opens up new opportunities for search and retrieval in 3D scenes, such as the recognition of objects by their specific surface properties as well as the distinction of different types of materials for improved scene understanding.

© Springer International Publishing Switzerland 2016
A. Bac and J.-L. Mari (Eds.): CTIC 2016, LNCS 9667, pp. 77–87, 2016.
DOI: 10.1007/978-3-319-39441-1_8

In this paper, we investigate the problem of describing and classifying 3D surfaces according to their geometric micro-structure. Two different types of approaches exist for this problem. Firstly, the dense processing of the surface in 3D space and secondly, the processing of the surface geometry in image-space based on depth maps derived from the surface.

For the representation of surface geometry in 3D, descriptors are required that capture the local geometry around a given point or mesh vertex. Different types of local 3D descriptors have been developed recently that are suitable for the description of the local geometry around a 3D point, such as spin images [12], 3D shape context [4], and persistent point feature histograms [24].

The dense extraction of surface geometry by local 3D descriptors, however, becomes a computationally demanding task when several millions of points need to be processed. A computationally more efficient approach is the analysis of 3D surfaces in image space. In such approaches a 3D surface is first mapped to a depth map which represents a height field of the surface. This processing step maps the 3D surface analysis problem to a 2D texture analysis task which can be approached by analyzing the surface by texture descriptors, such as HOG, GLCM, and Wavelet-based features [19,29,30].

The presented approach falls into the category of image-space approaches. We first map the surface to image-space by a depth projection. Next, we divide the resulting depth map into patches and describe them with traditional non-topological as well as with topological surface descriptors. For the classification of surface patches we use random undersampling boosting (RUSBoost) [25] due to its high accuracy for imbalanced class distributions [16].

2   Topological Approach

By mathematical standards topology, with its 120 years of history, is a relatively young discipline. It grew out of H. Poincares seminal work on the stability of the solar system as a qualitative tool to study the dynamics of differential equations without explicit formulas for solutions [20–22]. Due to the lack of useful analytic methods, topology soon became a purely theoretical discipline. However, in the last few years we observe a rapid development of topological data analysis tools, which open new applications for topology.

Topological spaces appearing in data analysis are typically constructed from small pieces or cells. A natural tool in the study of multidimensional images with topological methods are hypercubes (points, edges, squares, cubes etc.), e.g. a pixel in a 2 dimensional image is equivalent to a square, a voxel in a 3 dimensional volume is equivalent to a cube. Hypercubes are building blocks for structures called cubical complexes. Such representations give topology a combinatorial flavour and make it a natural tool in the study of multi-dimensional data sets.

Intuitively, the rank of the nth homology group, the so called nth Betti number denoted β_n, counts the number of n-dimensional holes in the topological space. In particular, β_0 counts the number of connected components. As an example consider the image of the digit 8. In this image there is one connected component and two holes, hence $\beta_0 = 1$ and $\beta_1 = 2$. For a hollow sphere we have $\beta_0 = 1$, $\beta_1 = 0$, $\beta_2 = 1$. For a tube in a tire we have $\beta_0 = 1$, $\beta_1 = 2$, $\beta_2 = 1$.

Betti numbers do not differentiate between small and large holes. In consequence, the holes resulting from the noise in the data cannot be distinguished from the holes indicative for the nature of the data. For instance, in a noisy image of the digit 8 one can get easily $\beta_0 > 1$. A remedy for this drawback is persistent homology, a tool invented at the beginning of the 20th century [7]. Persistent homology studies how the Betti numbers change when the topological space is gradually built by adding cubes in some prescribed order.

If X is a cubical complex, one can add cubes step by step. Typically, the construction goes through different scales, starting from the smallest pieces. However, in general an arbitrary function $f : X \rightarrow \mathbb{R}$, called the Morse function or measurement function, may be used to control the order in which the complex is built, starting from low values of f and increasing subsequently. This way we obtain a sequence of topological spaces, called a filtration,

$$\emptyset = X_{r_0} \subset X_{r_1} \subset X_{r_2} \subset \cdots \subset X_{r_n} = X,$$

where $X_r := f^{-1}((-\infty, r])$ and r_i is a growing sequence of values of f at which the complex changes. As the space is gradually constructed, holes are born, persist for some time and eventually may die. The length of the associated birth-death intervals (persistence intervals) indicates if the holes are relevant or merely noise. The lifetime of holes is usually visualized by the so called persistence diagram (PD). Persistence diagrams constitute the main tool of topological data analysis. They visualize geometrical properties of a multidimensional object X in a simple two dimensional diagram.

Figure 1(a) shows a 3D surface as a 2D depth map, where colors corresponds to depth (blue refers to low depth, yellow to high depth). In this case pixels are represented as 2-dimensional cells of a cubical complex. For the complex we can obtain a filtration X_r using a measuring function which has a value for a 2-dimensional cube equal to height (pixel color). For a lower dimensional cell (a vertex or an edge) we can set the function value as a maximum from the higher-dimensional neighborhoods of the cell. Figure 1(b) shows the persistence diagram for X_r.

There is still no specific answer on how and when the tools of computational topology and machine learning should be used together. A first attempt is to provide a descriptor of a topological space filtration based on elementary statistics of persistence intervals (or equivalently on persistence diagrams). Let

$$I := \{[b_1, e_1], [b_2, e_2], \ldots, [b_n, e_n]\}$$

be a set of persistence intervals. Let $D := \{d_i := (e_i - b_i)\}_{i=1}^{n}$ be a set of the interval lengths. We build an aggregated descriptor of D, denoted by PD_AGG, using following measures: number of elements, minimum, maximum, mean, standard deviation, variance, 1-quartile, median, 3-quartile, and norms $\sum \sqrt{d_i}$, $\sum d_i$, and $\sum (d_i)^2$.

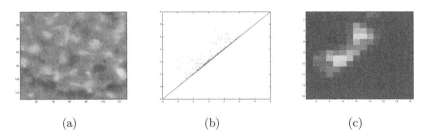

(a)   (b)   (c)

Fig. 1. Example patch: (a) the original 3D surface as a 2D depth map; (b) the corresponding persistent diagram; (c) and the persistent image with $\sigma = 0.001$ and resolution 16×16. (Color figure online)

Except the PD_AGG descriptor described above, which can be used with standard classification methods, there are also attempts to use PD directly with appropriately modified classifiers. Reininghaus et al. [23] proposed a multiscale kernel for PDs, which can be used with a support vector machine (SVM). While this kernel is well-defined in theory, in practice it becomes highly inefficient for a large number of training vectors (as the entire kernel matrix must be computed explicitly). As an alternative, Chepushtanova et al. [1] introduced a novel representation of a PD, called a persistence image (PI), which is faster and can be used with a broader range of machine learning (ML) techniques.

A PI is derived from mapping a PD to an integrable function $G_p : \mathbb{R}^2 \to \mathbb{R}$, which is a sum of Gaussian functions centered at each point of the PD. Taking a discretization of a subdomain of G_p defines a grid. An image can be created by computing the integral of G_p on each grid box, thus defining a matrix of pixel values. Formally, the value of each pixel p within a PI is defined by the following equation:

$$PI(p) = \iint_p \sum_{[b_i, e_i] \in I} g(b_i, e_i) \frac{1}{2\pi\sigma_x\sigma_y} e^{-\frac{1}{2}\left(\frac{(x-b_i)}{\sigma_x^2} + \frac{(y-e_i)}{\sigma_y^2}\right)} \, dy \, dx,$$

where $g(b_i, e_i)$ is a weighting function, which depends on the distance from the diagonal (points close to the diagonal are usually considered as noise, therefore they should have low weights), σ_x and σ_y are the standard deviations of the Gaussians in x and y direction. The resulting image (see Fig. 1c) is vectorized to achieve a standardized vectorial representation which is compatible to a broad range of ML techniques.

The advantage of PIs compared to PDs descriptor is a high classification accuracy and stability [1]. However, they require numerous parameters like the PI resolution, the weighting function g, as well as σ_x and σ_y.

3   Experimental Setup

In our experiments we investigate the robustness and expressiveness of the topological descriptors presented in Sect. 2 for 3D surface analysis and compare and combine them with traditional non-topological descriptors. For our experiments, we employ a dataset of high-resolution 3D reconstructions from the archaeological domain with a resolution below 0.1 mm [30]. The dimension of the scanned surfaces ranges from approx. 20×30 cm to 30×50 cm. The reconstructions represent natural rock surfaces that exhibit human-made engravings (so-called rock-art). The engravings represent symbols and figures (e.g. animals and humans) engraved by humans in ancient times. See Fig. 2 for an example surface. The engraved regions in the surface exhibit a different surface geometry than the surrounding natural rock surface. In our experiments we aim at automatically separating the engraved areas from the natural rock surface. The corresponding ground truth is depicted in Fig. 2c.

The employed dataset contains 4 surface reconstructions with a total number of 12.3 millions of points. For each surface a precise ground truth has been generated by domain experts that labels all engravings on the surface. The dataset contains two classes of surface topographies: class 1 represents engraved areas and class 2 represents the natural rock surface. Class priors are imbalanced. Class 1 represents 16.6 % of the data and is thus underrepresented.

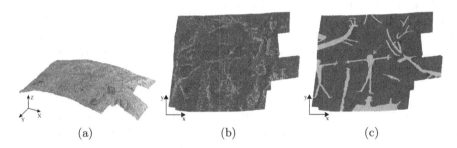

(a)   (b)   (c)

Fig. 2. Example data: (a) the 3D point cloud of the surface; (b) the depth projection of the surface with compensated global curvature; (c) ground truth labeling that specifies areas with different topography, such as the human-shaped figure in the center whose head is marked with an arrow.

For each scan we perform depth projection and preprocessing as described in [30]. The result is a depth map that reflects the geometric micro-structure of the surface, see Fig. 2b. This representation is the input to feature extraction.

From the depth map we extract a number of non-topological image descriptors in a block-based manner that serve as a baseline in our experiments. The block size is 128×128 pixels (i.e. 10.8×10.8 mm) and the step size between two blocks is 16 pixels (1.35 mm). The baseline features include: MPEG-7 Edge Histogram (EH) [11], Dense SIFT (DSIFT) [17], Local Binary Patterns (LBP) [18], Histogram of Oriented Gradients (HOG) [6], Gray-Level Co-occurrence Matrix (GLCM) [10], Global Histogram Shape (GHS), Spatial Depth Distribution (SDD), as well as manually modified enhanced versions of GHS and SDD (short EGHS and ESDD) that apply additional enhancements to the depth map described in [30].

Additionally to the baseline descriptors, we extract persistent homology descriptors in the same block-wise manner. For each patch, we compute a persistence diagram and derive the 12-dimensional aggregated descriptor (PD_AGG) as described in Sect. 2. Additionally, we extract persistence images (PIs) for different resolutions (8, 16, 32, 64) and standard deviations (0.00025, 0.0005, 0.001, 0.002) with and without weighting (see Sect. 2).

Alternatively, we first extract Completed LBP (CLBP) features [9] from the depth map as proposed in [15, 23] and then extract PD_AGG and PIs from the CLBP_S and CLBP_M maps.

For the discrimination of different surface topographies we employ supervised machine learning. All employed descriptors are represented by numerical vectors of fixed dimension and are thus suitable for statistical classification. As mentioned above, the class priors in our dataset are imbalanced. Skewed datasets pose problems to most classification techniques and often yield suboptimal models as one class dominates the other classes. A classifier expecially designed for imbalanced datasets is Random Undersampling Boosting (RUSBoost) [25]. RusBoost builds upon AdaBoost [8] which is an ensemble method that combines the weighted decisions of weak classifiers to obtain a final decision for a given input sample. RUSBoost extends this concept by a data sampling strategy that enforces similar class priors. During each training iteration the majority class in the training set is undersampled in a random fashion to balance the resulting class priors. In this manner, the weak classifiers can be learned from balanced datasets without being biased from the skewed class distribution.

For training the RUSBoost classifier we split the entire dataset into independent training and evaluation sets. The training set contains image patches from scans 1 and 2 from the dataset. Scans 3 and 4 make up the evaluation set. From the training set we randomly select 50 % of the blocks from class 1 (2962 blocks) and 30 % from class 2 (7592 blocks). From this subset of 9654 samples we train the RUSBoost classifier. For training we apply 5-fold cross-validation to estimate suitable classifier parameters (primarily the number of weak classifiers of the ensemble). The best parameters are used to train the classifier on the entire training set. The trained classifier is finally applied to the independent evaluation set of 27192 patches.

As a performance measure we employ the Dice Similarity Coefficient (DSC). DSC measures the mutual overlap between an automatic labeling X of an image and a manual (ground truth) labeling Y:

$$\mathrm{DSC}(X, Y) = \frac{2|X \cap Y|}{|X| + |Y|}.$$

DSC is between 0 and 1 where 1 means a perfect segmentation.

Each classification experiment is repeated 10 times with 10 different randomly selected subsets from the training set to reduce the dependency from the training data. From the 10 resulting DSC values we provide median and standard deviation as the final performance measures.

Aside from quantitative evaluations we investigate the following questions:

- Can persistent homology descriptors outperform descriptors like HOG, SIFT, and GLCM for surface classification?
- How does aggregation of the PD (PD_AGG) influence performance compared to non-aggregated representations like PI?
- Is CLBP a suitable input representation for persistent homology descriptors?
- How sensitive is PI to its parameters (resolution, sigma, weighting)?
- Do persistent homology descriptors add beneficial or even necessary information to the baseline descriptors in our classification task?

The experiment was implemented in Matlab. Most of the descriptors were extracted with VLFeat library [26], except PD_AGG and PI. We compute persistence intervals of the images using CAPD::RedHom library [13,14] with the PHAT [2,3] algorithm for persistence homology.

4   Results

We start our evaluation with the aggregated descriptor PD_AGG. The descriptor applied to our surfaces yields a DSC of 0.6528±0.0118 and represents a first baseline for further comparisons. Next, we apply PI with different resolutions, sigmas with and without weighting. Results are summarized in Table 1. All results for PI outperform that of PD_AGG. We assume the reason is that PD_AGG neglects the information about the points' localization, which is preserved in PI. The best result for PI is a DSC of 0.7335 ± 0.0024 without weighting. The difference between the best weighting and no weighting result is statistically significant[1] with $p - value = 0.006$. This result is surprising as it is contrary to the results of [1] where artificial datasets were used for evaluation. Results in Table 1 further show that PI has low sensitivity to different resolutions and sigmas.

Next, we evaluate the performance of PD_AGG and PI with CLBP as input representation, see Table 2. The best result for PD_AGG (0.6874 ± 0.0030) is

Table 1. DSC for PI descriptors depending on the sigma of the Gaussian function (σ) and resolution (*res*). Bold represents the best results for PI with and without weighting.

		$res = 8 \times 8$	$res = 16 \times 16$	$res = 32 \times 32$	$res = 64 \times 64$
weighting	$\sigma = 0.00025$	0.714 ± 0.005	0.718 ± 0.007	0.715 ± 0.007	0.709 ± 0.008
	$\sigma = 0.0005$	0.718 ± 0.005	0.715 ± 0.005	0.715 ± 0.006	0.714 ± 0.004
	$\sigma = 0.001$	0.715 ± 0.006	0.716 ± 0.005	0.718 ± 0.004	$\mathbf{0.718 \pm 0.005}$
	$\sigma = 0.002$	0.706 ± 0.003	0.719 ± 0.005	0.715 ± 0.004	0.710 ± 0.005
no wghting.	$\sigma = 0.001$	0.724 ± 0.004	$\mathbf{0.734 \pm 0.002}$	0.732 ± 0.004	0.733 ± 0.004

---

[1] Statistical significance is computed with the Wilcox signed rank test, as most of the samples do not pass the Shapiro-Wilk normality test.

Table 2. DSC for PD_AGG and PI descriptors extracted from the CLBP_S and CLBP_M maps. We consider two encodings for CLBP: rotation invariant uniform (riu2) and rotation invariant (ri) and vary radius r and the number of samples n. Bold numbers represent the best results for PD_AGG and PI.

Descriptor	CLBP type		$n = 8$	$n = 16$
PD_AGG	riu2	$r = 3$	0.613 ± 0.009	0.625 ± 0.005
		$r = 5$	0.654 ± 0.003	0.636 ± 0.010
	ri	$r = 3$	0.632 ± 0.009	0.666 ± 0.007
		$r = 5$	0.681 ± 0.004	$\mathbf{0.687 \pm 0.003}$
PI	riu2	$r = 3$	0.688 ± 0.005	0.702 ± 0.004
		$r = 5$	0.704 ± 0.002	0.717 ± 0.003
	ri	$r = 3$	0.699 ± 0.002	0.699 ± 0.002
		$r = 5$	0.703 ± 0.003	$\mathbf{0.703 \pm 0.003}$

obtained for the rotation invariant CLBP maps with radius 5 and number of samples 16. This improvement is statistically significant, with $p - value = 0.002$ (compared to the PD_AGG without CLBP). For PI we do not observe an improvement. This was confirmed by further experiments, where we combined PI obtained for the original depth map with PI on CLBP maps. The resulting DSC equals 0.7178 ± 0.0034. This shows not only that CLBP brings no additional information for PI, but further indicates that it can even be confusing for the classifier. The expressiveness of PI seems to be at a level where CLBP is not able to add additional information. Whereas PD_AGG is less expressive and thus benefits from the additional processing.

As a next step we investigate which locations of PI are the most important ones for classification. For this purpose we computed Gini importance measure for each location of the PI, see Fig. 3a. The most important pixels are located in the middle of the PI. It is worth noting that there are only few very important pixels, while the others are more than 10 time less important. Moreover, there are few important pixels near to the center of the diagonal. To get a more complete picture, we compute the Fisher discriminant for each location of the PI, see Fig. 3b. The result is to a large degree consistent with the Gini measure and confirms our observation.

Finally, we investigate the performance of topological vs. non-topological descriptors and their combinations. The DSC for baseline descriptors and for their combination with PD_AGG and PI are presented in Table 3. Our experiments show that both topological descriptors contribute additional valuable information to the baseline descriptors and improve the classification accuracy. All combinations with PD_AGG are significantly better than the baseline itself. Moreover, PI works significantly better than PD_AGG with all of the baseline descriptors (except for GHS, GHS+SDD, EGHS+ESDD where the improvement is not significant).

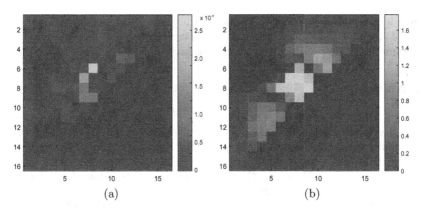

(a)   (b)

Fig. 3. Importance of the PI's pixels obtained with Gini importance measure and Fisher discriminant.

Table 3. DSC for baseline descriptors (B) and their combination with PD_AGG and PI descriptors (B + PD_AGG and B + PI, respectively). Asterisks (*) correspond to $p - values < 0.01$ when comparing B to B + PD_AGG and B + PD_AGG to B + PI.

Descriptor	Baseline (B)	B + PD_AGG	B + PI
EH	0.641 ± 0.007	$0.669 \pm 0.015^*$	$0.696 \pm 0.015^*$
LBP	0.452 ± 0.020	$0.531 \pm 0.023^*$	$0.587 \pm 0.027^*$
DSIFT	0.486 ± 0.003	$0.739 \pm 0.004^*$	$0.764 \pm 0.004^*$
HOG	0.503 ± 0.008	$0.712 \pm 0.007^*$	$0.732 \pm 0.003^*$
GLCM	0.645 ± 0.003	$0.706 \pm 0.002^*$	$0.732 \pm 0.002^*$
GHS	0.301 ± 0.048	$0.470 \pm 0.038^*$	0.476 ± 0.066
SDD	0.692 ± 0.003	$0.735 \pm 0.003^*$	$0.767 \pm 0.004^*$
GHS+SDD	0.399 ± 0.028	$0.426 \pm 0.027^*$	0.454 ± 0.029
EGHS	0.650 ± 0.008	$0.683 \pm 0.003^*$	$0.690 \pm 0.004^*$
ESDD	$\mathbf{0.743 \pm 0.002}$	$\mathbf{0.763 \pm 0.002^*}$	$\mathbf{0.790 \pm 0.002^*}$
EGHS+ESDD	0.728 ± 0.005	$0.740 \pm 0.003^*$	0.743 ± 0.005

5   Conclusion

We have presented an investigation of topological descriptors for 3D surface analysis. Our major conclusions are: (i) the aggregation of persistence diagrams removes important information which can be retained by using PI descriptors, (ii) PIs are expressive and robust descriptors that are well-suited to include topological information into ML pipelines, and (iii) topological descriptors are complementary to traditional image descriptors and represent necessary information to obtain peak performance in 3D surface classification. Furthermore, we observed that short intervals in the PD contribute more to classification accuracy than expected. This will be subject to future research.

Acknowledgements. Parts of the work for this paper has been carried out in the project 3D-Pitoti which is funded from the European Community's Seventh Framework Programme (FP7/2007-2013) under grant agreement no 600545; 2013-2016.

References

1. Adams, H., Chepushtanova, S., Emerson, T., Hanson, E., Kirby, M., Motta, F., Neville, R., Peterson, C., Shipman, P., Ziegelmeier, L.: Persistent images: A stable vector representation of persistent homology (2015). arXiv preprint arXiv:1507.06217
2. Bauer, U., Kerber, M., Reininghaus, J.: Phat - persistent homology algorithms toolbox (2013). https://code.google.com/p/phat/
3. Bauer, U., Kerber, M., Reininghaus, J., Wagner, H.: PHAT – Persistent homology algorithms toolbox. In: Hong, H., Yap, C. (eds.) ICMS 2014. LNCS, vol. 8592, pp. 137–143. Springer, Heidelberg (2014). http://dx.doi.org/10.1007/978-3-662-44199-2_24
4. Belongie, S., Malik, J., Puzicha, J.: Shape matching and object recognition using shape contexts. IEEE Trans. Pattern Anal. Mach. Intell. **24**(4), 509–522 (2002)
5. Crandall, D., Owens, A., Snavely, N., Huttenlocher, D.: Discrete-continuous optimization for large-scale structure from motion. In: 2011 IEEE Conference on Computer Vision and Pattern Recognition (CVPR), pp. 3001–3008. IEEE (2011)
6. Dalal, N., Triggs, B.: Histograms of oriented gradients for human detection. In: 2005 IEEE Computer Society Conference on Computer Vision and Pattern Recognition, CVPR 2005, vol. 1, pp. 886–893. IEEE (2005)
7. Edelsbrunner, H., Letscher, D., Zomorodian, A.: Topological persistence and simplification. Discrete Comput. Geom. **28**, 511–533 (2002)
8. Freund, Y., Schapire, R.E.: A decision-theoretic generalization of on-line learning and an application to boosting. J. Comput. Syst. Sci. **55**(1), 119–139 (1997)
9. Guo, Z., Zhang, L., Zhang, D.: A completed modeling of local binary pattern operator for texture classification. IEEE Trans. Image Process. **19**(6), 1657–1663 (2010)
10. Haralick, R.M., Shanmugam, K., Dinstein, I.H.: Textural features for image classification. IEEE Trans. Syst. Man Cybern. **6**, 610–621 (1973)
11. ISO-IEC: Information Technology - Multimedia Content Description Interface.15938, ISO/IEC, Moving Pictures Expert Group, 1st edn. (2002)
12. Johnson, A.E., Hebert, M.: Using spin images for efficient object recognition in cluttered 3D scenes. IEEE Trans. Pattern Anal. Mach. Intell. **21**(5), 433–449 (1999)
13. Juda, M., Mrozek, M., Brendel, P., Wagner, H., et al.: CAPD::RedHom (2010–2015). http://redhom.ii.uj.edu.pl
14. Juda, M., Mrozek, M.: CAPD:RedHom v2 - homology software based on reduction algorithms. In: Hong, H., Yap, C. (eds.) ICMS 2014. LNCS, vol. 8592, pp. 160–166. Springer, Heidelberg (2014)
15. Li, C., Ovsjanikov, M., Chazal, F.: Persistence-based structural recognition. In: IEEE Conference on Computer Vision and Pattern Recognition (CVPR), pp. 2003–2010. IEEE (2014)
16. López, V., Fernández, A., García, S., Palade, V., Herrera, F.: An insight into classification with imbalanced data: Empirical results and current trends on using data intrinsic characteristics. Inf. Sci. **250**, 113–141 (2013)
17. Lowe, D.G.: Distinctive image features from scale-invariant keypoints. Int. J. Comput. Vis. **60**(2), 91–110 (2004)

18. Ojala, T., Pietikäinen, M., Harwood, D.: A comparative study of texture measures with classification based on featured distributions. Pattern Recogn. **29**(1), 51–59 (1996)
19. Othmani, A., Lew Yan Voon, L., Stolz, C., Piboule, A.: Single tree species classification from terrestrial laser scanning data for forest inventory. Pattern Recogn. Lett. **34**(16), 2144–2150 (2013)
20. Poincaré, H.J.: Sur le probleme des trois corps et les équations de la dynamique. Acta Math. **13**, 1–270 (1890)
21. Poincaré, H.J.: Les méthodes nouvelles de la mécanique céleste. Gauthiers-Villars, Paris (1892, 1893, 1899)
22. Poincaré, H.J.: Analysis situs. J. Éc. Polytech., ser. 2 **1**, 1–123 (1895)
23. Reininghaus, J., Huber, S., Bauer, U., Kwitt, R.: A stable multi-scale kernel for topological machine learning (2014). arXiv preprint arXiv:1412.6821
24. Rusu, R.B., Marton, Z.C., Blodow, N., Beetz, M.: Persistent point feature histograms for 3D point clouds. In: Proceedings of the 10th International Conference on Intel Autonomous System (IAS-10), Baden-Baden, Germany, pp. 119–128 (2008)
25. Seiffert, C., Khoshgoftaar, T.M., Van Hulse, J., Napolitano, A.: Rusboost: A hybrid approach to alleviating class imbalance. IEEE Trans. Syst. Man Cybern. Part A: Syst. Hum. **40**(1), 185–197 (2010)
26. Vedaldi, A., Fulkerson, B.: Vlfeat: An open and portable library of computer vision algorithms. In: Proceedings of the International Conference on Multimedia, pp. 1469–1472. ACM (2010)
27. Wohlfeil, J., Strackenbrock, B., Kossyk, I.: Automated high resolution 3D reconstruction of cultural heritage using multi-scale sensor systems and semi-global matching. Int. Arch. Photogrammetry Remote Sens. Spat. Inf. Sci. XL-4 W **4**, 37–43 (2013)
28. Wu, C.: Towards linear-time incremental structure from motion. In: 2013 International Conference on 3DTV, pp. 127–134. IEEE (2013)
29. Zeppelzauer, M., Poier, G., Seidl, M., Reinbacher, C., Breiteneder, C., Bischof, H., Schulter, S.: Interactive segmentation of rock-art in high-resolution 3D reconstructions. In: 2015 Digital Heritage, vol. 2, pp. 37–44, September 2015. doi:10.1109/DigitalHeritage.2015.7419450
30. Zeppelzauer, M., Seidl, M.: Efficient image-space extraction and representation of 3D surface topography. In: Proceedings of the IEEE International Conference on Image Processing (ICIP). IEEE, Quebec, Canada (2015). http://arXiv.org/pdf/1504.08308v3.pdf

Towards a Topological Fingerprint of Music

Mattia G. Bergomi[1]([✉]), Adriano Baratè[2], and Barbara Di Fabio[3]

[1] Champalimaud Neuroscience Programme,
Champalimaud Centre for the Unknown, Lisbon, Portugal
`mattia.bergomi@neuro.fchampalimaud.org`
[2] Laboratorio di Informatica Musicale,
Università degli Studi di Milano, Milano, Italy
`barate@di.unimi.it`
[3] Dipartimento di Scienze e Metodi dell'Ingegneria,
Università di Modena e Reggio Emilia, Reggio Emilia, Italy
`barbara.difabio@unimore.it`

Abstract. Can music be represented as a meaningful geometric and topological object? In this paper, we propose a strategy to describe some music features as a polyhedral surface obtained by a simplicial interpretation of the *Tonnetz*. The *Tonnetz* is a graph largely used in computational musicology to describe the harmonic relationships of notes in equal tuning. In particular, we use persistent homology in order to describe the *persistent* properties of music encoded in the aforementioned model. Both the relevance and the characteristics of this approach are discussed by analyzing some paradigmatic compositional styles. Eventually, the task of automatic music style classification is addressed by computing the hierarchical clustering of the topological fingerprints associated with some collections of compositions.

Keywords: Music · Classification · Clustering · Tonnetz · Persistent homology

1 Introduction

Generally, the core of a piece of music consists of a small collection of strong, recognizable concepts, that are grasped by the majority of the listeners [13,17,29]. These *core concepts* are developed during the composition by varying levels of tension over time, drawing the attention of the listener to particular moments thanks to specific choices, frustrating his/her intuition through unexpected changes, or confirming his/her expectation with, for instance, a well-known cadence leading to resolution.

As the models for the analysis of audio signals take advantage of the strategies developed for image analysis [22,27,30], it is possible to borrow some tools from the topological analysis of shapes and data to tackle the problem of music analysis and classification. The main aim of this paper is the introduction of a low-dimensional geometric-topological model in order to describe, albeit in an extremely simplified form, music styles.

© Springer International Publishing Switzerland 2016
A. Bac and J.-L. Mari (Eds.): CTIC 2016, LNCS 9667, pp. 88–100, 2016.
DOI: 10.1007/978-3-319-39441-1_9

Loosely speaking, we introduce a metric representation of music as a planar polyhedral surface, whose vertices are then shifted along a third dimension in basis on a specific function. The shapes obtained via these deformations are fingerprinted by computing their *persistent homology* [15]. Afterwards, the musical meaning of this topological representation of music is discussed and applied to automatic style classification on three different datasets.

2   Background on Persistent Homology

In computational topology [14], persistent homology is actually considered an invaluable tool to describe both geometry and topology of a certain space, not only because of the simplicity of the method, but also because all the properties are ranked by importance, allowing us to choose the level of detail at which to perform such a description [15,18].

In more formal terms, given a topological space X, we define a continuous function $f : X \to \mathbb{R}$ to obtain a family of subspaces $X_u = f^{-1}((-\infty, u]), u \in \mathbb{R}$, nested by inclusion, i.e. a filtration of X. The map f, called therefore a *filtering function*, is chosen according to the geometrical properties of interest (e.g., height, distance from center of mass, curvature). Applying homology to the filtration, births and deaths of topological features can be algebraically detected and their lifetime measured. The scale at which a feature is significant is measured by its longevity. Formally, given $u \leq v \in \mathbb{R}$, we consider the inclusion of X_u into X_v. This inclusion induces a homomorphism of homology groups $H_k(X_u) \to H_k(X_v)$ for every $k \in \mathbb{Z}$. The image of such a homomorphism consists of the k-homology classes that live at least from u to v along the filtration, and is called the kth *persistent homology group* of the pair (X, f) at (u, v). When, for every $(u, v), u \leq v$, the kth persistent homology groups are finitely generated, we can compactly describe them using the so-called *persistence diagrams*. A persistence diagram $D_k(X, f)$ is the subset of $\{(u, v) \in \mathbb{R}^2 : u < v\}$ consisting of points (called *proper points*) and vertical lines (called *points at infinity*) encoding the levels of f at which the birth and the death of homological classes occur, union all the points belonging to the diagonal $u = v$. In particular, if there exists at least one k-homology class that is born at the level $\bar{u}$ and is dead at the level $\bar{v}$ along the filtration induced by f, then $p = (\bar{u}, \bar{v})$ is a proper point of $D_k(X, f)$; if there exists at least one k-homology class that is born at the level $\bar{u}$ and never dies along the filtration induced by f, then $p = (\bar{u}, +\infty)$ is a point at infinity of $D_k(X, f)$. A point at infinity is usually represented as the vertical line $u = \bar{u}$. Both points and lines are equipped with a multiplicity that depends on the number of classes with the same lifetime [19]. An example of persistence diagrams is displayed in Fig. 1. The surface $X \subset \mathbb{R}^3$ is endowed with the height function f. The associated persistence diagrams $D_0(X, f)$ and $D_1(X, f)$ are displayed on the right. $D_0(X, f)$ consists of one point at infinity r, whose abscissa u detects the absolute minimum of f, and one proper point p, whose abscissa and ordinate detect, respectively, the level at which the new connected component appears and merges with the existing one. $D_1(X, f)$ consists of one proper point, whose

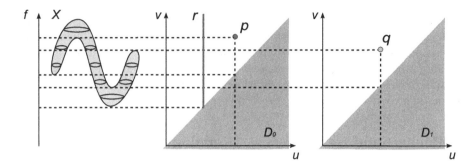

Fig. 1. Left: the height function on the topological space X. Right: the associated persistence diagrams $D_k(X, f)$, with $k = 0, 1$.

abscissa and ordinate detect, respectively, the level at which a new tunnel is created and disappears along the filtration.

One of the main reasons behind the usage of persistence diagrams in applications consists in the possibility of estimating the degree of dissimilarity of two spaces with respect to a certain geometrical property through an appropriate comparison of these shape descriptors. Because of its properties of stability [7] and optimality [10], the most used instrument to compare persistence diagrams is given by the so called *bottleneck distance* (a.k.a. *matching distance*) [9]. By stability, we mean that small changes of the filtering functions imply only small changes of the associated persistence diagrams in terms of this distance; by stability and optimality, we mean that, among all the stable distances between persistence diagrams, the bottleneck distance is the most discriminative one.

Definition 1. *The bottleneck distance between two persistence diagrams D and D' is defined as*

$$d_B(D, D') = \min_{\sigma} \max_{p \in D} d(p, \sigma(p)),$$

where σ varies among all the bijections between D and D' and

$$d\left((u, v), (u', v')\right) = \min\left\{\max\left\{|u - u'|, |v - v'|\right\}, \max\left\{\frac{v - u}{2}, \frac{v' - u'}{2}\right\}\right\}$$

for every $(u, v), (u', v') \in \{(x, y) \in \mathbb{R}^2 : x \leq y\}$.

3   Musical Setting

In order to safely introduce the main model presented in this paper, we start by defining some basic musical objects.

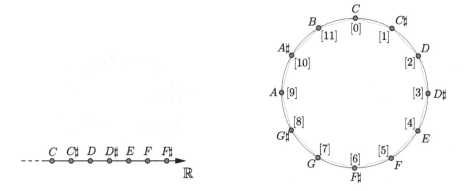

Fig. 2. Fundamental music representation spaces: the pitch space (left), and the pitch-class space (right).

We model a *note in equal tuning* $\mathfrak{n}$ as a pair $(p, d) \in \mathbb{R}^2$, where p is called the *pitch* of the note, and d is its *duration* in seconds. In particular, if ν denotes the fundamental frequency of $\mathfrak{n}$, the pitch $p(\nu)$ is defined as $p(\nu) = 69 + 12 \log_2\left(\frac{\nu}{440}\right)$, where 440 Hz is the fundamental frequency of the note A_4 (the *la* of the fourth octave of the piano). For further details on pitches, see, e.g., [11].

On a perceptual level, two notes an octave apart are really similar [4], thus, it is common to identify pitches modulo octave, by considering *pitch classes* $[p] = \{p + 12k : k \in \mathbb{Z}\} \cong \mathbb{R}/12\mathbb{Z}$. A representation of both the pitch and pitch-class spaces is depicted in Fig. 2.

3.1 The simplicial *Tonnetz*

The *Tonnetz* was originally introduced in [16] as a simple 3×4 matrix representing both the acoustical and harmonic relationships among pitch classes. Later, it has been largely generalized to several formalisms, see, e.g., [8,12,32]. We will focus on its interpretation as a simplicial complex [2]. In this setting, the *Tonnetz* is modeled as an infinite planar simplicial complex, whose 0-simplices are labeled with pitch classes in a way that 1-simplices form either perfect fifth, major, or minor third intervals, and 2-simplices correspond to either major or minor triads. A finite subcomplex of the *Tonnetz* T is depicted in Fig. 3a. We observe that the labels on its vertices are periodic with respect to the transposition of both minor and major third. This feature allows to work with the more comfortable toroidal representation $\mathbb{T}$ displayed in Fig. 3b.

It is possible to analyze and classify music by considering the subcomplexes of T generated by a sequence of pitch classes [2]. However, this approach does not allow to discriminate musical styles in a geometric or topological sense. In fact, as the example in Fig. 4 shows, two perceptively distinct sonorities (Fig. 4a and b) can be represented by isomorphic subcomplexes (Fig. 4c and d, respectively).

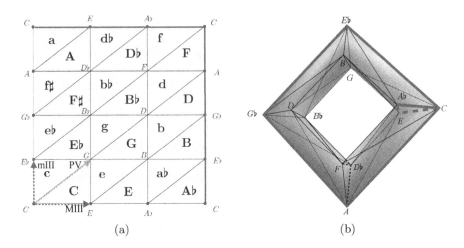

Fig. 3. (a) A finite subcomplex of the *Tonnetz*. (b) The *Tonnetz* torus $\mathbb{T}$ obtained by identifying vertices in (a) equipped with the same labels.

3.2 A Deformed *Tonnetz* for Music Analysis

In order to capture both the *temporal* and harmonic information encoded in a musical phrase, the associated *Tonnetz* should depend on both the pitches which are played and their duration. Our idea is based on the following observation: Let us consider again the example in Fig. 4. The musical phrases in Fig. 4a and b can be easily distinguished if we endow each vertex of the associated subcomplexes with a non-negative real number that detects for how long the associated pitches have been played during the execution of the phrase. Practically, we can replace the planar subcomplexes in Fig. 4c and d by the new subcomplexes Fig. 4e and f, respectively, obtained by shifting upward each vertex of a quantity equal to this weight.

In symbols, let V be the set of vertices of T and consider a finite collection of notes of a musical phrase, $\{\mathfrak{n}_1, \ldots, \mathfrak{n}_m\} = \{(p_1, d_1), \ldots, (p_m, d_m)\}$. Assume that $\{\mathfrak{n}_{i_1}, \ldots, \mathfrak{n}_{i_k}\}$ is the subset whose pitches $p_{i_1}, \ldots, p_{i_k}$ belong to $[p]$. We define a map that takes each vertex $v = (x_v, y_v, 0) \in \mathbb{R}^3$ labeled with $[p]$ to the point $(x_v, y_v, d_v) \in \mathbb{R}^3$, where $d_v = \sum_{j=1}^{k} d_{i_j}$, and then extend it linearly to all the simplices. The *Tonnetz* deformed under the action of this map will be denoted by $\mathcal{T}$, and will be used as the main object of our topological description of music style. A 3-dimensional interactive animation showing how the *Tonnetz* is deformed by a musical phrase is available at http://nami-lab.com/tonnetz/examples/deformed_tonnetz_int_sound_pers.html. It allows the user to play with its own keyboard to generate specific deformations or simply to analyze their evolution in real time during the execution of a certain song.

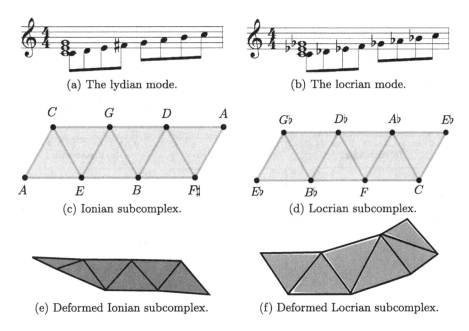

(a) The lydian mode.   (b) The locrian mode.

(c) Ionian subcomplex.   (d) Locrian subcomplex.

(e) Deformed Ionian subcomplex.   (f) Deformed Locrian subcomplex.

Fig. 4. Two modes (a) and (b) represented by isomorphic subcomplexes (c) and (d) of the *Tonnetz*. They can be distinguished when deforming the *Tonnetz* by taking into account the duration of the notes: (e) and (f) are deformed subcomplexes associated to the modes, observed from the same position.

4   A Topological Fingerprint of Music Styles

In order to describe the deformed *Tonnetz*, we use persistent homology.

We define the height function f on $\mathcal{T}$ to induce a lower level set filtration on the torus $\mathbb{T}$. The persistence diagrams obtained with this process are *descriptors* of the style characterizing the composition represented as a shape.

0th Persistence Diagrams. The connectedness of $\mathbb{T}$ is retrieved by the presence of only one point at infinity. Let $u = \bar{u}$ be its equation: $\bar{u}$ is the absolute minimum of f on the deformed *Tonnetz*. If $\bar{u} \approx 0$, then there exists at least one pitch-class set that does not have a relevant role in the composition, suggesting that it is based on a stable tonal or modal choice. On the contrary, if $\bar{u} >> 0$, then all the pitch classes have been used in the composition for a relevant time. This configuration suits a more atonal or chromatic style. The presence of proper points is due to the existence of minima of the height function, that are subcomplexes of the *Tonnetz* not connected by an edge, and hence, representing a dissonant interval [26]. Furthermore, the structure of the *Tonnetz* torus allows to retrieve a maximum of three connected components. To create this particular configuration, it is necessary to play a chromatic cluster: for instance, $C, C\sharp, D$, that is not usually used in a tonal or modal context.

1st Persistence Diagrams. The lifespan of 1-dimensional holes traversing the filtration provides symmetrical information with respect to the 0th persistence

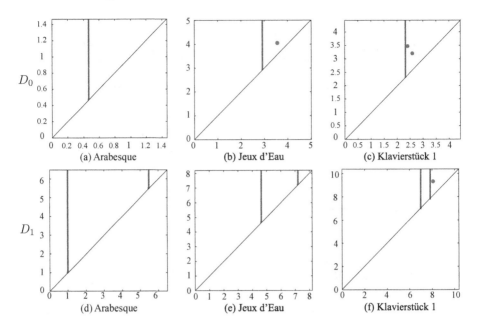

Fig. 5. The 0th (first row) and 1st persistence diagrams (second row) representing the topological fingerprints associated with three different compositions.

analysis. In this case, two points at infinity detect the two generators of the 1st homology group of the torus and, if there exists, proper points detect the presence of maxima of the height function, that are subcomplexes of the *Tonnetz* not connected by an edge.

As an example, we consider the persistence diagrams associated with Debussy's *Arabesque*, *Jeux d'Eau* by Ravel, and *Klavierstück 1* by Schönberg, shown in Fig. 5. In the 0th persistence diagram describing *Arabesque*, there are no proper points. This is an evidence of the pentatonic and diatonic/modal inspiration of the composition [28]. We also observe that the entire chromatic scale has been used, since $\bar{u} > 0$. The abscissa of the point at infinity in the 0th persistence diagram of *Jeux d'Eau* is characterized by a high value, thus the entire set of pitch classes has been largely used in the composition. Moreover, the presence of a proper point highlights the *ante-litteram* use of the *Petrushka chord*, a superposition of a major triad and its tritone substitute: for instance, $G = (G, B, D) + C\sharp = (C\sharp, E\sharp, G\sharp)$. Finally, the diagram associated with the *Klavierstück 1* has two relevant proper points: this last feature points out the atonal nature of the composition.

The second row of Fig. 5 shows the 1st persistence diagrams associated with the same compositions. The tonal nature of *Arabesque* is highlighted by the a large distance between the points at infinity and the absence of proper points. The chromatic style of *Jeux d'Eau* implies the reduction of the distance between the two points at infinity. This last feature appears also in the diagram describing the *Klavierstück 1*, whose atonal tendency is stressed by a proper point representing the relevant lifespan of a third non-connected subcomplex.

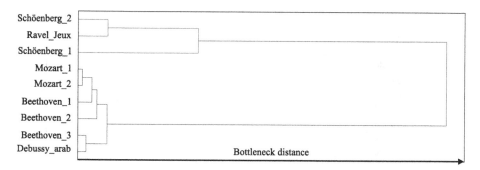

Fig. 6. Persistence-based clustering of nine classical and contemporary pieces.

4.1 Applications

In the following applications, we show how the persistence diagrams associated with a collection of compositions can effectively classify them according to their style. For $k = 0, 1$, let $\mathcal{D} = \{D_1, \ldots, D_n\}$ be the set of kth persistence diagrams. Let $M = m_{ij} = d_B(D_i, D_j)$, for $1 \leqslant i, j \leqslant n$, be their distance matrix. The hierarchical clustering analysis [25] allows us to describe the configuration of the diagrams $D_i \in \mathcal{D}$ with respect to the bottleneck distance. We will represent the organization of all their possible clusters as a dendrogram [21,23]. In this type of diagram, the abscissa of each splitting (vertical line) measures the distance between two clusters. Such distance is computed through elementary operations on the elements of M.

Tonal and Atonal Music. We consider a dataset composed by nine pieces selected among the compositions by Beethoven, Debussy, Mozart, Ravel and Schönberg available at http://nami-lab.com/tonnetz/examples/deformed_tonnetz_int_sound_pers.html.

The clustering computed using the 0th persistence of these pieces is depicted in Fig. 6. Data are organized in two main clusters, that segregate the two first pieces of Schönberg's *Drei Klavierstücke* and Ravel's *Jeux d'Eau*, from the ones by Mozart, Beethoven and Debussy. The association between *Klavierstück 2* and *Jeux d'Eau* mirrors the particular nature of this Schönberg's composition, that lies at the crossroad of tonal and atonal music, as it is proven by its disparate tonal interpretations [3,24,31]. The two movements of Mozart's *KV311* form immediately a cluster reached at an increasing distance by the two first movements of the *Sonata in C major* by Beethoven. The third movement of *Sonata in C major* is grouped with *Arabesque* because both are characterized by a generous use of the pentatonic scale.

Comparing Three Versions of *All the Things You Are*. The three interpretations are structured as follows:

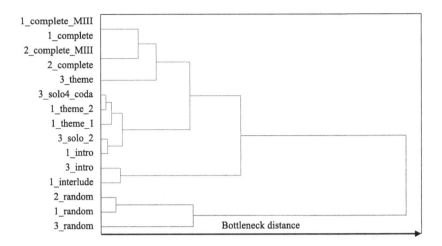

Fig. 7. Comparing three different versions of *All the Things You Are.*

1. Version 1 is played by a quartet in a standard way. Its main sections are a 3/4 introduction, a first exposition of the theme, and a 12 bars interlude introducing the last theme enriched by short improvisations.
2. Version 2 is performed by a piano solo, and it is characterized by a rich chromatic playing style of both hands. The main theme is executed twice.
3. Version 3 is performed by a duo (piano and bass). Its structure consists of an introduction, an exposition of the theme, and a piano improvisation.

The dataset is composed by the complete versions of the standard labeled as *i_complete* and a transposed version (*i_complete_interval*). Segments of each versions are included in the dataset and labeled as *i_segment*. In order to test the ability of the model to distinguish between a piece modulating in several tonalities, enriched with chromatic solos, and a non-structured sequence of pitches, a random version of each interpretation is also part of the dataset (*i_random*).

The resulting dendrogram is displayed in Fig. 7. We observe that the transposed versions have distance zero from the original ones, as an effect of the invariance of the filtration induced on the *Tonnetz* torus by the height function under uniform transposition. The randomized versions of the songs are well segregated. A small cluster groups the interlude of the first and the introduction of the third version, because both fragments share a very similar structure in terms of intervallic leaps and rhythm. Finally, in the top cluster, the two complete songs are linked to the fragment of the third version containing the theme. Hence, the 0th persistence homology retrieves the fragments containing the whole structure of the standard. This feature is surprising when taking into account the several modulations of the piece.

Big Pop Clustering. Figure 8 shows a simplification of the clustering resulting by the comparison of the 1st persistence diagrams associated with 58 pop
songs performed by 28 artists, spacing from Ray Charles to Lady Gaga. In order
to give a simplified representation of this dendrogram, we considered only the
three biggest clusters detected by the algorithm. On the left of each cluster, we
listed the artists whose songs belong to that group. Names written in black bold
characters indicate artists whose songs are entirely grouped in the cluster at
their right, while red bold characters identify the three artists whose songs are
spread among the three groups. We observe how the entire collection of songs
by Ringo Starr, Paul McCartney and Simon & Garfunkel are grouped together
in the blue cluster with Ray Charles, Stevie Wonder and George Benson. Moreover, the heterogeneity that characterizes Sting's compositions is mirrored by
the presence of one of his songs in each cluster. The second and third clusters
are less homogeneous, but promising, taking into account that so far each song
is identified by a single persistence diagram.

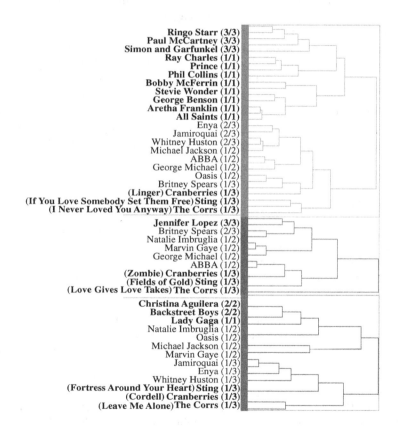

Fig. 8. A simplified representation of the clustering of 58 pop songs generated from
their 1st persistence diagrams.

5   Discussion and Future Works

We suggested a model describing music by taking into account the contribution of each pair (pitch class, duration) associated with the notes of a composition. The height function has been defined on the vertices of the simplicial *Tonnetz* to induce a lower level set filtration on the *Tonnetz* torus. The 0th and 1st persistence diagrams associated with different musical pieces have been interpreted in musical terms and their bottleneck distance has been used to classify them hierarchically. The possible clusterings have been represented as dendrograms, showing that 0th and 1st persistence can be used to analyze and classify music.

The analysis and classification of music we performed has been realized by considering datasets composed by MIDI files. However, the extension of this model to audio files is straightforward. Given an audio signal, the chroma analysis [20] retrieves the contribution in time of each pitch class. Using a chromagram to define the height function, it would surely be affected by the noisy data coming from the signal. The stability of persistence diagrams, when compared using the bottleneck distance, assures robustness against noise.

The model itself can be extended in several ways. For instance, it is possible to augment the dimensionality of the simplicial *Tonnetz*. This would result in losing the property of being easily visualizable, but it would give the possibility to encode more information. This could be done by associating with each pitch class of the *Tonnetz* a velocity, or by adding information concerning whatever pitch-class related feature. Moreover, topological persistence offers further tools to improve the strategies we suggested. A natural development is the study of the multidimensional persistent homology [5,6] of musical spaces and their time-varying nature [1].

References

1. Bergomi, M.G.: Dynamical and Topological Tools for (Modern) Music Analysis. Ph.D. thesis, Université Pierre et Marie Curie (2015)
2. Bigo, L., Andreatta, M., Giavitto, J.-L., Michel, O., Spicher, A.: Computation and visualization of musical structures in chord-based simplicial complexes. In: Yust, J., Wild, J., Burgoyne, J.A. (eds.) MCM 2013. LNCS, vol. 7937, pp. 38–51. Springer, Heidelberg (2013)
3. Brinkmann, R.: Arnold Schönberg, drei Klavierstücke Op. 11: Studien zur frühen Atonalität bei Schönberg. Franz Steiner Verlag (1969)
4. Burns, E.M., Ward, W.D.: Intervals, scales, and tuning. Psychol. Music **2**, 215–264 (1999)
5. Carlsson, G., Zomorodian, A.: The theory of multidimensional persistence. Discrete Comput. Geom. **42**(1), 71–93 (2009)
6. Cerri, A., Di Fabio, B., Ferri, M., Frosini, P., Landi, C.: Betti numbers in multidimensional persistent homology are stable functions. Math. Meth. Appl. Sci. **36**(12), 1543–1557 (2013)
7. Cohen-Steiner, D., Edelsbrunner, H., Harer, J.: Stability of persistence diagrams. Discrete Comput. Geom. **37**(1), 103–120 (2007)

8. Cohn, R.: Neo-riemannian operations, parsimonious trichords, and their "Tonnetz" representations. J. Music Theor. **41**, 1–66 (1997)

9. d'Amico, M., Frosini, P., Landi, C.: Using matching distance in size theory: a survey. Int. J. Imag. Syst. Tech. **16**(5), 154–161 (2006)

10. d'Amico, M., Frosini, P., Landi, C.: Natural pseudo-distance and optimal matching between reduced size functions. Acta Applicandae Mathematicae **109**(2), 527–554 (2010)

11. De Cheveigne, A.: Pitch perception models. In: Plack, C.J., Fay, R.R., Oxenham, A.J., Popper, A.N. (eds.) Pitch, pp. 169–233. Springer, New York (2005)

12. Douthett, J., Steinbach, P.: Parsimonious graphs: a study in parsimony, contextual transformations, and modes of limited transposition. J. Music Theor. **42**, 241–263 (1998)

13. Dowling, W.J.: Recognition of melodic transformations: inversion, retrograde, and retrograde inversion. Percept. Psychophys. **12**(5), 417–421 (1972)

14. Edelsbrunner, H., Harer, J.: Computational Topology: An Introduction. American Mathematical Society, Providence (2009)

15. Edelsbrunner, H., Harer, J.: Persistent homology-a survey. Contemp. Math. **453**, 257–282 (2008)

16. Euler, L.: De harmoniae veris principiis per speculum musicum repraesentatis. Opera Omnia **3**(2), 568–586 (1774)

17. Folgieri, R., Bergomi, M.G., Castellani, S.: EEG-based brain-computer interface for emotional involvement in games through music. In: Lee, N. (ed.) Digital Da Vinci. Computers in Music, pp. 205–236. Springer, New York (2014)

18. Frosini, P., Landi, C.: Size theory as a topological tool for computer vision. Pattern Recogn. Image Anal. **9**, 596–603 (1999)

19. Frosini, P., Landi, C.: Size functions and formal series. Appl. Algebra Eng. Comm. Comput. **12**(4), 327–349 (2001)

20. Harte, C., Sandler, M.: Automatic chord identifcation using a quantised chromagram. In: Audio Engineering Society Convention 118. Audio Engineering Society (2005)

21. Langfelder, P., Zhang, B., Horvath, S.: Defining clusters from a hierarchical cluster tree: the dynamic tree cut package for R. Bioinformatics **24**(5), 719–720 (2008)

22. Li, T.L., Chan, A.B., Chun, A.: Automatic musical pattern feature extraction using convolutional neural network. In: Proceedings of International Conference on Data Mining and Applications (2010)

23. Martinez, W.L., Martinez, A., Solka, J.: Exploratory Data Analysis with MATLAB. CRC Press, Boca Raton (2010)

24. Ogdon, W.: How tonality functions in Shoenberg Opus-11, Number-1. J. Arnold Schoenberg Inst. **5**(2), 169–181 (1981)

25. Ott, N.: Visualization of Hierarchical Clustering: Graph Types and Software Tools. GRIN Verlag, Munich (2009)

26. Plomp, R., Levelt, W.J.: Tonal consonance and critical bandwidth. J. Acoust. Soc. Am. **38**(4), 548–560 (1965)

27. Smaragdis, P., Brown, J.C.: Non-negative matrix factorization for polyphonic music transcription. In: 2003 IEEE Workshop on Applications of Signal Processing to Audio and Acoustics, pp. 177–180. IEEE (2003)

28. Trezise, S.: The Cambridge Companion to Debussy. Cambridge University Press, Cambridge (2003)

29. Tulipano, L., Bergomi, M.G.: Meaning, music and emotions: a neural activity analysis. In: NEA Science. pp. 105–108 (2015)

30. Wang, A., et al.: An Industrial Strength Audio Search Algorithm. In: ISMIR. pp. 7–13 (2003)
31. William, B.: Harmony in Radical European Music. In: Society of Music Theory (1984)
32. Žabka, M.: Generalized *Tonnetz* and well-formed GTS: a scale theory inspired by the neo-riemannians. In: Chew, E., Childs, A., Chuan, C.-H. (eds.) MCM 2009. CCIS, vol. 38, pp. 286–298. Springer, Heidelberg (2009)

Topological Comparisons of Fluvial Reservoir Rock Volumes Using Betti Numbers: Application to CO_2 Storage Uncertainty Analysis

Asmae Dahrabou[1], Sophie Viseur[2(✉)], Aldo Gonzalez-Lorenzo[3,4],
Jérémy Rohmer[5], Alexandra Bac[3], Pedro Real[4], Jean-Luc Mari[3],
and Pascal Audigane[5]

[1] Neuchâtel University, Neuchâte, Switzerland
[2] Aix-Marseille Université, CEREGE UM 34, CNRS, IRD, Marseille, France
viseur@cerege.fr
[3] Aix-Marseille Université, CNRS, LSIS UMR 7296, Marseille, France
[4] Department of Applied Mathematics I, University of Seville, Seville, Spain
[5] BRGM, Orléans, France

Abstract. To prevent the release of large quantities of CO_2 into the atmosphere, carbon capture and storage (CCS) represents a potential means of mitigating the contribution of fossil fuel emissions to global warming and ocean acidification. Fluvial saline aquifers are favourite targeted reservoirs for CO_2 storage. These reservoirs are very heterogeneous but their heterogeneities were rarely integrated into CO_2 reservoir models. Moreover, contrary to petroleum reservoirs, the available dataset is very limited and not supposed to be enriched. This leads to wide uncertainties on reservoir characteristics required for CSS management (injection location, CO_2 plume migration, etc.). Stochastic simulations are classical strategies in such under-constrained context. They aim at generating a wide number of models that all fit the available dataset. The generated models serve as support for computing the required reservoir characteristics and their uncertainties. A challenge is to optimize the uncertainty computations by selecting stochastic models that should have *a priori* very different flow behaviours. Fluid flows depend on the connectivity of reservoir rocks (channel deposits). In this paper, it is proposed to study the variability of the Betti numbers in function of different fluvial architectures. The aim is to quantify the impact of fluvial heterogeneities and their spatial distribution on reservoir rock topology and then on CO_2 storage capacities. Representative models of different scenarios of channel stacking and their internal heterogeneities are generated using geostatistical simulation approaches. The Betti numbers are computed on each generated models and statistically analysed to exhibit if fluvial architecture controls reservoir topology.

1 Introduction

The impact of the heterogeneities on the reservoir performances has always been a key topic in geosciences, especially in the case of fluvial reservoirs.

© Springer International Publishing Switzerland 2016
A. Bac and J.-L. Mari (Eds.): CTIC 2016, LNCS 9667, pp. 101–112, 2016.
DOI: 10.1007/978-3-319-39441-1_10

They consist of sediments deposited along rivers. They are very heterogeneous and characterized by a complex spatial organization of sedimentary entities.

Many studies have been developed for analyzing uncertainty of hydrocarbon fluvial reservoirs [1]. For instance, stochastic simulation methods have been proposed to generate plausible 3D models of fluvial sedimentary architectures, conditioned to available subsurface data (e.g., seismic, well). These sets of 3D models serve as support for risk assessments. An important issue is to characterize the uncertainties on dynamic reservoir characteristics because flow simulators are highly time and power consuming. As a consequence, only few 3D models of reservoir heterogeneity are used to perform flow simulations. For several years, optimization techniques [2] have been proposed to scan more rapidly the uncertainty space. It is based on the determination of distances between the generated stochastic simulations. Descriptors, i.e. variables estimated on the reservoir, are used to define distances between models and multivariate statistics are used to guide the selection of a limited subset of 3D reservoir models that are very different and that correspond to a representative subsampling of the generated plausible models.

In the CO_2 geological storage context, taking into account high resolution heterogeneity is still at the beginning even though some studies have shown the impact of heterogeneities on CO_2 reservoir performances and capacities [3,4]. Techniques proposed for petroleum reservoir analysis [1,2,5] may be used as basis for CO_2 applications. Several descriptors have been proposed to describe the geometry and the topology (i.e. connectivity) of reservoir rocks. However, differences exist between the problematics of petroleum and CO_2-storage reservoirs. First, the involved space and time scales are greater in the CO_2-storage domain than in the hydrocarbon industry. Second, the amount and availability of data are generally much more limited in CO_2 context. Finally, the use of flow simulation in CCS studies is not for predicting the flow path from an injection well to a producing one, but mainly for estimating the reservoir capacity and overpressure. This means that specific descriptors are needed.

In this paper, we focus on searching formal frameworks to define topological descriptors. Indeed, several geometrical characteristics have been already proposed [2,5,6], while only descriptors based on degraded flow simulations have been proposed for describing the reservoir connectivity. Even if these descriptors provide insights on rough flow behavior, they remain time consumming. The question is then to determine if the "static" topology of the reservoir rocks can also provide information about fluid flow behaviors. In this paper, we propose to use the formal framework of the Betti number to study the topology of the reservoir rock volume. However, we focus on determining if different fluvial architectures lead to significant differences of Betti number averages.

In a first section, basics on fluvial sediment architecture and the main modelling issues are presented. To compare the Betti numbers on different fluvial architectures, we propose to generate a synthetic data set of different 3D fluvial reservoirs. It is then possible to master the explanatory parameters of the models. Thus, the method used for generating the synthetic dataset and the explanatory parameters are described in a second section. In a third part, the mathematical background of the Betti numbers is briefly presented. The proposed approach for using the Betti numbers as descriptors is then described in a fourth section. The statistical approaches proposed to analyse the connectivity of these different reservoir models are also explained. Finally, the results are shown and discussed.

2   Fluvial Reservoir Modelling and Problematics

Rivers bring and deposit coarse sediments along their path (Fig. 1(A)) in a flood plain consisting in fine sediments. If the river becomes unstable due to internal or external (e.g. tectonics) factors, an *avulsion* occurs and the river path changes its location, potentially its orientation, in the flood plain. Coarse sediments are then deposited at the new river location. The sedimentary bodies deposited between two avulsions are often termed as *channels* even if the more convenient terminology is *channelbelts* [7,8]. Thus, fluvial reservoirs consist in the stackings of elongated tabular permeable bodies, the channelbelts, in an impermeable background. The fluvial architecture follows sedimentological laws such as: the channelbelt bodies pass entirely through the volume and their positions are correlated, in most cases [4,9] (Fig. 1(B)). The fluvial reservoirs are highly compartmentalized: the channelbelt bodies may be amalgamated in certain places and the connectivity between these bodies has then a major impact on the final reservoir rock organization, hence fluid flows.

For modelling purpose, it is common to only consider two lithologies ([4], Fig. 1(A)): the overbank deposits (mainly impermeable) and the channelbelts (reservoir rocks). Several algorithms have been proposed to stochastically reproduce the channelbelt stackings of fluvial architecture. Poisson point processes and simulated annealing are the most known techniques [1,10]. However, even if they assert to provide equiprobable models conditioned to subsurface data, both techniques consider channel positions as independent and do not assert that the generated objects totally pass through the volume. These drawbacks have been noticed by several authors [4,11,12]. In [4], the authors propose an algorithm to

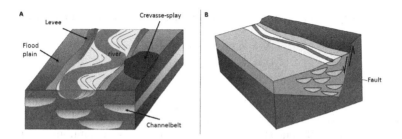

Fig. 1. Fluvial architecture: (A) channelbelt stackings and associated overbank deposits (levees, crevasse-splays, flood plain fines); (B) Conceptual model of a channelbelt stacking influenced by a fault activity (modified from [9]).

stochastically reproduce channel stackings that fit sedimentological laws. However, this technique is unconditional and can not then account for subsurface data. The integration of geological laws into stochastic conditioned models is still a sensitive issue in reservoir modelling [12] and a question is to determine which sedimentological laws really influence reservoir behaviors [5].

In this work, the objective is to determine if the connectivity of the reservoir rocks is in average influenced by different scenarios of channel stacking. An algorithm inspired from [4] was used to generate different families of stochastic models. The Betti numbers are proposed as descriptors for formelly compare the final topology, i.e. connectivity, of the reservoir rocks.

3   Synthetic Data Set

A grid of $100 \times 100 \times 100$ cells and $50 \times 50 \times 25$ scale was built. A channel-belt template is also constructed as a median polygonal line with perpendicular demi-ellipse cross-sections. Its length is twice the grid one to ensure that simulated objects totally pass through the volume. At a first glance, the object sizes have no importance as we only deal with topology. However, the channel sizes are chosen small enough compared to the grid in order to assert that the stochastic simulation is an ergodic process (i.e. statistics are reproduced on a realization [1]).

The principle of the algorithm is to simulate N channelbelt parametric objects within the volume. At each step, a channelbelt object is randomly generated and located in the volume using translation and rotation operations. Many parameters are required to define the dimensions, the sinuosity, the orientation and the location of a simulated channelbelt object. In this work, only two aspects have been considered as variable: the style of stacking and the orientations of the channelbelts. Indeed, the more parameters are variable, the more models are required for cross comparisons. Except the width and height of the channel sections, all the parameters are simulated using a Monte Carlo sampling on a density probability law. For a sake of simplicity, the probability laws are always considered as uniform and noticed $U(min, max)$ in the following. The minimum and maximum values are chosen so that it reproduces "realistic" shapes of channelbelts.

The sinuosity of the channel middle lines is simulated using a cosine equation as follows:

$$X = a \cdot cos(\phi + \tau \cdot \frac{Y+40}{280}$$
$$\phi = \frac{b \cdot \pi}{180}$$
$$\tau = c \cdot \pi$$
(1)

where $a \sim U(5, 20)$, $b \sim U(5, 30)$ and $c \sim U(1, 5)$. The width and height of the channel sections are constant and equal, respectively, to 10 and 3.

On one hand, two different styles of channel stacking were simulated (Fig. 2):

1. *unstructured*: the ith simulated object is randomly located in the volume, independently from the $i - 1$ already simulated objects. This is similar to Poisson point process as the events are considered as independent.

2. *structured*: the location of the ith object depends on the location of the $i-1^{th}$ object. Vertical and lateral translations are defined between them (Fig. 2(A)). These values are simulated using a Monte Carlo sampling on, respectively, $U(0,3)$ and $U(-10,10)$, except for the first simulated object for which the lateral translation is randomly chosen in the volume. The vertical translation asserts that there is no vertical gap between the $i-1^{th}$ and the ith objects. The values follow the sedimentological laws detailed in [4].

On the other hand, the orientations of the channel bodies are modified. In two model series, the channelbelt orientation θ varies from $-10°$ to $10°$, and in the two others from $-90°$ to $90°$ (Fig. 2(B)). In other words, the channelbelts are roughly oriented in the same directions (North-South) for the first two series, and in any orientations for the two others.

This leads to four series of models: (A) unstructured, orientations simulated on $U(-10°, 10°)$; (B) unstructured, $U(-90°, 90°)$; (C) structured, $U(-10°, 10°)$; and (D) structured, $U(-90°, 90°)$. Figure 3 shows top and cross-section views of two models corresponding to, respectively, A and D type.

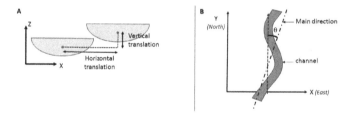

Fig. 2. Parameters for location simulation: (A) vertical and horizontal translations between objects $i-1$ and i; (B) Definition of the angle θ for simulating the main channel orientations.

Moreover, it may be easy to imagine that proportions can also control the Betti numbers, even if no specific study has been dedicated to that problem, to our knowledge. To illustrate this phenomenon, a purely random white noise of a binary variable was simulated for a proportion of black pixels from 0 to 100 %. It may be seen in Fig. 4 that the Betti numbers are influenced in average by the proportion. For this reason, we choose to generate 3D models of channel stacking for which the poportion of the reservoir rock (channel) ranges over $[0.35, 0.4]$ (in average, 20 objects are simulated in the volume). This proportion range is realistic and for a first study, corresponds to a balance between a too small reservoir rock proportion leading to isolated small bodies (unexploitable) and a high proportion (e.g. 80 %), for which it has been shown that the geometry and the organization of the channels have no more impact on reservoir behavior [5, 13].

Fifty simulations were provided for each model type $(A, B, C$ and $D)$ to have a sufficient sampling of models. Simulations were performed using a Gopy research plugin of the Gocad® software. The generated 3D models are firstly represented as a set of parametric objects (Fig. 3, top view). These models are secondly rasterized in the grid by storing the value 1 when a cell is intersected by a channel and 0, else. At the end, the 3D models represent binary volumes (3D grids), for which the value 1 corresponds to the reservoir rocks (channels) and 0 the background (impermeable).

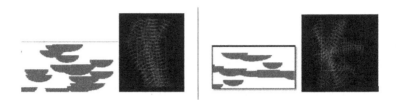

Fig. 3. Examples of 3D models: (left) the 8th simulation in the model A, section of the 3D grid and top view of the corresponding simulated parametric objects; (right) the 7th simulation in the model D, section of the 3D grid and top view of the corresponding simulated parametric objects.

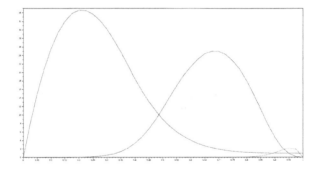

Fig. 4. Betti numbers (β_0 in blue, β_1 in green and β_2 in red) computed for each proportion of black pixels, over 10000 experiments on a $7 \times 7 \times 7$ grid. (Color figure online)

3.1   3D Cubical Complexes and Homology

The following section defines the Betti numbers for binary volumes. Let us first point out that these numbers cannot be directly obtained from the volume, so an intermediate structure, called cubical complex, must be considered.

An *elementary interval* is an interval of the form $[k, k + 1]$ or a degenerate interval $[k, k]$, where $k \in \mathbb{Z}$. An *elementary cube* is the Cartesian product of n elementary intervals, and the number of non-degenerate intervals in this product is its *dimension*. An elementary cube of dimension q will be called q-cube for short. Given two elementary cubes P and Q, we say that P is a *face* of Q if $P \subset Q$. A 3D cubical complex is a set of elementary cubes. The *boundary* of a q-cube is the collection of its $(q-1)$-dimensional faces.

Given a binary volume X, we can define a topologically equivalent 3D cubical complex $K(X)$ which contains a 3-cube $[x_1, x_1 + 1] \times [x_2, x_2 + 1] \times [x_3, x_3 + 1]$ for every voxel (x_1, x_2, x_3) of X.

A *chain complex* (C, d) is a sequence of $\mathfrak{R}$-modules $C_0, C_1, \ldots$ (called *chain groups* and homomorphisms $d_1 : C_1 \rightarrow C_0, d_2 : C_2 \rightarrow C_1, \ldots$ (called *differential* or *boundary operators*) such that $d_{q-1}d_q = 0$, for all $q > 0$, where $\mathfrak{R}$ is some ring, called the *ground ring* or *ring of coefficients*. In this paper we fix $\mathfrak{R} = \mathbb{Z}_2$ since we work in a three-dimensional space. The group chains are thus $\mathbb{Z}_2$-vector spaces.

A 3D cubical complex K induces a chain complex. $C_q(K)$ is the $\mathbb{Z}_2$-vector space of dimension $f_q(K)$, the number of q-cubes in K. Its elements (called *q-chains*) are formal sums of q-cubes with coefficients in $\mathbb{Z}_2$, so they can be interpreted as sets of q-cubes. The linear operator d_q maps each q-cube to the sum of its $(q-1)$-dimensional faces.

A q-chain x is a *cycle* if $d_q(x) = 0$, and a *boundary* if $x = d_{q+1}(y)$ for some $(q+1)$-chain y. By the property $d_{q-1}d_q = 0$, every boundary is a cycle, so we can define the q-th homology group of the chain complex (C, d):

$$H(C)_q = \ker(d_q)/\mathrm{im}(d_{q+1}). \tag{2}$$

This set is a group isomorphic to $(\mathbb{Z}_2)^b$ for some $b \geq 0$. The ranks of the homology groups are called the *Betti numbers*.

In the present work, the Betti numbers are scrutinized to determine if they can be used for describing differences on reservoir geobody network or connectivity.

4 Betti Numbers Used as Descriptor

4.1 Computations and Geological Meanings

This section aims at defining the "physical" meaning of each Betti number in the case where the targeted object is a reservoir 3D model. Indeed, we may consider two kinds of volume: the volume of reservoir rock (channelbelt sediments), which corresponds to fluid storage and drainage, and the volume of impermeable rocks (overbank deposits), which corresponds to flow barrier. We can then consider not only the Betti numbers β_0, β_1 and β_2 of the reservoir rock volumes but also the Betti numbers of the impermeable background, that we refer to as β_0^-, β_1^- and β_2^- in the following.

The Betti number of dimension β_0 represents the number of reservoirs contained in the background. By "reservoir", we mean the amalgamated channelbelt bodies that lead to a connected volume of reservoir rock. The Betti numbers β_1 and β_2 represent, respectively, the number of impermeable tunnels and cavities contained in the reservoir rock. These two parameters may have importance for

CO_2 storage application, as it has been shown that the presence of imperme-able lenses in a reservoir rock volume increases the capacity of CO_2 storage [14], because of the CO_2 accumulation under the lenses.

Concerning the complementary volume (i.e. the impermeable background), β_0^- is the number of the barriers that compartmentalize the reservoir. Obvi-ously, β_1^- and β_2^- are the number of tunnels and cavities in the background, respectively. The tunnels may have importance as they represent potential leaks for CO_2 in the impermeable covers. It is of paramount importance for storage efficiency and reliability that the CO_2 does not migrate inside the impermeable covers. Several studies have been conducted on the impact of thin permeable tunnels (e.g. faults) in impermeable sediments [15].

The Betti numbers were calculated using an add hoc C++/python code and the RedHom software [16]. The ad hoc code was developed to automate the model import into the RedHom software and to extract a convenient output format (i.e. column based file) from RedHom outputs for the statistical analysis over the 200 models.

4.2   Statistical Tests

As previously mentioned, the final objective is to determine if significant differ-ences in Betti number values can be observed between the four series of models. A common way to achieve this is the use of hypothesis testing. In this study, we propose to use the non-parametric Kruskal-Wallis test, whose purpose is to compare the mean of a random variable V between different samplings, that are assumed to be independent but that can have different sizes. This test does not require that V follows a particular parametric distribution law.

Let us consider N samplings on which a random variable V has been mea-sured and m_i the means of V on the ith sampling. The Kruskal-Wallis test is presented as follows:

$$\begin{cases} \text{The null hypothesis,} & H_0 : m_1 = m_2 = ... = m_N \\ \text{The alternative hypothesis,} & H_1 : \exists i, j / m_i \neq m_j \end{cases} \tag{3}$$

The underlying objective is to determine if the samplings are stemming from the same statistical population (the null hypothesis). Thus, the question is to reject or not H_0. As any hypothesis testing, it relies on the computation of a statistics t that should follow a given hypothetic law under H_0. Then, an observed value t_{obs} is computed from the samplings and compared to a critical value t_c. This critical value is defined by assuming a given risk α, which corresponds to the probability to reject H_0 although it is true. If t_{obs} is greater than t_c, H_0 is rejected, else it is accepted. Another way to interpret results is to compute the *p-value*, which corresponds to the probability of rejecting H_0 although it is true, computed from the samplings. If it is greater than α then H_0 is accepted, otherwise rejected.

In the case of H_0 rejection, it is a common practice to use post hoc tests that allows the determination of which means are different from the others.

In this study, the post hoc test "Kruskalmc" (multiple comparison) was chosen in order to evaluate differences in medians among the four models. It is important to notice that this test is less powerful than the Kruskal-Wallis test. This can lead to unconsitent results between the two tests. The power of an hypothesis testing corresponds to the probability to reject H_0 when H_1 is true. It characterizes its robustness.

Moreover, in order to scrutinize the values of the Betti numbers, we propose to use the boxplot (Fig. 5). The box plot displays the full range of variation (from min to max), the likely range of variation (the IQR), and a typical value (the median). They have also lines extending vertically from the boxes (whiskers) indicating variability outside the upper and lower quartiles. Outliers may be plotted as individual points. The spacing between the different parts of the box indicate the degree of dispersion (spread) and skewness in the data, and show outliers.

The statistical analysis was performed using the R software.

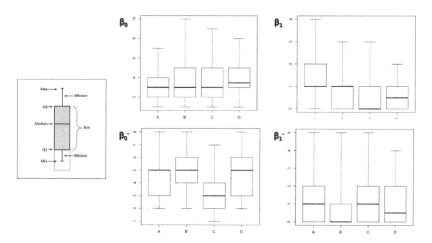

Fig. 5. Boxplots: left, diagram of the boxplot parameters; right, the four boxplots for each Betti number.

5   Results and Discussion

5.1   Kruskal-Wallis Tests

For the 200 generated simulations, no cavity was found: $\beta_2 = 0$ and $\beta_2^- = 0$. They are then not studied. Concerning the four other numbers, Table 1 summarizes the results found for the Kruskal-Wallis test. Except for β_0, all the p-values are under 5 %. Thus, it can be concluded that the mean of β_0 is not significantly different between the different series. On the contrary, for β_1, β_0^- and β_1^-, there is at least one mean that is significantly different from the others.

The post hoc test "Kruskalmc" was applied to evaluate differences among the four models for these three variables. Table 2 summarizes the results. For β_1, the models are all different. In terms of β_0^-, the models C is significantly different from the others. Finally, all the models are considered as similar in terms of β_1^-. This is unconsistent with previous results. However, as previously mentioned, the "kruskalmc" test is less powerful and "more easily accepts" H_0, which can explain these results. Moreover, the found p-values (0.03774) is very close to 5%. Similarly, the t_{obs} of the comparison between $A - B$ (27.3) is very close to the critical value (27.7123). This means that using a higher risk than 5%, the means could be considered as different and that the models B would be considered as different from A, C and D. In order to depict graphically the four groups of numerical data through their quartiles, boxplots were used (see Fig. 5). These plots confirm previous interpretations.

Table 1. Results of the Kruskal-Wallis test, p-values are always under 5%, except for β_0 (in bold). It may be noticed that the p-value of β_1^- (in italic) is close to 5%.

Betti number	β_0	β_1	β_0^-	β_1^-
p-value	**0.2334**	$7.5e^{-5}$	$1.33e^{-6}$	*0.03774*

Table 2. Results of the Kruskal-Wallis post hoc test, applied for β_1, β_0^-, and β_1^-.

β_1	t_{obs}	t_c	**Difference**
A-B	33.96	27.7123	TRUE
A-C	46.94	27.7123	TRUE
A-D	37.96	27.7123	TRUE
β_0^-	t_{obs}	t_c	**Difference**
A-B	11.94	27.7123	FALSE
A-C	46.15	27.7123	TRUE
A-D	1.59	27.7123	FALSE
β_1^-	t_{obs}	t_c	**Difference**
A-B	*27.3*	27.7123	FALSE
A-C	0.63	27.7123	FALSE
A-D	12.55	27.7123	FALSE

5.2   Interpretations and Discussions

Considering a proportion of reservoir rock between $[0.35; 0.4]$, it could be said that the number of reservoir geobodies does not depend on the stacking law and the channel orientation variability. On the contrary, both stacking and orientations modify significantly the number of impermeable tunnels in the reservoir rocks. Regarding Fig. 5, the unstructured models seem to increase the occurence of tunnels, in average. This means that this type of reservoir architecture is more tortuous and may facilitate dissolutions if these impermeable tunnels are preferentially horizontal [14]. Concerning the number of impermeable compartmentalization (β_0^-), it seems that unstructured models with sub-parallel channel orientation present smaller occurences of these features. Regarding Fig. 5, higher orientation variability leads to higher compartmentalization. On the contrary, the stacking correlation tends to decrease it. Both trends explain that the models C are different from the others, with a smallest mean. This result is suprising as we could imagine at a first glance that the stacking correlation tends to create vertical chimneys of reservoir geobodies that separate the background in different components. Finally, the number of permeable tunnel in background (β_1^-) seems to be positively influenced by the stacking correlation but negatively impacted by the channel orientation variability. This leads to differentiate models B from the others. However, the difference is not clearly significant as $p - value$ and observed statistics are close to the cutoffs.

6 Conclusion

This preliminary study allowed us to define a "physical meaning" to the Betti numbers computed on both permeable and impermeable volumes for reservoir characterizations. Six variables were defined, but only four were really used in the study because no cavity was found. This study also highlights that the type of channelbelt stackings influence the reservoir topology. These results have been only performed for a proportion of reservoir rocks ranged over $[0.35; 0.4]$. It could be interesting to test if the trends observed in this study remain similar for different proportions. It may be noticed that some Betti numbers correspond to entities that can be computed using more classical algorithms, such as β_0 and β_0^-. However, Betti numbers are known as powerful to characterize the number of tunnels, which generally remains the most difficult task. The presence of permeable tunnels in impermeable background caused by faults or lithologies have major impacts on CO_2 storage risks. Thus, the use of Betti numbers helps computing relevant topological indices for reservoir compactness characterization. In further studies, we will study the impact on proportions on the observed trends but also we will analyze more complex models having more than two lithologies. Finally, hydraulic behavior will be also studied to check if relationships exist between static (reservoir rock network) and dynamic (flow path) topology.

Acknowledgements. The authors would like to thank the ParadigmGeo company and the ASGA for its support in providing the Gocad software and its research plugins. This project belong to the ANR H-CUBE project and the authors would like to thank the ANR for funding this research.

References

1. Goovaerts, P.: Geostatistics for Natural Resources Evaluation. Oxford University Press, New York (1997)
2. Scheidt, C., Caers, J.: Representing spatial uncertainty using distances and kernels. Math. Geosci. **41**(4), 397–419 (2009)
3. Bouquet, S., Bruel, D., De Fouquet, C.: Influence of geological parameters on CO2 storage prediction in deep saline aquifer at industrial scale. In: TOUGH Symposium, Berkeley (2012)
4. Issautier, B., Viseur, S., Audigane, P., Le Nindre, Y.M.: Impacts of fluvial reservoir heterogeneity on connectivity: Implications in estimating geological storage capacity for CO2. Int. J. Greenhouse Gas Control **20**, 333–349 (2014)
5. Larue, D.K., Hovadik, J.: Why is reservoir architecture an insignificant uncertainty in many appraisal and development studies of clastic channelized reservoirs? J. Pet. Geol. **31**, 337–366 (2008)
6. Larue, D., Hovadik, J.: Connectivity of channelized reservoirs: a modelling approach. Pet. Geosci. **12**(4), 291–308 (2006)
7. Allen, J.: Studies in fluviatile sedimentation; an exploratory quantitative model for the architecture of avulsion-controlled alluvial sites. Sediment. Geol. **21**(2), 129–147 (1978)
8. Schumm, S., Mosley, M., Weaver, W.: Experimental Fluvial Geomorphology. Wiley, New York (1987)
9. Miall, A.: The Geology of Fluvial Deposits: Sedimentary Facies Basin Analysis and Petroleum Geology. Springer, Heidelberg (1996)
10. Lantuejoul, C.: Geostatistical Simulation: Models and Algorithms. Springer, Heidelberg (2001)
11. Bridge, J., Tye, R.: Interpreting the dimensions of ancient fluvial channelbars, channels, and channel belts from wireline-logs and cores. AAPG Bull. **84**(8), 1205–1228 (2000)
12. Larue, D.K., Friedmann, F.: The controversy concerning stratigraphic architecture of channelized reservoirs and recovery by waterflooding. Pet. Geosci. **11**, 131–146 (2005)
13. King, P.: The connectivity and conductivity of overlapping sand bodies. In: Buller, A., Berg, E., Hjelmeland, O., Kleppe, J., Torsaeter, O., Aasen, J. (eds.) North Sea Oil and Gas Reservoir II, pp. 353–362. Springer, Netherlands (1990)
14. Hesse, M.A., Woods, A.W.: Buoyant dispersal of CO2 during geological storage. Geophysical Research Letters **37**, 5 p. (2010)
15. Tsang, C.-F., Birkholzer, J., Rutqvist, J.: A comparative review of hydrologic issues involved in geologic storage of CO2 and injection disposal of liquid waste. Environ. Geol. **54**, 1723–1737 (2008)
16. Harker, S., Mischaikow, K., Mrozek, M., Nanda, V., Wagner, H., Juda, M., Dłotko, P.: The efficiency of a homology algorithm based on discrete Morse theory and coreductions. In: Rocio Gonzalez Diaz, P.R.J. (ed.): Proceedings of the 3rd International Workshop on Computational Topology in Image Context, Chipiona, Spain. Volume Image A, vol. 1. 41–47 (2010)

Topological Analysis of Amplicon Structure in Comparative Genomic Hybridization (CGH) Data: An Application to ERBB2/HER2/NEU Amplified Tumors

Sergio Ardanza-Trevijano[1], Georgina Gonzalez[2], Tyler Borrman[3], Juan Luis Garcia[4], and Javier Arsuaga[2,5(✉)]

[1] Department of Physics and Applied Mathematics, University of Navarra, 31080 Pamplona, Spain
sardanza@unav.es

[2] Department of Molecular and Cellular Biology, University of California Davis, One Shields Avenue, Davis, CA 95616, USA
gingonzalez@ucdavis.edu

[3] Medical School, University of Massachusetts, 368 Plantation Street, Worcester, MA 01605, USA
tyler.borrman@umassmed.edu

[4] Centro de Investigación del Cancer, Universidad de Salamanca, 37007 Salamanca, Spain
jlgarcia@usal.es

[5] Department of Mathematics, University of California Davis, One Shields Avenue, Davis, CA 95616, USA
jarsuaga@ucdavis.edu

Abstract. DNA copy number aberrations (CNAs) play an important role in cancer and can be experimentally detected using microarray comparative genomic hybridization (CGH) techniques. Amplicons, CNAs that extend over large sections of the genome, are difficult to study since they may contain multiple independent and dependent copy number changes. Here, we propose an algorithm to find the CNAs structure within a given amplicon. Our method relies on the observation that co-occurring CNAs can be encoded as 1-dimensional cycles. Applying this method to breast cancer patients known as ERBB2/HER2/NEU amplified we find three regions that can be co-occuring: the first region is in the cytoband 17q12, where the ERBB2 gene is located, the second region expands between 17q21.2 to 17q21.31 and includes the keratin genes, the third one is 17q21.33. We suggest that the first homology group helps uncovering the structure of amplicons.

Keywords: Copy number aberrations · Cancer · Computational homology · First homology group

S. Ardanza-Trevijano and G. Gonzalez contributed equally to this work.

A. Bac and J.-L. Mari (Eds.): CTIC 2016, LNCS 9667, pp. 113–129, 2016.
DOI: 10.1007/978-3-319-39441-1_11

1   Introduction

Cancer is a set of complex genetic diseases whose pathogenesis is not well under-stood. Initiation and progression of these diseases depend on the misregula-tion of key genes called *cancer/tumor genes*. Gene misregulation occurs through different mechanisms including the gain and losses of DNA chromosome frag-ments (e.g. [11,18,20,24]). These events are commonly termed DNA copy num-ber aberrations (CNAs) and are routinely detected in the laboratory through comparative genomic hybridization (CGH) arrays, single nucleotide polymor-phism (SNP) arrays and sequencing (e.g. [12–14,17,22,36,47]). However not all detected CNAS are relevant for tumor initiation and/or progression. It is cur-rently believed that CNAs that contain tumor genes are those that are relevant for tumor progression. These CNAs are called *drivers* while those which appear to have no biological implications are called passengers. Determining which CNAs are driving tumor progression and which ones are just passengers remains an open problem. Certain CNAs expand over large fragments of the genome and are sometimes termed *Amplicons*. These regions are important because contain multiple tumor genes and the presence or absence of certain CNAs within an amplicon has been associated with patient's prognosis (e.g. [23,41]). Examples include 9p in breast cancer, colon and glioblastoma tumors and lymphomas [5,19], $11q$ in head and neck, breast, oral and liver tumors (reviewed in [46]) and $17q$ in ERBB2/HER2/NEU (ERBB2+, thereafter) positive breast cancer [4]. The detailed structure of amplicons is complex and difficult to investigate using traditional statistical methods since some amplifications appear to occur simultaneously, hence they are not significant as independent CNAs, and have synergistic effects [1,28,43]. In this work we will call *co-occurring CNAs* those that occur simultaneously independently of their functional effects. One poten-tial approach to study the structure of an amplicon and identify potential co-occurring CNAs is to encode combinations of CNAs as a single predictor variable and perform association studies between these new predictor variables and phe-notypes of interest.

Here we extend our previously reported supervised approach, termed Topo-logical Analysis of array CGH (TAaCGH), to study the structure of an ampli-con. In TAaCGH, we associate a point cloud to each CGH profile (or section of a CGH profile) through a sliding window algorithm [15], build a Vietoris-Rips (VR) simplicial complex [31] and perform an association study between the topo-logical properties of the VR complex and the chosen phenotype. The difference between TAaCGH and other current association studies is that TAaCGH uses the topological properties of the point cloud, instead of the probes, as predic-tor variables. The advantage of using topological properties as predictors is that they can encode relationships between probes. In previous works we showed that using the rank of the zero homology group (β_0) as a predictor variable in asso-ciation studies of breast cancer is comparable to other statistical methods [3]. Here we hypothesize that performing association between the rank of the first homology group β_1 and a specific phenotype helps analyze the underlying struc-ture of amplicons. This hypothesis is based on recent analytical and numerical results that shows that β_1 encodes for periodic patterns [34] and by our own observations that show that neighboring (not-necessarily periodic) regions of amplifications are mirrored by β_1 [10,38].

To test our hypothesis and to illustrate our methodology we analyze the amplicon on $17q$ in ERBB2/HER2/NEU (ERBB2+, thereafter) positive breast cancer samples. ERBB2+ breast cancer is an aggressive form of the disease that comprises 25 % of all breast tumors diagnosed (reviewed in [35]). The ERBB2 gene is located in the region of the genome labeled as cytoband $17q12$ (where 17 is the chromosome arm, q denotes the long arm of the chromosome and 12 denotes a specific band that can be detected by chromosome staining). Misregulation of ERBB2 in ERBB2+ tumors commonly occurs through copy number gains of $17q12$. In many patients, this amplification is accompanied by gains of other regions in the same chromosome arm. This includes amplifications of $17q21.2$ that encompasses the $Top2A$ gene [32], chromosome regions $17q21.1$, $17q22$ [27] and $17q21.33 - q25.1$ which is predictive of early recurrence [9] and contains TANC2 ($17q23$) and PPM1D genes [29,37], two independent co-amplified regions have also been reported in $17q23$ [4,39].

To test whether TAaCGH can detect these events, we analyzed two independently published data sets [13,20]. We first confirmed the presence of the amplicon in $17q$ in both data sets using β_0, we then identified specific regions within this arm using β_1 analysis. This study revealed two regions of significance delimited by $17q12$ and $17q12 - 17q21.33$. To further localize the regions of the genome that contributed to the significance of β_1 we calculated the generators of the first homology group and the correspondence between the probes and the generators. Statistical analysis quantifying the over-representations of genomic regions in the generators allowed us to further subdivide the region $17q12 - 17q21.33$. A first amplification was detected in between the neighboring regions $17q21.2 - 17q21.31$ (extending from base pairs 40,884,763-41,826,877) and the region $17q21.33$ (from base-pairs 46,603,678-49,075570). Using the UCSC genome browser we observed that the first region contains the keratin cluster (e.g. [30]) and the second contains, among others the HOXB cluster (see [8] for a review). Both of these clusters have been previously reported in breast cancer studies. Whether their functionality is synergistic in some patients remains to be determined.

2   Data Sets and Methods

2.1   CGH Data

CNAs are defined as gains or losses of genome fragments and can be detected using microarray technologies. Through Comparative Genomic Hybridization (CGH), DNA probes (i.e., fragments of DNA sequences) are spotted on a platform. Tumor DNA, labeled with Cy3, and control DNA, labeled with Cy5, are co-hybridized in a 1:1 ratio. The intensity of the hybridized samples is captured and transformed into a red-green ratio value called the $\log_2$ ratio. Since the physical position of each probe is known, these $\log_2$ ratios can be mapped to the original genome producing a CGH profile (Fig. 1). In traditional statistical approaches each CGH profile is normalized and segmented, and significant copy number aberrations are then identified [6, 33, 45].

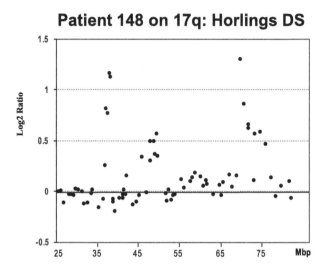

Fig. 1. A CGH profile for chromosome arm 17q. The x-axis indicates the genomic position and the y-axis the $\log_2$ ratio of the intensity of the tumor and control samples co-hybridized to the same array.

2.2   Simulation Data Set

We simulated single and co-occurring aberrations. A detailed description of the simulation methods for a single aberration can be found in [3, 25, 26]. In brief, each simulation consisted of 200 profiles, 100 in the control set and 100 in the test set. Each simulated profile contained 100 aCGH probes. The value of the copy number along the profile was determined by three parameters: the mean value of the aberration μ, the length of the aberration λ, and the standard deviation associated with noise σ. Probes outside the aberration and in the control set had $\mu = 0$, whereas for those probes inside the aberration was $\mu = 0.6$ or 1. Aberration length λ was equal to 5 and 10 probes. Noise was implemented by drawing samples from a Gaussian distribution of mean 0 and standard deviation σ of values 0.2, 0.6 or 1. The control set for single aberrations was made of profiles without aberrations (i.e. only noise).

Co-occurring aberrations were represented by two aberrations of different lengths. In the first aberration $\mu = 0.6$ or 1 and in the second $\mu = 1$. The control set was made of profiles with no aberrations or with only one aberration.

2.3   Horlings Data Set

This dataset analyzed was published by Horlings and colleagues [20] and was obtained from the supplementary data [21]. Measurements of copy number variations were performed on microarrays containing 3.5 k BAC, PAC-derived DNA segments covering the entire genome with a spacing average of 1 Mb. Each BAC clone was spotted and triplicated on every slide (Code Link Activated Slides, Amersham Biosciences). Our own preprocessing of the data can be found in [3]. This study contained 14 ERBB2+ patients determined by clinical diagnosis. The control set consisted of the patients belonging to the remaining subtypes.

2.4   Climent Data Set

This data set was used as a validation set. In [13] genome-wide measurements of copy number variations were performed by array CGH (UCSF Hum. Array 2.0) with an average spacing between probes of 1Mb. The study contained 180 patients diagnosed with a stage I/II lymph node-negative breast cancer. The data set was downloaded from the GEO data base with accession number GSE6448. Arrays were preprocessed by averaging/removing probes as follows: 18 clones mapping to chromosome Y or missing genomic location information were removed, 80 probes mapping to identical genomic regions were averaged and represented as single values, 179 probes missing entries for 30 % or more patients were removed, and missing values were imputed using the lowess regression method in the aCGH package for R [16]. This resulted in 2,168 unique clones from the original 2,445 printed in the array. We classified as ERBB2+ tumors the subset of 9 patients that showed a copy number change >1 (in log scale) at the clone DMPC-HFF#1-61H8 which contains the ERBB2 gene.

2.5   Multidimensional Analysis of CGH Profiles Using Computational Algebraic Topology

We previously reported a new method to analyze CGH data called topological analysis of array CGH (TAaCGH) [3,15]. Our method uses a sliding window algorithm that associates a point cloud to a given CGH profile (or section of a CGH profile). The dimension of the point cloud is determined by the size of the sliding window. In this study and based on our previous work [3] we considered windows of size $n = 2$. TAaCGH assigns a β_0 curve to each CGH profile, computes the average $\langle \beta_0 \rangle$ curve for each population of patients (test and control) and performs statistical analysis to determine differences between them (see below). Here we extended TAaCGH by incorporating a similar analysis using $\langle \beta_1 \rangle$ curves. We used the program JavaPlex to perform the calculation of β_1 and its generators [40]. As in the case of β_0, we generated the function $\beta_1(\epsilon)$ for each patient. In this case ϵ took values between 0 and the value at which $\beta_0 = 1$. Given the $\beta_1(\epsilon)$ for each patient, we computed the average $\langle \beta_1 \rangle$ for the ERBB2 set and the control set (consisting of the reminder of the patients) and test for statistically significant differences between the two $\langle \beta_1 \rangle$ curves.

2.6   Testing for Statistical Differences

To test for statistically significant differences between $\langle \beta_i \rangle$ curves associated to different patient groups, we assumed the null hypothesis that $\langle \beta_i \rangle$ curves for a sample of patients was independent of the cancer subtype. We quantified deviations from the null distribution by the statistic S_{exp}, which was defined as the sum of the squares of the differences between the average $\langle \beta_i \rangle$ curves across all radii, i.e.

$$S_{exp,i} = \sum (a_{ij} - b_{ij})^2 \quad \text{for} \quad j = 1, \ldots, N$$

where a_{ij} and b_{ij} are the $\langle \beta_i(j) \rangle$ value for each population under study and for the value of the filtration parameter $\epsilon = j$.

2.7   Finding Co-Occurring Aberrations

In order to determine the regions of the genome that contributed to the first homology we found the CGH probes that were mapped to each of the vertices of the generators. First, generators for each patient and value of the filtration coefficient were calculated using JavaPlex [40]. Second, the probes of the CGH profile that mapped to the vertices of the generators were identified. Third, since generators were not necessary minimal and, due to the noise of the data, some generators mapped to different areas of the genome we determined a CNA by measuring the concentration of the probes. Regions with higher concentration of probes than the control set were called CNAs.

2.8   Software for Visualization of Generators

We created an exploratory tool using Shiny app to visualize the generators in the point cloud together with their corresponding probes in the CGH profile. The app highlights the probes and generators as the values of the filtration coefficient changes. The software allows to visualize the dispersion of the probes associated with the probes through the CGH profile. An example is shown in Fig. 5. The software is available from the authors upon request.

3   Results

3.1   Computer Simulations

To better interpret our results we performed computer simulations. Since the analysis of β_0 has been performed elsewhere [3,15], we focused on simulations concerning the detection of CNAs using β_1. Figure 2 shows an example of two simulated profiles, one with no aberrations as control (Fig. 2) and a second one with two co-occurring aberrations (Fig. 2B). In both Fig. 2A and B, the x-axis represents the position along the chromosome and the y-axis the log_2 ratio of the copy number values. The $\langle \beta_1 \rangle$ curves (Fig. 2C) obtained from the curves above help understand the growth and disappearance of the first homology. In the case of no amplification (red), the $\langle \beta_1 \rangle$ curve starts at $\langle \beta_1 \rangle = 0$, since for very small values of ϵ there is no 1-dimensional homology. $\langle \beta_1 \rangle$ rapidly increases due to the structure of the noise until it reaches a maximum after which it decays to 0.

The graph for $\langle \beta_1 \rangle$ is different when two aberrations are present (blue). For small values of the filtration parameter the graph behaves similarly to the graph without aberrations, however in this case the graph shows more than one local maximum and a lower log_2 ratio of copy number values at the first maximum.

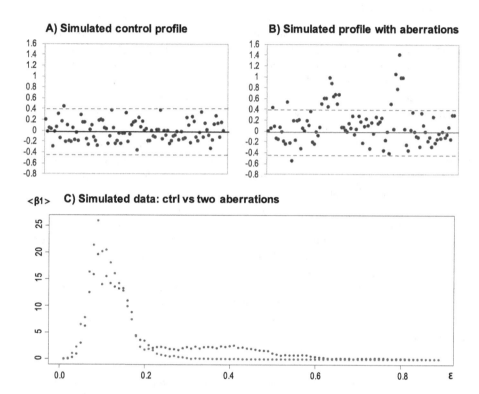

Fig. 2. Examples of simulated aberration profiles and $\langle \beta_1 \rangle$ curve. (A) shows a control profile with no aberrations with σ=0.2. (B) shows a profile with two aberrations with parameters $\lambda = 10$ and 5, $\mu = 0.6$ and 1 and $\sigma = 0.2$ for both. The blue dashed lines represent two standard deviations. The bottom graph shows in red the $\langle \beta_1 \rangle$ for the control group with no aberrations and in blue the $\langle \beta_1 \rangle$ curve for a pair of aberrations with $\lambda = 10$, $\mu = 0.6$ and 1 and $\sigma = 0.2$ (Color figure online).

We tested our method by performing a sensitivity and specificity analysis in three different simulation experiments. Each experiment consisted of 200 profiles (100 tests and 100 controls) and all possible combinations of parameters were considered. A successful identification of an aberration was scored when the obtained P-value was less than 0.05 after correcting by FDR. First we considered the case of one single amplification (test set) taking as control set a population with no aberrations. In this case sensitivity was 87.5 %. In the second experiment we used profiles with two amplifications as a test set and no amplifications as the control set. In this experiment we got average sensitivity of 95 %. In the third experiment we compared double amplifications with single (as control). Results showed 82.5 % in sensitivity. Specificity was measured by comparing two control data sets resulting in 97.5 %. Our method has bigger chances to fail when the length of the aberration is small (5 or less) and $\mu = \sigma$.

β_0 Significance of 17q

As discussed elsewhere [3,15], $\langle \beta_0 \rangle$ curves can detect chromosome aberrations. Since we are interested in the entire amplicon in 17q, we applied TAaCGH to full chromosome arms. The chromosome arm 17q was significant in both data sets. In the Horlings data set we found significance on $\langle \beta_0 \rangle$ curves when comparing chromosome arm 1q (P-value = 0.021) and 17q (P-value = 0.004). The graph for chromosome 1q however showed that the control curve was above the test set indicating that the control set (ERBB-) had more CNAs that the test set (ERBB+). Therefore was not relevant in this study. In our validation data set, we found only 17q to be significant with a corresponding P-value after FDR correction of 0.0037. Figure 3 shows examples of $\langle \beta_0 \rangle$ curves for both chromosomes. Since β_0 is the number of connected components of the simplicial complex, $\langle \beta_0 \rangle$ curves start at the value of the number of probes in each chromosome arm for $\epsilon = 0$ and gradually decays with increasing ϵ until a single connected component remains. All blue curves shown in Fig. 3 represent the ERBB2+ population and all red curves represent the ERBB2- population. Results shown in Fig. 3A and B include $\langle \beta_0 \rangle$ curves associated to 17q for the Climent and Horlings data sets respectively; Fig. 3C shows $\langle \beta_0 \rangle$ curves associated to 1q and Fig. 3D $\langle \beta_0 \rangle$ curves associated to the negative control 19q. Chromosome arm 17q showed, as expected, a higher number of chromosome aberrations in the ERBB2+ patients than in the ERBB2- patients.

β_1 Significance of 17q

Next, we analyzed the significance of β_1 in chromosome arm 17q. We considered two approaches. First we tested for β_1 significance of the entire chromosome arm 17q and then for overlapping sections of the chromosome arm. We found important to use both approaches since co-occurring CNAs may be local or spread over the entire arm. Analysis using the whole arm showed 17q to be significant in the Climent data set (with a P-value of 0.040), but not in the Horlings data set (P-value 0.172). Figure 4 shows the corresponding $\langle \beta_1 \rangle$ curves for both studies suggesting that any amplicon structure, if present, would be local.

Following our previous work [3] we subdivided chromosome arm 17q in the Horlings data set into 6 sections, which corresponded to 5 sections in the Climent data set. Each section containing 20 CGH probes with 10 overlapping probes. Results are shown in Table 1. Column 1 shows the section analyzed; columns 2 and 5 the cytogenetic band, columns 3 and 6 the location in base pairs, and columns 4 and 7 the p-values [7]. Both data sets showed some significant sections. In the Horlings data set, Sects. 2 and 3 significant after correction for multiple testing (column 4). In the Climent data set all sections except Sect. 4 were significant (column 7). Based on the reproducibility of these results we concluded

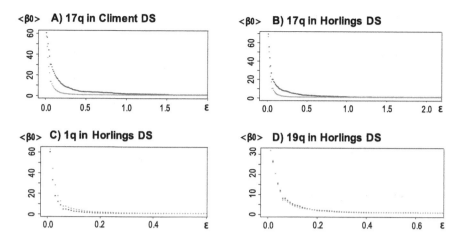

Fig. 3. Examples of $\langle \beta_0 \rangle$ curves in dimension 20. Blue indicates the ERBB2+ population and red the ERBB2-. (A) Arm 17q arm in Climent; (B) Arm 17q in Horlings, (C) Arm 1q in Horlings and (D) Arm 19q in Horlings (Color figure online).

that sections containing cytobands 17q12 to 17q21.33 had co-occurring CNAs and are therefore good candidates for uncovering the underlying structure of the amplicon.

To further identify the regions within 17q12 and 17q21.31 − 17q21.33 we identified the generators of the first homology group for each patient and mapped the probes to the vertices of the corresponding generators. Before we discuss the statistical results we highlight some interesting properties of the generators: (1) probes that made up the generators may be distributed throughout the entire

Table 1. Chromosome Sections. Correspondence between sections, cytobands and base pairs range for each of the sections used to analyze chromosome 17q.

Section	Cytoband (Horlings *et al.*)	Basepair	$(P-$value) FDR correction	Cytoband (Climent *et al.*)	Basepair	$(P-$value) FDR Correction
17q.s1	q11.1-q12	25440972- 37812853	(0.043) 0.08640	q11.1-q21.2	25530227- 40615955	0.0088
17q.s2	q12-q21.31	32489785- 43339849	(0.0008) 0.00480	q12-q21.33	35669421- 47644854	0.0016
17q.s3	q21.2-q21.33	38428492- 49075570	(0.0116) 0.03480	q21.31-q22	42170022- 55594526	0.0378
17q.s4	q21.31-q22	44084882- 57340119	(0.471) 0.47170	q21.33-q24.3	47968636- 70573094	0.100
17q.s5	q22-q24.2	51080264- 66108804	(0.253) 0.30432	q23.1-q25.3	58025830- 78774742	0.009
17q.s6	q23.1-q25.3	57996713- 80780814	(0.237) 0.30432			

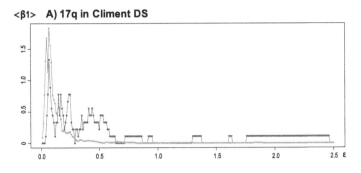

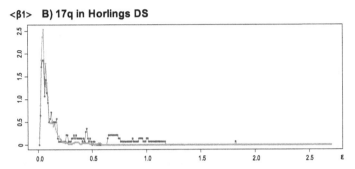

Fig. 4. $\langle \beta_1 \rangle$ **Significance of** $17q$ **in the climent and the horlings data sets.**
(A)The figure shows the $\langle \beta_1 \rangle$ curves for ERBB2+ (blue) and ERBB2-(red) in the
Climent data set (significant). (B) Here we show the $\langle \beta_1 \rangle$ curves for both categories
for the Horlings data set (non-significant) (Color figure online).

arm or localized in a specific region (2) unlike β_0 generators do not necessarily
detect the global maximum in the profile but different regions that contribute
to several local maxima (3) neighboring maxima or even sections of the same
maximum are detected at different values of the filtration parameter. Figure 5
shows the profile of a patient for $17q$ and the point cloud. Probes in blue are
those that were mapped to the generators at two different filtration coefficient
values. The corresponding 2D point cloud (with edges included) and with the
vertices in each cycle highlighted in blue are also shown.

   These inherent variability of the generators and the noise of the data motivated us to use a statistical approach. As detailed in the methods sections for
each patient and value of the filtration parameter we computed the cycles and
the probes that defined those cycles. The frequency at which a probe was mapped
to a particular region of the genome is represented by a histogram (see Fig. 6).
The top graphs show the histograms for the Horlings data set and the bottom
ones the histograms for the Climent data set. The histograms on the left are
the control and the ones on the right correspond to the ERBB2+. The most
remarkable feature is the difference between the control and the ERBB2 data
sets. While the control show no significant concentration of the probes that

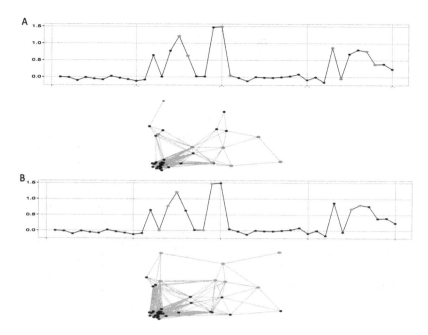

Fig. 5. Correspondence between CGH probes and generators. Different values of the filtration parameter detects different generators which corresponds to different probes in the genome. Panel A shows the profile of one patient and its associated point cloud. The probes highlighted in blue correspond to the vertices of the single generator, also in blue. The filtration coefficient was $\epsilon = 0.78$. Panel B shows the same patient and point cloud for a different value of the filtration coefficient $\epsilon = 0.83$

belong to cycles the ERBB2+ clearly show three regions of interest. $17q12$ has a significant concentration of cycle elements and corresponds to the position of the gene $ERBB2$. Two regions extend beyond the position of $ERBB2$ The first one is in the boundary between $17q21.2$ and $17q21.31$. The Horlings data set suggests that the region of interest is more localized in $17q21.31$ while the Climent data set suggest a region contained in $17q21.2$. The last region is located at $17q21.33$ and is common to both studies.

Since our simulations show that the first homology group can also identify single amplifications one may argue that the found amplifications correspond to single independent events. To address this problem we analyzed the distribution of the cycles-forming-probes. Figure 7 show some examples of the distribution of cycles in the genome for specific patients. Each plate corresponds to one patient, the x-axis is the position along the genome and the y-axis the "life" of the cycle. Each color represents a different cycle. If the amplifications were independent events one would expect to see single colors concentrated at specific regions. However we see cycles dispersed over the entire profile indicating the presence of co-occurring CNAs.

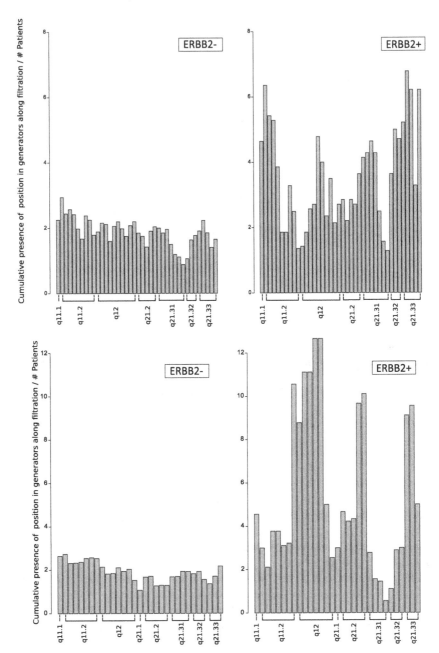

Fig. 6. Comparison of ERBB2- (left) and ERBB2+ (right) patients at the generator level. The top histograms correspond to the Horlings data set and the bottom to the Climent data set. Each bar in the histogram represents a probe. Its height represents the cumulative presence of that probe on the generators of the first homology group divided by the number of patients. The cumulative presence is calculated by counting the number of cycles in which the probe is part of the generator for each value of the filtration parameter (multiplied by the number of generators if they were more than one).

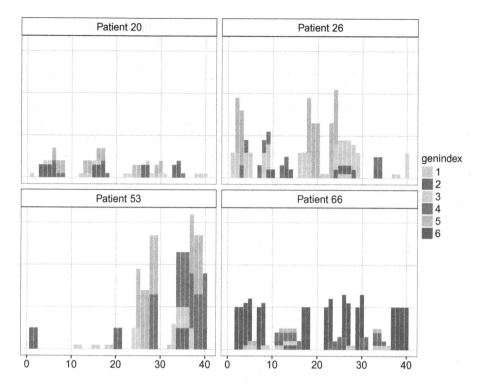

Fig. 7. Distribution of cycles in CGH profiles. Each plate corresponds to the CGH profile of a patient and how the vertices of the cycles are mapped back to the profile. Different colors indicate different cycles and do not represent the same cycle in each plate. The height of the bars represent the life of the cycle (Color figure online).

4   Discussion

Copy number measurements provide an unparalleled opportunity to identify the underlying mechanisms of cancer. Previous efforts in analyzing copy number data have mainly focused on the identification of single, independent chromosome copy number aberrations. These approaches however are known to be deficient in the identification of co-occurring copy number changes since there is a large number of combinations of probes that one needs to interrogate. In this study, we have presented a methodology that helps circumvent the search for simultaneously occurring CNAs by encoding copy number data as topological objects. In particular we have used the rank of the first homology group to perform this association. To test this hypothesis, we searched for co-occurring aberrations in ERBB2+ breast cancer patients. Our results show β_1 significance in chromosome cytobands that extend from 17q12 to 17q21.33. By identifying the probes that form the generators and measuring their concentration along the CGH profiles we were able to further narrow this significant region to three amplifications. The first is 17q12 which contains the ERBB2 gene. The second and the third have also been reported in ERBB2+ patients. The second amplification is in the boundary between $17q21.2$ and $17q21.31$ and according to our estimation is delimited by the Top2A and BRCA1 genes (base pairs $40,884763 - 41,826,877$). This region encompasses the type I keratin gene cluster. Finally we identified $17q21.33$ (base pairs $47,400,368 - 49,075570$) a large region that contain multiple tumor associated genes including the HOXB cluster [42], Prohibitin [44] and amplification of this region has been associated with poor prognosis [41]. Unfortunately at this point, due to the small sample size, we cannot determine how common these co-occurring CNAs are in the general population of ERBB2+ patients or whether they form subtypes within the ERBB2+ subtype. Nevertheless the fact that these regions are significant in two independent data sets is encouraging. It is therefore our immediate plan to scale up this study on larger data sets.

Our work presents also new tools for the topological analysis of time series. We and others [34] independently introduced the concept of using the sliding window algorithm to analyze time series. In our previous work we noted that: (1) the overall shape of the point cloud already provides information of the data [2,3,15], (2) The point cloud can be seen as the reconstruction set of the dynamical system induced by the sliding window algorithm [2], (3) the zero homology group identifies large step increments between consecutive measurements [15]. Our contributions in this work is the development of algorithms that (1) detect the single and co-occuring maxima in the data in non-necessarily periodic signals using the first homology group (2) Identify local maxima by computing the concentration of the pre-images (by the sliding window algorithm) of the vertices that form the cycles. It is our belief that the use of topological methods for the analysis of signals using simple construction techniques, such as the commonly used sliding window algorithm, can provide new insights in the analysis of time series.

Acknowledgments. We would like to thank H. Bengtsson and T. Speed for very helpful comments during the development of this methodology. T.B and J.A. were partially supported by NSF grant 1217324 and by NIH-RIMI (Research Infrastructure in Minority Institutions) grant 2P20MD000544-06. SA was partially supported by the *Ministerio de Economía y competitividad* grant MTM2013-42486-P.

References

1. Arriola, E., Marchio, C., Tan, D.S., et al.: Genomic analysis of the HER2/TOP2A amplicon in breast cancer and breast cancer cell lines. Lab Invest. 88(5), 491–503
2. Arsuaga, J., Baas, N.A., DeWoskin, D., et al.: Topological analysis of gene expression arrays identifies high risk molecular subtypes in breast cancer. Appl. Algebra Eng. Commun. Comput. **23**(1), 3–15 (2012)
3. Arsuaga, J., Borrman, T., Cavalcante, R., Gonzalez, G., Park, C.: Identification of copy number aberrations in breast cancer subtypes using persistence topology. Microarrays **4**(3), 339–369 (2015)

4. Barlund, M., Tirkkonen, M., Forozan, F., Tanner, M.M., Kallioniemi, O., Kallioniemi, A.: Increased copy number at 17q22-q24 by CGH in breast cancer is due to high-level amplification of two separate regions. Genes Chromosom. Cancer. **20**(4), 372–376 (1997)
5. Barrett, M.T., Anderson, K.S., Lenkiewicz, E., et al.: Genomic amplification of 9p24.1 targeting JAK2, PD-L1, and PD-L2 is enriched in high-risk triple negative breast cancer. Oncotarget **6**(28), 26483–26493 (2015)
6. Bengtsson, H., Ray, A., Spellman, P., Speed, T.P.: A single-sample method for normalizing and combining full-resolution copy numbers from multiple platforms, labs and analysis methods. Bioinformatics **25**(7), 861–867 (2009)
7. Benjamini, Y., Hochberg, Y.: Controlling the false discovery rate: a practical and powerful approach to multiple testing. J. Roy. Statist. Soc. Ser. B **57**(1), 289–300 (1995)
8. Bhatlekar, S., Fields, J.Z., Boman, B.M.: HOX genes and their role in the development of human cancers. J. Mol. Med. (Berl) **92**(8), 811–823 (2014)
9. Bilal, E., Vassallo, K., Toppmeyer, D., et al.: Amplified loci on chromosomes 8 and 17 predict early relapse in ER-positive breast cancers. PLoS One **7**(6), e38575 (2012)
10. Cavalcante, R.: Using Homology and networks to locate copy number aberrations associated to recurrence in breast cancer. MA Thesis, San Francisco State University (2012)
11. Chin, K., DeVries, S., Fridlyand, J., Spellman, P.T., Roydasgupta, R., et al.: Genomic and transcriptional aberrations linked to breast cancer pathophysiologies. Cancer Cell **10**, 529–541 (2006)
12. Ching, H.C., Naidu, R., Seong, M.K., Har, Y.C., Taib, N.A.: Integrated analysis of copy number and loss of heterozygosity in primary breast carcinomas using high-density SNP array. Int. J. Oncol. **39**(3), 621–633 (2011)
13. Climent, J., Garcia, J.L., Mao, J.H., Arsuaga, J., Perez-Losada, J.: Characterization of breast cancer by array comparative genomic hybridization. Biochem Cell Biol. **85**(4), 497–508 (2007)
14. Desmedt, C., Voet, T., Sotiriou, C., Campbell, P.J.: Next-generation sequencing in breast cancer: first take home messages. Curr Opin. Oncol. **24**(6), 597–604 (2012)
15. DeWoskin, D., Climent, J., Cruz-White, I., Vazquez, M., Park, C., et al.: Applications of computational homology to prediction of treatment response in breast cancer patients. Topology Appl. **157**, 157–164 (2010)
16. Fridlyand, J., Dimitrov, P.: aCGH: Classes and functions for Array Comparative GenomicHybridization data. R package version 1.34.0
17. Fridlyand, J., Snijders, A.M., Pinkel, D., Albertson, D.G., Jain, A.N.: Hidden Markov models approach to the analysis of array CGH data. J. Multivar. Anal. **90**, 132–153 (2004)
18. Fridlyand, J., Snijders, A.M., Ylstra, B., Li, H., Olshen, A., et al.: Breast tumor copy number aberration phenotypes and genomic instability. BMC Cancer **6**, 96 (2006)
19. Green, M.R., Monti, S., Rodig, S.J., et al.: Integrative analysis reveals selective 9p24.1 amplification, increased PD-1 ligand expression, and further induction via JAK2 in nodular sclerosing Hodgkin lymphoma and primary mediastinal large B-cell lymphoma. Blood 116(17), 3268–3277
20. Horlings, H.M., Lai, C., Nuyten, D.S.A., et al.: Integration of DNA copy number alterations and prognostic gene expression signatures in breast cancer patients. Clin Cancer Res. **16**(2), 651–663 (2010)

21. Horlings, H.M., Lai, C., Nuyten, D.S.A., et al.: Supplementary Data. Clin. Cancer Res. 16(2), 651–663 (2010b). http://clincancerres.aacrjournals.org/content/16/2/651/suppl/DC1

22. Hupe, P., Stransky, N., Thiery, J.P., Radvanyi, F., Barillot, E.: Analysis of array CGH data: from signal ratio to gain and loss of DNA regions. Bioinformatics 20(18), 3413–3422 (2004)

23. Jacot, W., Fiche, M., Zaman, K., Wolfer, A., Lamy, P.J.: (2013) The HER2 amplicon in breast cancer: Topoisomerase IIA and beyond. Biochim. Biophys. Acta. 1, 146–157 (1836)

24. Jonsson, G., Staaf, J., Vallon-Christersson, J., Ringner, M., Holm, K., et al.: Genomic subtypes of breast cancer identified by array comparative genomic hybridization display distinct molecular and clinical characteristics. Breast Cancer Res. 12(3), R42 (2010)

25. Lai, W.R., Johnson, M.D., Kucherlapati, R., Park, P.J.: Comparative analysis of algorithms for identifying amplifications and deletions in array CGH data. Bioinformatics (2005). doi:10.1093/bioinformatics/bti611

26. Lai, C., Horlings, H., van de Vijver, M.J., et al.: SIRAC: supervised identification of regions of aberration in aCGH datasets. BMC Bioinform. 8, 422 (2007)

27. Latham, C., Zhang, A., Nalbanti, A., et al.: Frequent co-amplification of two different regions on 17q in aneuploid breast carcinomas. Cancer Genet. Cytogenet. 127(1), 16–23 (2001)

28. Leiserson, M.D., Vandin, F., H-T, Wu, et al.: Pan-cancer network analysis identifies combinations of rare somatic mutations across pathways and protein complexes. Nat. Genet. 47, 106–114 (2015)

29. Mahmood, S.F., Gruel, N., Chapeaublanc, E., et al.: A siRNA screen identifies RAD21, EIF3H, CHRAC1 and TANC2 as driver genes within the 8q23, 8q24.3 and 17q23 amplicons in breast cancer with effects on cell growth, survival and transformation. Carcinogenesis 35(3), 670–682 (2014)

30. Martin-Castillo, B., Lopez-Bonet, E., Bux, M., et al.: Cytokeratin 5/6 fingerprinting in HER2-positive tumors identifies a poor prognosis and trastuzumab-resistant basal-HER2 subtype of breast cancer. Oncotarget 6(9), 7104–22 (2015)

31. Niyogi, P., Smale, S., Weinberger, S.: Finding the homology of submanifolds with high confidence from random samples. Discrete Comput. Geom. 39, 419–441 (2008)

32. Nielsen, K.V., Muller, S., Mller, S., Schonau, A., Balslev, E., Knoop, A.S., Ejlertsen, B.: Aberrations of ERBB2 and TOP2A genes in breast cancer. Mol. Oncol. 4(2), 161–168 (2010)

33. Olshen, A.B., Venkatraman, E.S., Lucito, R., Wigler, M.: Circular binary segmentation for the analysis of array-based DNA copy number data. Biostatistics 5(4), 557–572 (2004)

34. Perea, J., Harer, J.: Sliding windows and persistence: An application of topological methods to signal analysis. Found. Computat. Math. 15(3), 799–838

35. Perou, C., Borresen-Dale, A.L.: Systems biology and genomics of breast cancer. Cold Spring Harbor Perspect. Biol. 3, a003293 (2011)

36. Pinkel, D., Albertson, D.G.: Array comparative genomic hybridization and its applications in cancer. Nat. Genet. 37(Suppl), S11–S17 (2005)

37. Rauta, J., Alarmo, E.L., Kauraniemi, P., et al.: The serine-threonine protein phosphatase PPM1D is frequently activated through amplification in aggressive primary breast tumours. Breast Cancer Res. Treat. 95(3), 257–263 (2006)

38. Rebouh: Exploring topological methods to study topological imbalance in breast cancer. San Francisco State University MA thesis (2012)

39. Sinclair, C.S., Rowley, M., Naderi, A., Couch, F.J.: The 17q23 amplicon and breast cancer. Breast Cancer Res. Treat. **78**(3), 313–322 (2003)
40. Tausz, A., Vejdemo-Johansson, M., Adams, H.: JavaPlex: A research software package for persistent (co)homology. In: Hong, H., Yap, C. (eds.) Mathematical Software – ICMS 2014. LNCS, vol. 8592, pp. 129–136. Springer, Heidelberg (2014)
41. Thompson, P.A., Brewster, A.M., Kim-Anh, D.: Selective genomic copy number imbalances and probability of recurrence in early-stage breast cancer. PLoS One **6**(8), e23543 (2010)
42. Torresan, C., Oliveira, M.M., Pereira, S.R., et al.: Increased copy number of the DLX4 homeobox gene in breast axillary lymph node metastasis. Cancer Genet. **207**(5), 177–187 (2014)
43. Ulz, P., Heitzer, E., Speicher, M.: Co-occurrence of MYC amplification and TP53 mutations in human cancer. Nat. Genet. **48**(2), 104–106 (2016)
44. Webster, L.R., Provan, P.J., Graham, D.J., et al.: Prohibitin expression is associated with high grade breast cancer but is not a driver of amplification at 17q21.33. Pathology **45**(7), 629–636 (2013). doi:10.1097/PAT.0000000000000004
45. Willenbrock, H., Fridlyand, J.: A comparison study: applying segmentation to array CGH data for downstream analyses. Bioinformatics **21**(22), 4084–4091 (2005)
46. Wilkerson, P.M., Reis-Filho, J.S.: The 11q13-q14 amplicon: clinicopathological correlations and potential drivers. Genes Chromosom. Cancer **52**(4), 333–355 (2013)
47. Zhou, X., Rao, N.P., Cole, S.W., Mok, S.C., Chen, Z., Wong, D.T.: Progress in concurrent analysis of loss of heterozygosity and comparative genomic hybridization utilizing high density single nucleotide polymorphism arrays. Cancer Genet. Cytogenet **159**(1), 53–57 (2005)

Fast, Simple and Separable Computation of Betti Numbers on Three-Dimensional Cubical Complexes

Aldo Gonzalez-Lorenzo[1,2]([⊠]), Mateusz Juda[3], Alexandra Bac[1], Jean-Luc Mari[1], and Pedro Real[2]

[1] Aix-Marseille Université, CNRS, LSIS UMR 7296, Marseille, France
`aldo.gonzalez-lorenzo@univ-amu.fr`
[2] Institute of Mathematics IMUS, University of Seville, Seville, Spain
[3] Institute of Computer Science and Computational Mathematics, Jagiellonian University, Krakow, Poland

Abstract. Betti numbers are topological invariants that count the number of holes of each dimension in a space. Cubical complexes are a class of CW complex whose cells are cubes of different dimensions such as points, segments, squares, cubes, etc. They are particularly useful for modeling structured data such as binary volumes.

We introduce a fast and simple method for computing the Betti numbers of a three-dimensional cubical complex that takes advantage on its regular structure, which is not possible with other types of CW complexes such as simplicial or polyhedral complexes. This algorithm is also restricted to three-dimensional spaces since it exploits the Euler-Poincaré formula and the Alexander duality in order to avoid any matrix manipulation. The method runs in linear time on a single core CPU. Moreover, the regular cubical structure allows us to obtain an efficient implementation for a multi-core architecture.

Keywords: Cubical complex · Betti numbers · 3D · Separable · Computational topology · Homology

1 Introduction

Understanding a discrete volume can be addressed by determining its volume, its convexity, its diameter or any other geometrical descriptor. A higher level analysis can be made through topology, which tolerates continuous deformations. This could be seen as a less interesting approach, as we could not distinguish a sphere from a cube, but it actually furnishes a more essential information of the object. Homology is a powerful tool as its formalizes the concept of hole.

M. Juda—This research is supported by the Polish National Science Center under grant 2012/05/N/ST6/03621.

© Springer International Publishing Switzerland 2016
A. Bac and J.-L. Mari (Eds.): CTIC 2016, LNCS 9667, pp. 130–139, 2016.
DOI: 10.1007/978-3-319-39441-1_12

Holes of dimension 0, or 0-holes, correspond to connected components. 1-holes are tunnels or handles, which are particularly difficult to count in a volume depending on their shape. 2-holes correspond to voids in a volume. These notions can be generalized to higher dimensions, but they do not have an intuitive interpretation. We can compute the number of holes in each dimension or even draw them on the volume, though this is not useful with a complex shape.

Homology can be used for understanding an object without visualizing it, or to compare objects in a flexible way. It has been applied to dynamical systems [13,15], material science [4,18], electromagnetism [7,8], image understanding [1,14] and sensor networks [6].

In this article we aim at counting the number of holes (the Betti numbers) of a cubical complex embedded in a three-dimensional space. This is far from being an abstract work, as binary volumes (3D binary images, with voxels instead of pixels) can be transformed into equivalent cubical complexes. Our algorithm has a very specific input, since it cannot treat meshes or higher dimension cubical complexes, but it benefits from a good time complexity (linear) and a wide range of applications where data is structured in a lattice.

There have been a lot of works in computational homology in the last decades. Many of them [9,16,17] can compute the homology groups of more general spaces in cubical time. Computing only the Betti numbers (number of holes), which are the ranks of these groups, should be faster, but this has not been algorithmically proved. Delfinado and Edelsbrunner [5] introduce an algorithm with almost linear time complexity that computes the Betti numbers of a simplicial complex which is a subcomplex of a triangulation of S^3. The software library RedHom [12] is optimized for computing the homology in the context of cubical complexes. Wagner [19] also proposes an adapted algorithm for computing persistent homology on a cubical complex.

We propose an algorithm that is based on the computation of connected components and avoids any matrix manipulation. This is possible due to the Euler-Poincaré formula and the Alexander duality, which turn to be extraordinarily useful in the context of three-dimensional cubical complexes.

A simple description of the algorithm is given in Sect. 3. Then, we explain in Sect. 4 how to parallelize the computation by considering a different method for counting the connected components which is more adapted to the input data. Sections 5 and 6 explain the implementation of the algorithm and compare it with a previous software respectively.

2   Preliminaries

2.1 nD Cubical Complex

An *elementary interval* is an interval of the form $[k, k + 1]$ or a degenerate interval $[k, k]$, where $k \in \mathbb{Z}$. An *elementary cube* is the Cartesian product of n elementary intervals, and the number of non-degenerate intervals in this product is its *dimension*. An elementary cube of dimension d will be called d-cube for short. Given two elementary cubes p and q, we say that p is a *face* of q if $p \subset q$.

The *Khalimsky coordinates* of an elementary cube $\prod_{i=1}^{n} [a_i, b_i]$ are $(a_1 + b_1, \cdots, a_n + b_n)$. The dimension of an elementary cube and its faces can be easily deduced from its Khalimsky coordinates. For a cube q we denote its Khalimsky coordinates by $q[]$ and its ith component by $q[i]$.

An nD *cubical complex* is a set of elementary cubes. The *boundary* of a d-cube is the collection of its $(d-1)$-dimensional faces. By virtue of its regular structure, an nD cubical complex can be represented as an n-dimensional array (called CubeMap in [19]), where the cubes are represented by their Khalimsky coordinates.

From now on we assume that cubes of a given nD cubical complex K have all positive coordinates bounded by integers w_i $(1 \leq i \leq n)$. A_K is the binary n-dimensional array of size $L := \prod_{i=1}^{n}(2w_i + 1)$ where elementary cubes are represented by a Boolean equal to true associated to their Khalimsky coordinates. An element of the array with coordinates $x = (x_1, \ldots, x_n)$ is denoted by $A_K[x_1] \ldots [x_n]$ or $A[x]$ for short. The element $A_K[q[]]$ associated to the cube q is denoted by $A_K[q]$.

It is straightforward to provide an enumeration of Khalimsky coordinates in $\prod_{i=1}^{n} [0, 2w_i]$. Namely, there exists a bijection $I : \prod_{i=1}^{n} [0, 2w_i] \to [0, L-1]$. Such bijection I will be referred to as the *index map* and its image as the *index set*. For a cube q, $I(q)$ means $I(q[]) = I(q[1], \ldots, q[n])$.

The *support* of K, denoted by $\mathrm{supp}(K)$, is the nD cubical complex containing all the elementary cubes in $\prod_{i=1}^{n} [0, w_i]$. Thus, A_K encodes both K and $\mathrm{supp}(K) \setminus K$.

2.2   Homology

A *chain complex* (C, d) is a sequence of $\mathfrak{R}$-modules $C_0, C_1, \ldots$ (called *chain groups*) and homomorphisms $d_1 : C_1 \to C_0, d_2 : C_2 \to C_1, \ldots$ (called *differential* or *boundary operators*) such that $d_{q-1}d_q = 0$, for all $q > 0$, where $\mathfrak{R}$ is some ring, called the *ground ring* or *ring of coefficients*. In this paper we will fix $\mathfrak{R} = \mathbb{Z}_2$.

An nD cubical complex K induces a chain complex. C_q is the free $\mathfrak{R}$-module generated by the q-cubes of K. Its elements (called *q-chains*) are formal sums of q-cubes with coefficients in $\mathbb{Z}_2$, so they can be interpreted as sets of q-cubes. The linear operator d_q maps each q-cube to the sum of its $(q-1)$-dimensional faces.

A q-chain x is a *cycle* if $d_q(x) = 0$, and a *boundary* if $x = d_{q+1}(y)$ for some $(q+1)$-chain y. By the property $d_{q-1}d_q = 0$, every boundary is a cycle, but the reverse is not true: a cycle which is not a boundary contains a "hole". The qth homology group of the chain complex (C, d) contains the q-dimensional "holes": $H(C)_q = \ker(d_q)/\mathrm{im}(d_{q+1})$. This set is a finite-dimensional vector space, so there is a basis typically formed by the holes of the complex, whose elements are called *homology generators*. The ranks of the homology groups are called the *Betti numbers*, which count the number of holes in each dimension.

There is a slightly different homology theory called *reduced homology* where d_0 is defined otherwise. Thus, the zeroth Betti number β_0 is decremented by one. This avoids exceptional cases in several theorems.

3   The Algorithm

In this section we give a first presentation of our algorithm. It considers a restricted class of complexes: 3D cubical complexes. We explain in the following how we obtain each Betti number.

0th Betti number — It is well known that $\beta_0(K)$ is the number of connected components of K. This is easy to compute with a traversal of the complex.

2nd Betti number — Alexander duality relates the homology of a complex K of dimension 3 to the homology of its complementary in the three-dimensional sphere $S^3 \setminus K$.

Proposition 1 (Alexander Duality). *Let K be a 3D cubical complex. Then $H_q(K)$ and $H^{2-q}(S^3 \setminus K)$ are isomorphic for reduced homology and cohomology.*

As a consequence, $\beta_2(K) = \beta_0(S^3 \setminus K) - 1$. That is, the number of voids in K is the number of connected components in the complementary minus one.

This result, which holds for more general spaces, is computationally interesting in the context of cubical complexes. First, the sphere S^n is easy to build. Figure 1 shows the spheres S^1 and S^2 as cubical complexes.

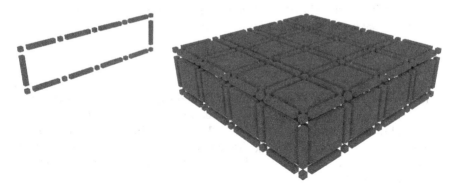

Fig. 1. Cubical complexes homeomorphic to S^1 and S^2.

Also, the complementary of a cubical complex is obvious to compute given its regular structure. Figure 2 illustrates the complementary of a cubical complex.

We want to obtain the number of connected components (minus one) of $S^3 \setminus K$ for deducing $\beta_2(K)$. Nevertheless, we do not need to build $S^3 \setminus K$. It suffices to count the connected components in $\text{supp}(K) \setminus K$ and consider only those which do not contain a cube in the boundary of $\text{supp}(K)$. These connected components are connected to $S^3 \setminus \text{supp}(K)$, thus making only one connected component in $S^3 \setminus K$. Note that this fact is far easier to understand for a 1D or a 2D cubical complex.

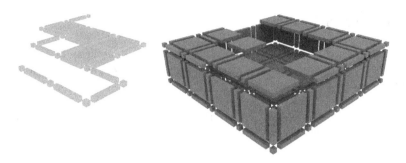

Fig. 2. A two-dimensional cubical complex K and its complementary $S^2 \setminus K$

1st Betti number — Once $\beta_0(K)$ and $\beta_2(K)$ are known, $\beta_1(K)$ is easy to obtain via the Euler-Poincaré formula. The *Euler-Poincaré characteristic* of a 3D cubical complex K is the alternating sum of its cubes. Formally,

$$\chi(K) = k_0 - k_1 + k_2 - k_3,$$

where k_q denotes the number of cubes of dimension q in K. This number, which is easy to compute, is a topological invariant.

Proposition 2 (Euler-Poincaré Formula). *Let K be a 3D cubical complex. Then $\chi(K) = \beta_0(K) - \beta_1(K) + \beta_2(K)$.*

Therefore, $\beta_1(K) = \beta_0(K) + \beta_2(K) - \chi(K)$.

Algorithm 1 combines these three ideas. It passes by all the elements of A_K and traverses the connected components of K and $\text{supp}(K) \setminus K$. For the sake of simplicity we do not explicitly describe the computation of $\chi(K)$ in Algorithm 1. It can be obtained by adding $\chi \leftarrow \chi + (-1)^{\dim(p)}$ to line 13. As each cube is connected to six other cubes in A_K (except for the cubes in the boundary of A_K), the complexity of the algorithm is $O(n+6n) = O(n)$ where n is the number of cubes in $\text{supp}(K)$.

4   Recursive Version of the Algorithm

The core of the previous algorithm is the computation of connected components through a traversal of the three-dimensional array A_K. This is difficult to parallelize because it uses a queue data structure. In this section we describe an algorithm for computing connected components of an nD cubical complex K in parallel. The algorithm total CPU utilization (i.e. work) is almost linear. It significantly uses the representation of a cubical complex as a multidimensional array A_K with an index map I.

In Sect. 3 we count connected components by traversing the connectivity graph of the cubical complex. Another well known approach to compute connected components is to use disjoint set data structure. The data structure

---

Algorithm 1. BettiViaCC

Input: K a 3D cubical complex; A_K its associated binary array
Output: The Betti numbers of K: β_0, β_1, β_2

1 $\beta_0 \leftarrow 0$, $\beta_2 \leftarrow 0$;
2 **foreach** $p \in A_K$ *not marked* **do**
3 | $b \leftarrow$ false;
4 | $Q \leftarrow$ an empty queue;
5 | $Q.push(p)$; mark p;
6 | **while** Q *not empty* **do**
7 | | $q \leftarrow Q.pop()$;
8 | | **if** q *belongs to the boundary of* A_K **then**
9 | | | $b \leftarrow$ true;
10 | | **foreach** q' *6-neighbor of* q, $A_K[q'] = A_K[q]$, q' *not marked* **do**
11 | | | $Q.push(q')$; mark q';
12 | **if** $A_K[p] = true$ **then**
13 | | $\beta_0 \leftarrow \beta_0 + 1$;
14 | **else if** $b = false$ **then**
15 | | $\beta_2 \leftarrow \beta_2 + 1$;
16 $\beta_1 \leftarrow \beta_0 + \beta_2 - \chi(K)$;
17 **return** $(\beta_0, \beta_1, \beta_2)$;

---

maintains a collection $S = \{ S_1, \ldots, S_k \}$ of disjoint sets. Each set in S is identified by a representative, which is a member of the set (see [3, Chap. 21]). The following operations may be performed on the disjoint set data structure C:

- $C.\text{makeSet}(x)$ - creates a new set whose only member (and thus representative) is x.
- $C.\text{find}(x)$ - returns a pointer to the representative of the (unique) set containing x.
- $C.\text{union}(x, y)$ - merges the sets that contain x and y into a new set that is the union of these two sets.

To compute connected components of a cubical complex it is enough to call $C.\text{union}(x, y)$ for each pair x, y of adjacent cubes. A parallel version of such algorithm requires synchronization, so in practice it cannot be implemented efficiently. However, the regular structure of a cubical complex allows us to propose a different approach where synchronization is not needed. The idea is to recursively cut the complex in two halves, find the connected components in each half and then merge them.

Let K be a cubical complex and I the index map of Khalimsky coordinates. Let J be a subset of the index set associated with K. We define $K_J := \{ q \in K \mid I(q) \in J \}$. We also define the *left slice*, *right slice* and *middle slice* of J in dimension d by x respectively as

$$S(J, x_-, d) := \{\, y \in J \mid I^{-1}(y)[d] < x \,\}$$
$$S(J, x^+, d) := \{\, y \in J \mid x \leq I^{-1}(y)[d] \,\}$$
$$S(J, x, d) := \{\, y \in J \mid x - 1 \leq I^{-1}(y)[d] \leq x \,\}.$$

For a $j \in J$ we denote by $cc_J(j)$ the connected component of K_J to which j belongs. Algorithm 2 computes recursively connected components of a cubical complex. Observe that at each step of the recursion the set J is split following some rule. We do not give an explicit description of the rule, but it should divide J into two sets of similar size by separating K_J along alternate axes. We thus obtain three subsets that cover J, one of them intersecting the other two so we can merge the connected components computed on each side. The first two recursive steps (lines 4 and 5) work on independent data, so they can be executed in parallel. The third recursive step at line 6 always jumps to the line 8 (since $J \not> \epsilon = \infty$) and it depends on the previous two steps.

Algorithm 2. RecursiveCC

Input: K a 3D cubical complex; I its associated index map; $J \subset I$
Input: C a disjoint set data structure on the index set of K, such that
      $C.\text{find}(i) \neq C.\text{find}(j)$ for all $i, j \in J$
Input: Parameters: $d \in \mathbb{Z}$ and $\epsilon > 0$
Output: For each pair $i, j \in J$ we have $cc_J(i) = cc_J(j)$ if and only if
      $C.\text{find}(i) = C.\text{find}(j)$

1   **if** *size of $J > \epsilon$* **then**
2      $d \leftarrow$ using d choose dimension for next slicing;
3      $x \leftarrow$ choose slicing value in dimension d;
4   RecursiveCC$(K, I, S(J, x_-, d), d, \epsilon, C)$;
5   RecursiveCC$(K, I, S(J, x^+, d), d, \epsilon, C)$;
6   RecursiveCC$(K, I, S(J, x, d), d, \infty, C)$;

7   **else**
8   **foreach** $p \in K_J$ **do**
9   **foreach** q *2n-neighbor of p in K_J* **do**
10          $C.\text{union}(I(p), I(q))$;

Algorithm 3 computes the Betti numbers of a 3D cubical complex K. It computes the connected components of K and $\text{supp}(K) \setminus K$ in two calls to Algorithm 2. Again, $\chi(K)$ can be computed during the traversal of the complex.

5 Implementation

Algorithm 3 is implemented as a part of the CAPD::RedHom project [11]. Our parallel version of the implementation uses Threading Building Blocks library [10]. A crucial part of the implementation is a data structure for efficient

---

Algorithm 3. RecursiveBetti

Input: K a 3D cubical complex; I its associated index map
Input: Parameter $\epsilon > 0$
Output: The Betti numbers of K: $\beta_0, \beta_1, \beta_2$

1 $C_1 \leftarrow$ a disjoint set for im I;
2 **foreach** $q \in K$ **do**
3 $\quad \lfloor \; C_1.\text{makeSet}(I(q));$

4 RecursiveCC$(K, I, \text{im } I, 0, \epsilon, C_1)$;
5 $\beta_0 \leftarrow$ number of sets in C_1;

6 $K_0 \leftarrow \text{supp}(K) \setminus K$;
7 $C_0 \leftarrow$ a disjoint set for im I;
8 **foreach** $q \in K_0$ **do**
9 $\quad \lfloor \; C_0.\text{makeSet}(I(q));$

10 RecursiveCC$(K_0, I, \text{im } I, 0, \epsilon, C_0)$;
11 $r \leftarrow$ number of sets in C_0 containing a cube in the boundary of $\text{supp}(K)$;
12 $\beta_2 \leftarrow$ number of sets in C_0 minus r;
13 $\beta_1 \leftarrow \beta_0 + \beta_2 - \chi(K)$;
14 **return** $(\beta_0, \beta_1, \beta_2)$;

---

slicing of the index set. For this we use Boost.MultiArray, a library from Boost Project [2]. It is an implementation of a multidimensional array container. In our case the data structure contains the index set. It provides an efficient slicing operation implemented as views to the original container. We use it to implement the operation S from the algorithm. At each recursion step we take a direction an cut the multidimensional array in the middle of the direction.

The data structure provides a mapping from multidimensional indices (in our case Khalimsky coordinates) to the index set. Technically it is enough to implement a mapping from the set of indices to a linear space of memory $[0, L-1]$ containing the value i at the ith position. Taking advantage of this fact, features of the C++ language, and Boost.MultiArray, we do not have to allocate memory for the index set. We get the index set and the slicing operation without any additional cost. Of course we can achieve it in many ways, however with our approach we can reuse well tested code.

6   Validation

Table 1 shows results of numerical experiments with the algorithm implementation. We compare also with standard approach for Betti numbers computations using elementary reductions, coreduction, and Morse decomposition from CAPD::RedHom [11]. All the computations were performed using one data structure, only algorithms vary.

Data sets N0001 and P0001 come from computer assisted proofs in dynamics. Data sets rand_pP_S were generated randomly, where S is the size of the grid and each 3-cube (together with its faces) is included with probability P.

The data sets are in binary format, thus reading time can be omitted. Computations were performed on a 2,3 GHz Intel Core i7 (4 real cores, 8 virtual) with 16 GB RAM. The results show that the parallel implementation is around 4 times faster than the sequential one. It suggest a perfect scalability with the number of real cores. Also, we see that for the new algorithm only grid size matters.

Table 1. CPU time (format [h:]mm:ss) usage for cubical complexes. Computations with following algorithms from CAPD::RedHom: Algorithm 3 parallel, Algorithm 3 sequential, standard

Data set	Grid size	Number of cells	Parallel CPU	sequential CPU	standard CPU
N0001	256^3	75357994	0:23	1:18	1:31
P0001	256^3	75559573	0:23	1:18	1:39
rand_p25_256	256^3	75897341	0:22	1:13	3:35:22
rand_p50_256	256^3	110450571	0:23	1:15	
rand_p75_256	256^3	127326478	0:23	1:17	
rand_p25_384	384^3	256006045	1:21	4:12	> 4h
rand_p50_384	384^3	372383238	1:18	4:17	
rand_p75_384	384^3	429007477	1:17	4:15	

7   Conclusion

This paper introduces a linear algorithm that computes the Betti numbers of a 3D cubical complex. It counts the connected components of the complex and its complementary in S^3 and uses the Euler-Poincaré formula. The algorithm is specially conceived for cubical complex as it takes advantage of its regular structure both in a theoretical and a practical manner. It cannot be extended to 4D cubical complexes since the Euler-Poincaré formula does not suffices to obtain all the Betti numbers.

An interesting issue that should be addressed in the near future is how to adapt this algorithm for simplicial complexes. The main problem is that we need a triangulation of the complementary of the complex in S^3, which is not as easy as for cubical complexes.

The current implementation outperforms the existing software for computing Betti numbers on cubical complexes. It is available as a part of the CAPD::RedHom [11] project. A more detailed comparison will be done in a forthcoming paper.

References

1. Allili, M., Corriveau, D.: Topological analysis of shapes using Morse theory. Comput. Vis. Image Underst. **105**(3), 188–199 (2007)
2. BoostCommunity. Boost Project (2016). http://www.boost.org/
3. Cormen, T.H., Stein, C., Rivest, R.L., Leiserson, C.E.: Introduction to Algorithms. McGraw-Hill Higher Education, New York (2001)
4. Day, S., Kalies, W.D., Wanner, T.: Verified homology computations for nodal domains. Multiscale Model. Simul. **7**(4), 1695–1726 (2009)
5. Cecil, J.A., Delfinado, H.E.: An incremental algorithm for Betti numbers of simplicial complexes on the 3-sphere. Comput. Aided Geom. Des. **12**(7), 771–784 (1995)
6. Dlotko, P., Ghrist, R., Juda, M., Mrozek, M.: Distributed computation of coverage in sensor networks by homological methods. Appl. Algebra Eng. Commun. Comput. **23**(1–2), 29–58 (2012)
7. Dłotko, P., Specogna, R.: Efficient cohomology computation for electromagnetic modeling. CMES: Comput. Model. Eng. Sci. **60**(3), 247–278 (2010)
8. Gross, P.W., Robert Kotiuga, P.: Electromagn. Theory Comput. Cambridge University Press, Cambridge (2004). Cambridge Books Online
9. Harker, S., Mischaikow, K., Mrozek, M., Nanda, V.: Discrete morse theoretic algorithms for computing homology of complexes and maps. Found. Comput. Math. **14**(1), 151–184 (2013)
10. Intel. Threading Building Blocks (2016). https://www.threadingbuildingblocks.org/
11. Juda, M., Mrozek, M., Brendel, P., Wagner, H., et al.: CAPD: : RedHom (2010–2016). http://redhom.ii.uj.edu.pl
12. Juda, M., Mrozek, M.: CAPD:RedHom v2 - homology software based on reduction algorithms. In: Hong, H., Yap, C. (eds.) ICMS 2014. LNCS, vol. 8592, pp. 160–166. Springer, Heidelberg (2014)
13. Mischaikow, K.: Conley index theory. In: Johnson, R. (ed.) Dynamical Systems. Lecture Notes in Mathematics, vol. 1609, pp. 119–207. Springer, Heidelberg (1995)
14. Mrozek, M., Zelawski, M., Gryglewski, A., Han, S., Krajniak, A.: Homological methods for extraction and analysis of linear features in multidimensional images. Pattern Recogn. **45**(1), 285–298 (2012)
15. Mrozek, M.: Index pairs algorithms. Found. Comput. Math. **6**, 457–493 (2006)
16. Munkres, J.R.: Elements of Algebraic Topology. Addison-Wesley, Reading (1984)
17. Peltier, S., Alayrangues, S., Fuchs, L., Lachaud, J.-O.: Computation of homology groups and generators. In: Andrès, É., Damiand, G., Lienhardt, P. (eds.) DGCI 2005. LNCS, vol. 3429, pp. 195–205. Springer, Heidelberg (2005)
18. Teramoto, T., Nishiura, Y.: Morphological characterization of the diblock copolymer problem with topological computation. Jpn. J. Ind. Appl. Math. **27**(2), 175–190 (2010)
19. Wagner, H., Chen, C., Vuçini, E.: Efficient computation of persistent homology for cubical data. In: Peikert, R., Hauser, H., Carr, H., Fuchs, R. (eds.) Topological Methods in Data Analysis and Visualization II. Mathematics and Visualization, pp. 91–106. Springer, Berlin Heidelberg (2012)

Computation of Cubical Steenrod Squares

Marek Krčál and Paweł Pilarczyk[✉]

Institute of Science and Technology Austria,
Am Campus 1, 3400 Klosterneuburg, Austria
{marek.krcal,pawel.pilarczyk}@ist.ac.at

Abstract. Bitmap images of arbitrary dimension may be formally perceived as unions of m-dimensional boxes aligned with respect to a rectangular grid in $\mathbb{R}^m$. Cohomology and homology groups are well known topological invariants of such sets. Cohomological operations, such as the cup product, provide higher-order algebraic topological invariants, especially important for digital images of dimension higher than 3. If such an operation is determined at the level of simplicial chains [see e.g. González-Díaz, Real, Homology, Homotopy Appl, 2003, 83–93], then it is effectively computable. However, decomposing a cubical complex into a simplicial one deleteriously affects the efficiency of such an approach. In order to avoid this overhead, a direct cubical approach was applied in [Pilarczyk, Real, Adv. Comput. Math., 2015, 253–275] for the cup product in cohomology, and implemented in the ChainCon software package [http://www.pawelpilarczyk.com/chaincon/].

We establish a formula for the Steenrod square operations [see Steenrod, Annals of Mathematics. Second Series, 1947, 290–320] directly at the level of cubical chains, and we prove the correctness of this formula. An implementation of this formula is programmed in C++ within the ChainCon software framework. We provide a few examples and discuss the effectiveness of this approach.

One specific application follows from the fact that Steenrod squares yield tests for the topological extension problem: Can a given map $A \to S^d$ to a sphere S^d be extended to a given super-complex X of A? In particular, the *ROB-SAT* problem, which is to decide for a given function $f \colon X \to \mathbb{R}^m$ and a value $r > 0$ whether every $g \colon X \to \mathbb{R}^m$ with $\|g - f\|_\infty \leq r$ has a root, reduces to the extension problem.

Keywords: Cohomology operation · Cubical complex · Cup product · Chain contraction

1 Introduction

Binary images (or bitmaps) appear in various contexts, not only image processing. One can perceive a 2-dimensional bitmap image as a finite collection of squares (black pixels) aligned with respect to a fixed grid in $\mathbb{R}^2$, and indexed in both directions by the integers. A generalization to $\mathbb{R}^m$ may be called an m-dimensional binary image. For example, a 3-dimensional binary image corresponds to a collection of voxels that represent a 3-dimensional object embedded

© Springer International Publishing Switzerland 2016
A. Bac and J.-L. Mari (Eds.): CTIC 2016, LNCS 9667, pp. 140–151, 2016.
DOI: 10.1007/978-3-319-39441-1_13

in $\mathbb{R}^3$. This definition of an m-dimensional bitmap does not limit the area of applications to image processing alone. For example, a rectangular lattice in $\mathbb{R}^m$ is often used for numerical simulations of PDEs or simply for approximating bounded sets in $\mathbb{R}^m$, e.g., an outer bound for a set of solutions to some equation.

Cohomology and homology groups are well known topological invariants that can be used to describe or classify the rough shape defined by an m-dimensional bitmap. There exist theory and software that allow one to efficiently compute these invariants. For example, the monograph [11] and the CHomP [2] and CAPD [3] software projects contain algorithms aimed specifically at efficient homology computation of m-dimensional bitmaps described in terms of full cubical sets and cubical complexes (see Sect. 2 for precise definitions of these terms).

Cohomological operations, such as the cup product, provide higher-order algebraic topological invariants than (co)homology groups alone. This is especially important for bitmaps of dimension higher than 3, where the natural human intuition may easily fail. One way to compute the operations effectively is to use an approach to homology computation known as "effective homology" [20, 21]. In this approach, instead of *reducing* the topological information to a minimal linear system that describes the degree of connectivity of the objects, one computes an algebraic skeleton, further called an algebraic-topological model (or an AT model for short), for *representing* these objects. In particular, an AT model contains homomorphisms that allow to instantly obtain representative cycles for each homology generator, and to efficiently compute, for an arbitrary cycle, the combination of the corresponding homology generators. We refer to [9] for an explanation of the philosophy behind this approach, and to [17] for an application in the context of cubical sets.

Steenrod squares [23] are cohomology operations which, roughly speaking, enhance the cohomology ring structure and thus help discriminating between topologically different spaces which still might have isomorphic cohomology rings. Steenrod squares, as well as any cohomology operations, are by definition *natural*: They are compatible with the induced homomorphisms, and thus can discriminate between non-homotopic maps. Moreover, due to the naturality, Steenrod squares can provide tests and sometimes even complete characterization for problems in homotopy theory. The primary example of such and also an important motivation for our work is the *topological extension problem (for maps into a sphere)*: Can a given map $A \to S^d$ to a sphere S^d be extended to a given super-complex X of A? Here the Steenrod squares yield stronger test than the one obtained by plain cohomology and this test is complete if $\dim X \le d + 2$. In fact, this has been the original motivation for Steenrod to introduce his squares, which has been a major breakthrough in homotopy theory of the end of the first half of the twentieth century.

We are interested in the extension problem mainly because another computational problem – called *robust satisfiability* – reduces to it [5]. In that problem, given a continuous function $f \colon X \to \mathbb{R}^n$ and a value $r > 0$, one has to decide whether every $g \colon X \to \mathbb{R}^n$ with $\|g - f\|_\infty \le r$ has a root. In plain words, we ask for solvability of the system of equations $f(x) = 0$ under uncertainty about the function f quantified by the value of r. Robust satisfiability and the topological extension problem are essentially computationally equivalent [5]. In particular, theoretical complexity study of the extension problem [1] shows that robust satisfiability is decidable in polynomial time when $\dim X \leq 2n - 3$ or $n = 1, 2$, and undecidable when $\dim X \geq 2n - 2$ and n is odd. (In the remaining cases of n even and arbitrary dimension of X, the problem is decidable [25].) More practical point of view and an actual implementation is presented in [15] where the Steenrod squares are used in the context of cubical complexes.

Determining cohomological operations at the level of cochains allows one to use an AT model to compute them effectively. The idea is to take the cocycles corresponding to cohomology generators, apply the operation to them, and to determine the combination of the cohomology generators that defines the cohomology class represented by the resulting cocycle; see [8] for an in-depth description of this approach. In particular, the simplicial formula provided implicitly in the first Steenrod's paper on the topic [23] (see also [7] for an alternative approach), computes the Steenrod squares at the level of representing simplicial cocycles. This yields an effective method for computing the Steenrod squares on the level of cohomology of simplicial complexes, if combined with the computation of an AT model. Unfortunately, decomposing a cubical set into a simplicial complex deleteriously affects the efficiency of such an approach. A direct cubical approach was proposed in [17] for the cup product in cohomology (see also [12]), and implemented in the ChainCon software package [16].

The purpose of our paper is to establish a formula for the Steenrod square operations directly at the level of cubical chains. We emphasize the fact that our result is not based upon the formula provided by Real [19] for the case of simplicial sets; rather, we develop a direct cubical formula, following Steenrod's original approach [23]; the correctness of this formula follows from the axiomatic approach to Steenrod squares, as explained at the beginning of Sect. 3. An implementation of this formula is programmed in C++ and put within the ChainCon software framework [16], and serves the purpose of proof-of-concept and benchmarking.

2   Topological Preliminaries

Cubical Complexes. An *(abstract) cubical complex* X is a family of sets $X = (X_0, X_1, \ldots)$ equipped with *face operators*

$$\partial_i^s \colon X_n \to X_{n-1} \text{ for each } i \in \{1, \ldots, n\} \text{ and } s \in \{+, -\}$$

satisfying the relation

$$\partial_i^s \partial_j^t = \partial_j^t \partial_{i+1}^s \text{ for } i \geq j \text{ and } s, t \in \{+, -\}. \tag{1}$$

The elements of each X_n are called n-*cubes* and the face operators $\partial_i^\pm$ can be thought of as an abstract counterpart of obtaining a facet of the cube $[-1,1]^n$ by fixing the ith coordinate to ± 1, that is,

$$\partial_i^\pm [-1,1]^n = [-1,1]^{i-1} \times \{\pm 1\} \times [-1,1]^{n-i}.$$

The simplest example is indeed the cubical complex I^m where each set of n-cubes I_n^m consists of all n-faces of the geometric cube $[-1,1]^m$ and the face operators are defined in the obvious sense. An important example will be cubical complexes derived from *cubical sets* as explained below.

A cubical map $f\colon X \to Y$ from a cubical complex X to a cubical complex Y is a family of maps $f_n\colon X_n \to Y_n$ that commute with all the face operators.

Cubical Sets. We follow the terminology and notation based upon [11,14].

Let m be a positive integer. An *elementary cube* is the cartesian product of m intervals of length 1 (the non-degenerate case) or 0 (the degenerate case) with integer coordinates; formally:

$$[a_1, b_1] \times \cdots \times [a_m, b_m],$$

where $a_i, b_i \in \mathbb{Z}$, and either $b_i = a_i + 1$, or $a_i = b_i$ (and then $[a_i, b_i]$ denotes the singleton $\{a_i\}$). If all the intervals in this product are non-degenerate then the elementary cube is called a *full cube*.

A set $A \subset \mathbb{R}^m$ is called a *cubical set* if it is a finite union of elementary cubes; Note that cubical sets are obviously compact ENRs. The cubical set A is called a *full cubical set* if it is a finite union of full cubes. For example, an m-dimensional binary image (or a bitmap) can be perceived as a full cubical set in $\mathbb{R}^m$ for the purpose of topological analysis.

Since the face of an elementary cube is also an elementary cube, sets of elementary cubes yield a natural cubical complex structure. The homological properties of this cubical complex agree with the (singular) homology of the corresponding cubical set.

Cubical Chain Complexes. To each cubical complex X and an Abelian group G we assign a *cubical chain complex*

$$\dots \xrightarrow{d_3} C_2(X;G) \xrightarrow{d_2} C_1(X;G) \xrightarrow{d_1} C_0(X;G),$$

where each $C_n(X;G)$ is the group of formal sums $\sum_{\sigma \in X_n} g_\sigma \cdot \sigma$ with coefficients in G and each *boundary operator* d_n is the homomorphism defined by

$$d_n(g \cdot \sigma) = \sum_{i=1}^n (-1)^i (g \cdot \partial_i^+ \sigma - g \cdot \partial_i^- \sigma).$$

At Models. Let us recall the notion of an AT model, which helps us to compute Steenrod squares effectively at the cohomology level using the formula defined at the level of cochains.

A *chain map* between two chain complexes is a homomorphism that commutes with the boundary operator. A *chain contraction* from a chain complex C_* to another chain complex C'_* is a triple (π, ι, ϕ) of chain maps $\pi\colon C_* \to C'_*$ (*projection*), $\iota\colon C'_* \to C_*$ (*inclusion*) and $\phi\colon C_* \to C_{*+1}$ (*chain homotopy*) that satisfy the following conditions: (a) $\mathrm{Id}_C - \iota\pi = \partial\phi + \phi\partial$; (b) $\pi\iota = \mathrm{Id}_{C'}$; (c) $\pi\phi = 0$; (d) $\phi\iota = 0$; (e) $\phi\phi = 0$. See e.g. [4, §12] for the motivation of this definition, and [8, p. 86] for comments on the terminology and applications. Note that the existence of a chain contraction from C_* to C'_* implies the fact that the homology and cohomology modules of both chain complexes are isomorphic.

An *algebraic topological model* (introduced in [6]), or an *AT model* for short, of a cubical complex K, is a chain contraction from $C_*(K)$ to some free chain complex M_* with null differential. Note that M_* is isomorphic to the homology module of K. In particular, an AT model of K exists if $H_*(K)$ has no torsion. In what follows, we work with coefficients in $\mathbb{Z}_2$, so this condition is satisfied.

We use an AT model for representing the homology of K in the following way. The image of each element of M_* by the inclusion ι is a cycle that represents the corresponding homology class. Additionally, the image of each cycle in $C_*(K)$ by the projection map π is the homology class that contains the cycle. In this way, the homomorphisms ι and π are used to go back and forth between homology generators and the corresponding cycles in $C_*(K)$.

3   The Cubical Formulas for the Steenrod Operations

We follow the general scheme of many standard textbooks and sources addressing Steenrod operations such as [10, 24], but the particular notation is very close to [18]. The (non-algorithmic) construction there is based on the existence of chain maps $D_*^k\colon C_*(X; \mathbb{Z}_2) \to \big(C_*(X; \mathbb{Z}_2) \otimes C_*(X; \mathbb{Z}_2)\big)_{*+k}$ for $k = 0, 1, \ldots$ satisfying the following: D_*^0 is a diagonal approximation[1] and, all the chain maps D_*^k satisfy the relation

$$D_*^k - TD_*^k = (d \otimes d)D_*^{k+1} + D_*^{k+1}d, \tag{2}$$

where $T\colon C_*(X; \mathbb{Z}_2) \otimes C_*(X; \mathbb{Z}_2) \to C_*(X; \mathbb{Z}_2) \otimes C_*(X; \mathbb{Z}_2)$ is defined as follows: $T(\sigma \otimes \tau) = \tau \otimes \sigma$. Each chain map D_*^k is called the *kth higher diagonal approximation*.

---

[1] A diagonal approximation for X is any chain map $C_*(X; G) \to C_*(X; G) \otimes C_*(X; G)$ which induces the map $\Delta_*\colon H_*(X; G) \to H_*(X \times X; G)$ where $\Delta\colon X \to X \times X$ is the diagonal map $x \mapsto (x, x)$. In the case of cubical chain complexes, the explicit formula was given by Serre [22].

Definition 1 ([18, pp. 186,187]). *The Steenrod square* $\mathrm{Sq}^j\colon H^n(X;\mathbb{Z}_2) \to H^{n+j}(X;\mathbb{Z}_2)$ *is induced by the composition*

$$Z^n(X;\mathbb{Z}_2) \xrightarrow{\;\Delta\;} Z^n(X;\mathbb{Z}_2) \otimes Z^n(X;\mathbb{Z}_2) \xrightarrow{(D^{n-j}_{n+j})^*} Z^{n+j}(X;\mathbb{Z}_2)\,,$$

where $\Delta(z) := z \otimes z$ *is the diagonal map.*

Our goal here is to give formulas for the chain maps D^k_* in the special case when X is a cubical complex – see Definition 2 and Theorem 3 below.

There is a subtle difference between our definition above and the definition provided in [18]: The chain maps D^k_* are defined in [18] for chain complexes with integral coefficients. Indeed, their existence can be proved in this stronger sense; however, for the definition of the Steenrod operations alone the "modulo 2" version is only relevant. The proof of [18, Theorem 3.60] gives a chain homotopy between any two choices of higher diagonal approximations (no matter whether over $\mathbb{Z}$ or $\mathbb{Z}_2$). Thus the higher diagonal approximations over $\mathbb{Z}_2$ necessarily lead to the identical cohomology operations – Steenrod squares.

Definition 2. *For given integers* $n, k \geq 0$, *let us define the set*

$$\mathcal{F}^k_n := \{(A, B) \mid A, B \subseteq [n], A \cap B = \emptyset \text{ and } |A| + |B| = n - k\}.$$

Let X be a cubical complex. We define the homomorphisms $D^k_n\colon C_n (X;\mathbb{Z}_2) \to \big(C_*(X;\mathbb{Z}_2) \otimes C_*(X;\mathbb{Z}_2)\big)_{n+k}$ of degree k by the formula

$$D^k_n(\sigma) := \sum_{(A,B)\in\mathcal{F}^k_n} \partial^{-*}_A\sigma \otimes \partial^*_B\sigma\,,$$

where for $A = \{a_1 < a_2 < \ldots < a_p\}$ *and* $B = \{b_1 < b_2 < \ldots < b_q\}$ *we define* $\partial^{-*}_A = \partial^{-s(a_1)}_{a_1} \ldots \partial^{-s(a_p)}_{a_p}$ *and* $\partial^*_B = \partial^{s(b_1)}_{b_1} \ldots \partial^{s(b_q)}_{b_q}$ *where*

$$s(x) = (-1)^{|[x]\setminus(A\cup B)|}\,.$$

Theorem 3. *The homomorphisms* D^k_n *defined above satisfy relation* (2).

Proof. Over $\mathbb{Z}_2$, relation (2) is equivalent to

$$D^k_n + TD^k_n + D^{k+1}_{n-1}d_n = (d \otimes d)_{n+k+1}D^{k+1}_n\,. \tag{3}$$

The right-hand side of (3) evaluates on a given generator $1 \cdot \sigma \in C_n(X;\mathbb{Z}_2)$ as follows (we will denote $1 \cdot \sigma$ simply by σ):

$$\sum_{(A,B)\in\mathcal{F}^{k+1}_n} \left(\sum_{\substack{i\in[n-|A|]\\ sigma\in\{+,-\}}} \partial^s_i\partial^{-*}_A\sigma \otimes \partial^*_B\sigma + \sum_{\substack{i\in[n-|B|]\\ sigma\in\{+,-\}}} \partial^{-*}_A\sigma \otimes \partial^s_i\partial^*_B\sigma \right)$$

$$= \sum_{(A,B)\in\mathcal{F}^{k+1}_n} \left(\sum_{\substack{j\in[n]\setminus A\\ sigma\in\{+,-\}}} \partial^{-*,s}_{A,j}\sigma \otimes \partial^*_B\sigma + \sum_{\substack{j\in[n]\setminus B\\ sigma\in\{+,-\}}} \partial^{-*}_A\sigma \otimes \partial^{*,s}_{B,j}\sigma \right),$$

where the operator $\partial_{B,j}^{*,s}$ is equal to ∂_B^* with ∂_j^s inserted at the correct position (that is, $\partial_{B,j}^{*,s} = \partial_{b1}^{s(b_1)} \ldots \partial_{b_r}^{s(b_r)} \partial_j^s \partial_{b_{r+1}}^{s(b_{r+1})} \ldots \partial_{b_q}^{s(b_q)}$ for $b_r < j < b_{r+1}$ and $s(x) = (-1)^{|[x] \setminus (A \cup B)|}$) and similarly for $\partial_{A,j}^{-*,s}$. This equality follows by applying relation (1). Each term of the sum above can be rewritten into one of the following forms according to whether the face operator $\partial_j^\pm$ is present on both sides of the tensor product:

1. When the face operator $\partial_j^\pm$ is present on both sides of the tensor product, each term can be rewritten in one of the two types according to whether the signs of the operators $\partial_j^\pm$ on the left and on the right agree or not, explicitly

$$\partial_{A'}^{-*} \partial_j^t \sigma \otimes \partial_{B'}^* \partial_j^{-t} \sigma \text{ and } \partial_{A'}^{-*} \partial_j^t \sigma \otimes \partial_{B'}^* \partial_j^t \sigma$$

   for unique $(A', B') \in \mathcal{F}_{n-1}^{k+1}$ and $t \in \{+, -\}$ determined by A, B, j and s. It is not difficult to see that for any fixed A', B', t and j, the term of the first type appears either twice (when $t = s(j)$) or never (when $t \neq s(j)$). In the second type, the term appears exactly once (either $\partial_{A'}^{-*} \partial_j^t = \partial_A^*$ or $\partial_{B'}^* \partial_j^t = \partial_B^*$ for some A and B).

2. When the face operator $\partial_j^\pm$ is present on one side of the tensor product only, we set up the following labeling:
   - $\partial_{A,j}^{-*,s} \sigma \otimes \partial_B^* \sigma$ will be called an $(A \cup \{j\}, B, j, -)$-term when $s = s(j) = (-1)^{|[j] \setminus (A \cup B)|}$.
   - $\partial_{A,j}^{-*,s} \sigma \otimes \partial_B^* \sigma$ will be called an $(A \cup \{j\}, B, j, +)$-term when $s = -s(j)$.
   - $\partial_A^{-*} \sigma \otimes \partial_{B,j}^{*,s} \sigma$ will be called an $(A, B \cup \{j\}, j, +)$-term when $s = s(j)$.
   - $\partial_A^{-*} \sigma \otimes \partial_{B,j}^{*,s} \sigma$ will be called an $(A, B \cup \{j\}, j, -)$-term when $s = -s(j)$.

   It follows that for each $(A', B') \in \mathcal{F}_n^k$, each $j \in A' \cup B'$, and each sign $t \in \{+, -\}$, there is exactly one (A', B', j, t)-term in the sum above. We define the following pairing on the set of all such (A', B', j, t)-terms: We pair each $(A', B', j, +)$-term with the $(A', B', j', -)$-term for $j' = \min\left((A \cup B) \setminus [j]\right)$ when the minimum exists.[2] Note that the paired terms are equal. The unpaired
   - $(A', B', j, +)$-terms for $j = \max(A' \cup B')$ and
   - $(A', B', j, -)$-terms for $j = \min(A' \cup B')$

   are equal to
   - $\partial_{A'}^{-*} \sigma \otimes \partial_{B'}^* \sigma$ and
   - $\partial_{A'}^* \sigma \otimes \partial_{B'}^{-*} \sigma$,

   respectively.

Summing up what has been said above, the right-hand side of (3) equals to

$$\sum_{(A',B') \in \mathcal{F}_n^k} \partial_{A'}^{-*} \sigma \otimes \partial_{B'}^* \sigma + \partial_{A'}^* \sigma \otimes \partial_{B'}^{-*} \sigma \quad + \sum_{(A',B') \in \mathcal{F}_{n-1}^{k+1}} \sum_{\substack{j \in [n] \\ t \in \{+,-\}}} \partial_{A'}^{-*} \partial_j^t \sigma \otimes \partial_{B'}^* \partial_j^t \sigma,$$

which is exactly the left-hand side of (3).   □

---

[2] Or, equivalently, we pair each $(A', B', j, -)$-term with the $(A', B', j', +)$-term for $j' = \max\left((A \cup B) \cap [j]\right)$ when the maximum exists.

Corollary 4. *Definition 2 yields an explicit cubical formula for the Steenrod squares as follows:*

$$\langle \mathrm{Sq}^j(z^n), \sigma \rangle := \sum_{\substack{A,B \subseteq [n+j] \\ A \cap B = \emptyset \\ |A| = |B| = j}} \langle z^n, \partial_A^{-*}\sigma \rangle \langle z^n, \partial_B^*\sigma \rangle. \qquad (4)$$

4 The Algorithm, Software, and Examples

Algorithm for Computing Steenrod Squares. In order to compute all the nontrivial Steenrod squares in a cubical complex K, we first compute an AT model of K, using the algorithm provided in [17]. The AT model consists of a finitely generated free chain complex M_* with null differential, and a chain contraction (π, ι, ϕ) from $C_*(K)$ to M_*. In particular, M_* is represented by a finite collection $\mathcal{M}_*$ of its generators. In this algorithm, for a chain $z \in C_n(K)$, its dual cochain is denoted as z^n.

Algorithm 5.
INPUT:
    (M_*, π, ι, ϕ) – an AT model of K;
OUTPUT:
    $\mathcal{P} = \{(z_p, j_p, \sigma_p) : p = 1, \ldots, P\}$ for some $P \in \mathbb{Z}$, where $z_p \in \mathcal{M}_{n_p}$,
    $n_p \in \mathbb{Z}$, $j_p \in \mathbb{Z}$, $\sigma_p \in \mathcal{M}_{j_p+n_p}$, and the elements of $\mathcal{P}$ represent
    all the nontrivial Steenrod squares in K, that is, $\langle \mathrm{Sq}^{j_p}(z_p^{n_p}), \sigma_p \rangle = 1$;
CODE:
    $\mathcal{P} := \emptyset$;
    $d :=$ the dimension of M_*;
    for each $n \in \{0, \ldots, d\}$
        for each $j \in \{0, \ldots, \max(n, d-n)\}$
            for each generator $z \in \mathcal{M}_n$
                for each generator $\sigma \in \mathcal{M}_{n+j}$
                    $\alpha(z, j, \sigma) := 0$;
                    for each $A, B \subseteq [n+j]$, $A \cap B = \emptyset$, $|A| = |B| = j$
                        for each s in $\iota(\sigma)$
                            if z appears in $\pi(\partial_A^{-*}s)$ and in $\pi(\partial_B^* s)$ then
                                $\alpha(z, j, \sigma) := \alpha(z, j, \sigma) + 1$;
                  if $\alpha(z, j, \sigma) \neq 0$ then
                    $\mathcal{P} := \mathcal{P} \cup \{(z, j, \sigma)\}$;
    return $\mathcal{P}$.

Software Implementation. The software publicly available at [16] under the GNU General Public License is written in the C++ programming language using the technique of generic programming. In particular, the type of cells in a cellular complex is a template parameter, so the same software applies to simplicial and cubical complexes alike, provided that the cell-specific operations (like the boundary) have been defined. In addition to a programming library that is accessible from within a program written in C++ and is the most efficient way of using this software, there are a few command-line programs provided that read definitions of cellular complexes saved in human-readable text files, and output the results in text format to the console. These programs thus constitute an easy to use interface to the main features of the software. Simplicial and cubical cells are defined, and several algorithms are implemented, including the computation of an AT model, the cohomology ring, and the Steenrod squares. The main program written especially for this paper is called `ssqcub`, and computes the Steenrod squares of a cubical complex. Additional programs that may be used for comparison and for gathering additional information, are: `ssqsim` (computation of Steenrod squares for simplicial complexes), `ssqcubs` (computation of Steenrod squares for cubical complexes using simplicial subdivision), `cringcub` (computation of the cohomology ring for cubical complexes), and `cringsim` (computation of the cohomology ring for simplicial complexes). We refer to the website [16] and instructions provided there for further information.

Approximations of Sample Manifolds. Given a finite set $X \subset \mathbb{R}^n$ that roughly approximates a bounded set $M \subset \mathbb{R}^n$ whose homological information we would like to demonstrate, we approximate it by means of a full cubical set A as follows. For each point $x = (x_1, \ldots, x_n) \in X$, we take the point $a = (a_1, \ldots, a_n) \in \mathbb{Z}^n$, where $a_i := \lfloor x_i \rfloor$ is the largest integer that does not exceed x_i, and is effectively computed by truncating the coordinates of x down to the nearest integers. Then we take the union of all the full cubes whose minimal vertices are given in this way. More precisely:

$$A := \bigcup \left\{ \left[\lfloor x_1 \rfloor, \lfloor x_1 \rfloor + 1 \right] \times \cdots \times \left[\lfloor x_n \rfloor, \lfloor x_n \rfloor + 1 \right] : x \in X \right\}.$$

In order to reduce a full cubical set A to a cubical set $A' \subset A$ which has the same homological properties, we apply the reduction techniques introduced in [14], which include removal of full cubes at the boundary of the set, and then a sequence of free face collapses. The examples of full cubical sets and (general) cubical sets discussed in this section are available at [16], and were obtained as described above, with the application of the reductions. In particular, the inclusion $A' \to A$ induces an isomorphism in (co)homology.

The parametrization given by $(\alpha, \beta) \mapsto (R \cos \alpha + r \cos \alpha \cos \frac{\alpha}{2} \cos \beta, R \sin \alpha + r \sin \alpha \cos \frac{\alpha}{2} \cos \beta, r \sin \frac{\alpha}{2} \cos \beta, r \sin \beta)$, with $R = 4$, $r = 2$, and $\alpha, \beta \in [0, 2\pi]$, was used for the Klein bottle embedded in $\mathbb{R}^4$, and the parametrizations provided in [13] were used for $\mathbb{R}P^2$ embedded in $\mathbb{R}^5$ and for $\mathbb{C}P^2$ embedded in $\mathbb{R}^8$ (the third coordinate in both original formulas was dropped).

Sample Computation of the Steenrod Squares. We consider three representative examples which exhibit nontrivial Steenrod squares, in addition to the obvious $Sq^0 \cong$ id. A summary of these examples is gathered in Table 1. The cohomology over $\mathbb{Z}_2$ (in terms of Betti numbers) and the nontrivial cup products, as well as the nontrivial Steenrod squares are listed in Table 2.

Table 1. A list of sample cubical sets that approximate selected manifolds with nontrivial Steenrod squares.

Name of the example	Embedding dimension	Number of full cubes	Number of elementary cubes
K^2 – Klein bottle	4	111	406
$\mathbb{RP}^2$ – real projective plane	5	38	288
$\mathbb{CP}^2$ – complex projective plane	8	281	16,915

Table 2. The nontrivial cup products and the nontrivial Steenrod squares. Cohomology generators are denoted by consecutive alphabetic letters for each dimension (e.g., a for dimension 0, e for dimension 4) with appended indices starting from 1 within each dimension separately.

Example	Betti numbers	Nontrivial cup products	Nontrivial Steenrod squares
K^2	$(1,2,1)$	$b_1 \smile b_1 = c_1$, $b_1 \smile b_2 = c_1$	$\mathrm{Sq}^1(b_1) = c_1$
$\mathbb{RP}^2$	$(1,1,1)$	$b_1 \smile b_1 = c_1$	$\mathrm{Sq}^1(b_1) = c_1$
$\mathbb{CP}^2$	$(1,0,1,0,1)$	$c_1 \smile c_1 = e_1$	$\mathrm{Sq}^2(c_1) = e_1$

Using Simplicial Subdivision. In the previous approach for the computation of the Steenrod squares of digital images [7,8], one would have to first compute a simplicial subdivision of a cubical set, and then compute the simplicial Steenrod squares, e.g., using the efficient formulas provided in [7]. This approach is considerably less efficient than using the direct cubical formula, especially in higher dimensions, where it takes a considerable number of simplices to fill a full cube. For example, our approximation of Klein bottle consisting of 111 full cubes of dimension 4 (see Table 1) yields a simplicial complex with 27,404 simplices, as opposed to the cubical complex containing 5,724 cubical cells. The 38 full cubes of dimension 5 that approximate $\mathbb{RP}^2$ yield 76,475 simplicial cells and 7,113 cubical cells, respectively. Obviously, the high numbers of simplicial cells detrimentally affect the computation speed (which is not linear!), and the memory usage as well.

Computation for the Suspension. For each of the sample cubical sets, we construct a cubical counterpart of its suspension, as follows. Given a cubical set $A \subset \mathbb{R}^m$, let B be a contractible cubical set in $\mathbb{R}^m$ such that $A \subset B$. For example, if $[m_i, M_i]$ is the range of the i-th coordinate of all the points in A then the cartesian product $\Pi_{i=1}^m [m_i, M_i]$ is a cubical set that contains A and is obviously contractible. Then we take

$$S^c(A) := (A \times [0,1]) \cup (B \times \{0,1\}).$$

It is immediate to see that $S^c(A)$ is homotopically equivalent to the suspension of A, defined as

$$S(A) := (A \times [0,1])/\{(x_1, 0) \sim (x_2, 0) \text{ and } (x_1, 1) \sim (x_2, 1) \text{ for all } x_1, x_2 \in A\}.$$

As expected, computations show that the nontrivial Steenrod squares remain in the suspension, although are shifted by one dimension (see Table 3), but the nontrivial cup products disappear in the suspension (and are thus not shown in the table).

Table 3. Nontrivial Steenrod squares for suspensions. Cohomology generators are denoted by consecutive alphabetic letters for each dimension (e.g., a for dimension 0, e for dimension 4) with appended indices starting from 1 within each dimension separately.

Example	Betti numbers	Nontrivial Steenrod squares
$S^c(\mathrm{K}^2)$	$(1, 0, 2, 1)$	$\mathrm{Sq}^1(c_1) = d_1$
$S^c(\mathbb{RP}^2)$	$(1, 0, 1, 1)$	$\mathrm{Sq}^1(c_1) = d_1$
$S^c(\mathbb{CP}^2)$	$(1, 0, 0, 1, 0, 1)$	$\mathrm{Sq}^2(d_1) = f_1$

Time Complexity. Provided that an AT model of a chain complex has been already computed, computing all the Steenrod squares involves the computation of the inclusion map on selected homology generators, applying the formulas for the Steenrod squares at the level of chains, and checking if specific homology generators appear in the projections of faces that appear in the formula. Let s denote the number of cells in the chain complex, let g be the number of homology generators. An upper bound for the number of how many times (4) is applied is at most $O(g^2)$. Assume the dimension is fixed, and then (4) evaluates in constant time times the cost of checking the projections, which is at most $O(g)$. Since the chains are not longer than $O(s)$, the overall pessimistic time complexity of the computation of all the Steenrod squares is $O(g^3 s^2)$. Note that the time complexity of computing an AT model is $O(s^3)$, and in typical applications the numbers of homology generators are very small; therefore, the cost of computing Steenrod squares is neglibigle if computation of complete homological information of a cubical complex is taken into consideration.

Acknowledgements. The research conducted by both authors has received funding from the People Programme (Marie Curie Actions) of the European Union's Seventh Framework Programme (FP7/2007-2013) under REA grant agreements no. 291734 (for M. K.) and no. 622033 (for P. P.).

References

1. Čadek, M., Krčál, M., Matoušek, J., Vokřínek, L., Wagner, U.: Polynomial-time computation of homotopy groups and Postnikov systems in fixed dimension. Siam J. Comput. **43**(5), 1728–1780 (2014)

2. Computational Homology Project software. http://chomp.rutgers.edu/software/
3. Computer Assisted Proofs in Dynamics group. http://capd.ii.uj.edu.pl/
4. Eilenberg, S., Mac Lane, S.: On the groups $H(\Pi, n)$, I. Ann. Math. **58**, 55–106 (1953)
5. Franek, P., Krčál, M.: obust satisfiability of systems ofequations. J. ACM **62**(4), 26:1–26:19 (2015). http://doi.acm.org/10.1145/2751524
6. Gonzalez-Díaz, R., Medrano, B., Sánchez-Peláez, J., Real, P.: Simplicial perturbation techniques and effective homology. In: Ganzha, V.G., Mayr, E.W., Vorozhtsov, E.V. (eds.) CASC 2006. LNCS, vol. 4194, pp. 166–177. Springer, Heidelberg (2006)
7. González-Díaz, R., Real, P.: A combinatorial method for computing Steenrod squares. J. Pure Appl. Algebra **139**(1–3), 89–108 (1999)
8. González-Díaz, R., Real, P.: Computation of cohomology operations on finite simplicial complexes. Homology Homotopy Appl. **5**(2), 83–93 (2003)
9. González-Díaz, R., Real, P.: HPT and cocyclic operations. Homology Homotopy Appl. **7**(2), 95–108 (2005)
10. Hatcher, A.: Algebraic Topology. Cambridge University Press, Cambridge (2001). http://www.math.cornell.edu/ hatcher/AT/ATpage.html
11. Kaczynski, T., Mischaikow, K., Mrozek, M.: Computational homology, Applied Mathematical Sciences, vol. 157. Springer-Verlag, New York (2004)
12. Kaczynski, T., Mrozek, M.: The cubical cohomology ring: An algorithmic approach. Found. Comput. Math. **13**(5), 789–818 (2013)
13. Kühnel, W., Banchoff, T.F.: The 9-vertex complex projective plane. Math. Intelligencer **5**(3), 11–22 (1983)
14. Mischaikow, K., Mrozek, M., Pilarczyk, P.: Graph approach to the computation of the homology of continuous maps. Found. Comput. Math. **5**, 199–229 (2005)
15. Franek, P., Krčál, M., Wagner, H.: Robustness of zero sets: Implementation, submitted
16. Pilarczyk, P.: The ChainCon software. Chain contractions,homology and cohomology software and examples. http://www.pawelpilarczyk.com/chaincon/
17. Pilarczyk, P., Real, P.: Computation of cubical homology, cohomology, and (co)homological operations via chain contraction. Adv. Comput. Math. **41**(1), 253–275 (2015)
18. Prasolov, V.V.: Elements of Homology Theory. Graduate Studies in Mathematics, American Mathematical Society (2007)
19. Real, P.: On the computability of the Steenrod squares. Ann. Univ. Ferrara, Nuova Ser., Sez. VII, Sc. Mat. **42**, 57–63 (1996)
20. Sergeraert, F.: Effective homology, a survey (1992). http://www-fourier.ujf-grenoble.fr/~sergerar/Papers/Survey.pdf
21. Sergeraert, F.: The computability problem in algebraic topology. Adv. Math. **104**(1), 1–29 (1994)
22. Serre, J.P.: Homologie singulière des espaces fibrés. Ann. Math. **54**(3), 425–505 (1951)
23. Steenrod, N.E.: Products of cocycles and extensions of mappings. Ann. Math. **48**(2), 290–320 (1947)
24. Steenrod, N.E.: Cohomology operations, and obstructions to extending continuous functions. Adv. Math. **8**, 371–416 (1972)
25. Vokřínek, L.: Decidability of the extension problem for maps into odd-dimensional spheres. [math.AT] (2014). arXiv:1401.3758

On Homotopy Continuation
for Speech Restoration

Darian M. Onchis[1(✉)] and Pedro Real[2]

[1] Faculty of Mathematics, University of Vienna, Vienna, Austria
darian.onchis@univie.ac.at
[2] Department of Applied Mathematics I, University of Seville, Seville, Spain
real@us.es

Abstract. In this paper, a homotopy-based method is employed for the recovery of speech recordings from missing or corrupted samples taken in a noisy environment. The model for the acquisition device is a compressed sensing scenario using Gabor frames. To recover an approximation of the speech file, we used the basis pursuit denoising method with the homotopy continuation algorithm. We tested the proposed method with various speech recordings.

Keywords: Homotopy continuation · Speech restoration · Basis pursuit · ℓ^1 regularization · Gabor frames · Numerical algorithm

1 Introduction

The reconstruction of an audio signal with missing sampled or clipped, is a classical problem in signal processing and it was largely discussed in the specialized scientific literature, see [1–3].

In this paper, we report the experiments performed with a method inspired from computational topology, namely the homotopy continuation method in order to enhance the typical recovery of audio speech recordings based on ℓ^1- minimization, [4–6].

To precisely formulate the problem, we consider the following non parametric model with observations:

$$y = \Theta s + e \in \mathbb{R}^P$$

where $s \in \mathbb{R}^N$ is the speech signal to recover, $e \in \mathbb{R}^P$ is a noise vector, and $\Theta \in \mathbb{R}^{P \times N}$ models the acquisition device. This device is nowadays equipped with the additional assumption of sparsity, which refers to the circumstance that many natural signals can be expanded (using a suited dictionary Θ) with only few non zero coefficients. We assume a compressed sensing scenario where the operator Θ could be the realization of a random Gaussian, Bernoulli, or partial Fourier matrix satisfying the restricted isometry property (RIP) [7]. But given the special characteristics of nature signals as the speech recordings, which usually consist of sets of distinct components as transients and harmonics with orientation in time and frequency, we have used for the proposed method a Gabor frame generated by the Alltop sequences as proposed in [8,9].

© Springer International Publishing Switzerland 2016
A. Bac and J.-L. Mari (Eds.): CTIC 2016, LNCS 9667, pp. 152–156, 2016.
DOI: 10.1007/978-3-319-39441-1_14

2   Gabor Frames and the ℓ^1−minimization

Frames $(g_i)_{i \in I}$ generalize the idea of a basis in a Hilbert space $\boldsymbol{H}$ and consist of the indexed families such that the so-called frame operator S

$$Sf = \sum_{i \in I} \langle f, g_i \rangle g_i \tag{1}$$

is invertible. The main tool for time-frequency analysis is the Short-Time Fourier Transform, defined for functions $f, g \in \boldsymbol{L}^2(\mathbb{R}^d)$ at $\lambda = (\alpha, \beta) \in \mathbb{R}^{2d}$ by

$$V_g f(\lambda) = V_g f(\alpha, \beta) = \langle f, M_\beta T_\alpha g \rangle = \langle f, \pi(\lambda) g \rangle \tag{2}$$

where $T_\alpha f(t) = f(t - \alpha)$ is the translation (time shift) and $M_\beta f(t) = e^{2\pi i \beta \cdot t} f(t)$ is the modulation (frequency shift). The operators $\pi(\lambda) := M_\beta T_\alpha$ are called time-frequency shifts and the set $\Lambda = \{\lambda; \lambda = (\alpha, \beta) \in \mathbb{R}^d \times \widehat{\mathbb{R}}^d\}$ is a lattice, [11]. The Gabor system $\mathcal{G}(g, \Lambda) = \{\pi(\lambda) g; \lambda \in \Lambda\}$ over the lattice Λ consisting of the translated and modulated versions of one atom g, is a frame for the space $L^2(\mathbb{R}^d)$, if and only if there exist $0 < A \le B < \infty$ (frame bounds) with

$$A||f||^2 \le \sum_{\lambda \in \Lambda} |\langle f, \pi(\lambda) g \rangle|^2 \le B||f||^2 \quad \text{for every} f \in L^2(\mathbb{R}^d), \tag{3}$$

We will use in the construction of Gabor frames the Alltop sequences as proposed in [8].

To recover an approximation of the signal s, a standard method is the basis pursuit denoising or ℓ^1-minimization [10]. This method is based on using the ℓ^1 norm as a sparsity enforcing penalty. That turns into an optimization problem and allows us to recover the signal minimizing the expression:

$$s_\rho \in argmin_{s \in \mathbb{R}^N} \frac{1}{2} ||y - \Theta s||^2 + \rho ||s||_1 \tag{4}$$

where the ℓ^1 norm is defined as $||s||_1 = \sum_i |s_i|$.

The parameter ρ should be set in accordance to the noise level $||e||$.

In the case where there is no noise, $e = 0$, we let $\lambda \to 0^+$ and solve the basis pursuit constrained optimization $s_{0+} \in argmin_{\Theta s = y} ||s||_1$.

In order to avoid technical difficulties, we could further assume that Θ is such that s_ρ is uniquely defined.

In the following, for some index set $I \subset \{1, \ldots, N\}$, we denote by

$$\Theta_I = (\theta_i)_{i \in I} \in \mathbb{R}^{P \times |I|}$$

the sub-matrix obtained by extracting the columns $\theta_i \in \mathbb{R}^P$ of Θ indexed by I. The support of a vector is $supp(x) = \{i \in \{1, \ldots, N\} : x_i \ne 0\}$.

Using results from the convex analysis, we obtain that s_ρ is a solution of (4) if and only if

$$\begin{cases} (C1) & \Theta_I^*(y - \Theta_I s_{\rho,I}) = \rho sign(s_{\rho,I}), \\ (C2) & ||\Theta_J^*(y - \Theta_I s_{\rho,I})||_\infty \le \rho \end{cases}$$

where $I = supp(s_\rho)$ and $J = I^c$ is the complementary.

3   The Homotopy-Continuation Algorithm and Experiments

Topology helps to understand the different degrees of connectivity a geometric object has. To deal with topological isomorphisms or homeomorphisms between continuous geometric objects is a hard task and discretization strategies, such as triangulations, are employed for reducing the computational complexity of the topological interrogation. While homology considers the notion of hole in linear algebra terms, the homotopy is dealing with the same issues in a purely combinatorial terms. Therefore, homotopy computation is much more harder in general than homology computation, but in combination with numerical methods it can be proven to be a useful tool for signal recovery but also in image recognition.

The proposed homotopy-based method for speech recovery is gradually deforming a trivial initialization of the speech vector into the original speech vector through the process of path-tracking. The numerical homotopy procedure is based on the fact that the objective function undergoes a homotopy from the ℓ^2 to the ℓ^1 optimization as the algorithm progresses. The homotopy algorithm proceeds by computing iteratively the value s_ρ.

We sum below the complete algorithm:

Homotopy-speech restoration algorithm

```
Input: y-noisy speech file,
  Θ-Gabor frame compressed sensing operator,
Initialization: Corr = Θ' * y, ρ = max(Corr), s_ρ = 0,
I_sparsity = supp(s_ρ)
Output: s_ρ-restored speech file, ρ,
I_sparsity
Begin iteration
  Compute the correlations   Corr = Θ' * (y − Θ * s_ρ);
  Update direction dir = Θ' * Θ sign(Corr);
  Compute J the complementary support of I_sparsity
  Compute minimum α for condition (C1) and (C2)
  Update solution s_ρ = s_ρ + α * dir; ρ = ρ − α; I_sparsity = supp(s_ρ).
End Iteration
```

(ℓ^1−minimization with homotopy deformation)

For the numerical experiments, we have used 5 speech data s of 2 to 5 s, recorded by a microphone and sampled at 16 kHz. All signals were normalized, and after that the following noise level $\sigma = 0.05 * norm(\Theta * s)/sqrt(P)$ was applied. We used $P = round(N/4)$ where $N = size(s)$. The distorted measurements where defined by the expression $y = \Theta * s + \sigma * randn(P, 1)$ as in [5]. These measurements were the input for our algorithm.

In Fig. 1, we displayed 6 iterations of the algorithm to visualize the homotopic progression towards the correct restoration. For clarity reasons, only the first 2000 samples of the speech signal are shown.

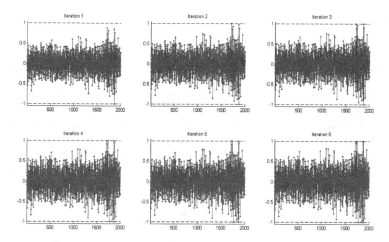

Fig. 1. First 6 iterations of the homotopy algorithm (orginal in red, recovered in blue) (Color figure online)

Even though the application of the algorithm provides a complete recovery of the original speech recording, a drawback is the large number of iterations. In our experiments, we managed to recover the 5 speech data, with a number of iterations proportional with almost half the size of the signal, depending on the distortion applied. In comparison with other ℓ^1−minimization methods like the iterative shrinkage-thresholding, proximal gradient or augmented Lagrange multiplier, the homotopy achieves the best accuracy, even though, as mentioned before, in terms of speed, the homotopy takes longer time to converge when the distortion is high. But since speech recognition is usually a sensitive issue, the accuracy degree of the reconstruction made us confident in the utility of the proposed algorithm.

4   Conclusions

In this report, we presented the results of speech restoration using the basis pursuit algorithm in a sparse Gabor frames scenario, enhanced with a topology-inspired procedure entitled the homotopy-continuation method. The method allows a complete recovery of a speech recording with missing samples or clipped but with a high computational cost given by a large number of iteration necessary. Further parallelization of the algorithm are considered by the authors.

Acknowledgments. The first author gratefully acknowledge the support of the Austrian Science Fund (FWF): project number P27516.

References

1. Abel, J.S., Smith III., J.O.: Restoring a clipped signal. In: Proceedings of the International Conference on Acoustics, Speech, and Signal Processing. IEEE, pp. 1745–1748 (1991)
2. Godsill, S.J., Rayner, P.J.: A Bayesian approach to the restoration of degraded audio signals. IEEE Trans. Speech Audio Process. **3**(4), 267–278 (1995)
3. Adler, A., Emiya, V., Jafari, M., Elad, M., Gribonval, R., Plumbley, M.D.: Audio inpainting. IEEE Trans. Audio Speech Lang. Process. **20**(3), 922–932 (2012)
4. Emmanuel Candes. http://statweb.stanford.edu/~candes/l1magic/
5. Numerical Tours of Signal Processing. http://www.numerical-tours.com/matlab/optim8homotopy/
6. Malioutov, D.M., Cetin, M., Willsky, A.S.: Homotopy continuation for sparse signal representation. In: IEEE International Conference on Acoustics, Speech and Signal Processing, Philadelphia, PA, vol. 5, pp. 733–736, March 2005
7. Candes, E.J., Tao, T.: Decoding by linear programming. IEEE Trans. Inf. Theor. **51**(12), 4203–4215 (2005)
8. Herman, M.A., Strohmer, T.: High-resolution radar via compressed sensing. IEEE Trans. Signal Process. **57**(6), 2275–2284 (2009)
9. Strohmer, T., Heath, R.: Grassmanian frames with applications to coding and communication. Appl. Comput. Harmon. Anal. **14**(3), 257–275 (2003)
10. Gill, P.R., Wang, A., Molnar, A.: The in-crowd algorithm for fast basis pursuit denoising. IEEE Trans. Signal Process. **59**(10), 4595–4605 (2011)
11. Ricaud, B., Stempfel, G., Torresani, B., Wiesmeyr, C., Lachambre, H., Onchis, D.: An optimally concentrated Gabor transform for localized time-frequency components. Adv. Comput. Math. **40**(3), 683–702 (2014)

Finding Largest Rectangle Inside a Digital Object

Apurba Sarkar[1]([⊠]), Arindam Biswas[2], Mousumi Dutt[3],
and Arnab Bhattacharya[4]

[1] Department of Computer Science and Technology,
Indian Institute of Engineering Science and Technology, Howrah, India
as.besu@gmail.com
[2] Department of Information Technology,
Indian Institute of Engineering Science and Technology, Howrah, India
barindam@gmail.com
[3] Department of Computer Science and Engineering,
International Institute of Information Technology, Naya Raipur, India
duttmousumi@gmail.com
[4] Department of Computer Science and Engineering,
Indian Institute of Technology, Kanpur, India
arnabb@iitk.ac.in

Abstract. We present a combinatorial algorithm which runs in $O(n \log n)$ time to find largest rectangle (LR) inside a given digital object without holes, n being the number of pixels on the contour of digital object. The object is imposed on background isothetic grid and inner isothetic cover is obtained for a particular grid size, g, which tightly inscribes the digital object. Certain combinatorial rules are applied on the isothetic cover to obtain the largest rectangle. The largest rectangle is useful for shape analysis of digital objects by varying grid size, by rotating the object, etc. Experimental results on different digital objects are also presented.

Keywords: Digital object · Isothetic grid · Rectangle · Inner isothetic cover · Shape analysis

1 Introduction

The problem of finding the Largest area axis-parallel Rectangle (LR) inside a general polygon of n vertices is a geometric optimization problem in the class of polygon inclusion problem [4]. There are many solutions for this problem in various scenarios (e.g. in convex polygon, in orthogonal polygon, etc.) because of the practical importance of the problem. Chazelle et al. [5,6] proposed an algorithm to find largest area rectangle with sides parallel to the given rectangle containing n points and reported that their algorithm runs in $O(n \log^3 n)$ time and $O(n \log n)$ space. Aggarwal et al. [1] simplifies that algorithm by Chazelle et al. [5,6] and proposed an algorithm that takes same $O(n \log^3 n)$ time but

© Springer International Publishing Switzerland 2016
A. Bac and J.-L. Mari (Eds.): CTIC 2016, LNCS 9667, pp. 157–169, 2016.
DOI: 10.1007/978-3-319-39441-1_15

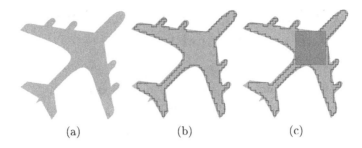

(a)   (b)   (c)

Fig. 1. (a) The digital object, A, (b) Inner isothetic cover ($g = 8$), (c) Largest
Rectangle.

$O(n)$ space. They proposed another algorithm that runs in $O(n \log^2 n)$ time and
$O(n)$ space. Daniels et al. [7] considered a geometric optimization problem of
finding maximum area axis parallel rectangle from a n-vertex general polygon.
They characterized the largest area rectangle problem by considering different
cases based on the types of contacts between the rectangle and the polygon.
They also proposed a framework that can transform an algorithm for orthogonal
polygons into an algorithm for non-orthogonal polygons and showed that the
running time of their algorithm for general polygons to be $O(n \log^2 n)$. They
have established lower bound for finding the largest empty rectangles in both self-
intersecting polygons and general polygons with holes to be $O(n \log n)$. McKenna
et al. [9] use a divide-and-conquer approach to find the LR in an orthogonal
polygon in $O(n \log^5 n)$ time. For the merge step at the first level of divide-and-
conquer, they obtain an orthogonal, vertically separated, horizontally convex
polygon. At the second level, their merge step produces an orthogonally convex
polygon, for which they solve the LR problem in $O(n \log^3 n)$ time. They also
establish a lower bound of time in $\Omega(n \log n)$ for finding the LR in orthogonal
polygons with degenerate holes, which implies the same lower bound for general
polygons with degenerate holes.

LR problem has many applications in electronic design automation, design
and verification of physical layout of integrated circuits [10,11]. Largest area
rectangle problem has many interesting industrial applications also, e.g., consider
a sheet of fabric or a rectangular piece metal with certain number of flaws in it.
This problem can be salvaged to find a maximum area rectangular sheet that
does not contain any flaws.

In this paper we present another flavor of the same problem - finding largest
rectangle in a digital object which is useful for shape analysis of the object. It is to
be noted here that the resulting largest rectangle may not be unique. The digital
object (Fig. 1(a)) is imposed on a background grid (grid size may vary depending
on the shape and size of the object). Inner isothetic cover which tightly inscribes
the digital object is shown in Fig. 1(b). The corresponding largest rectangle
is shown in Fig. 1(c) for grid size $g = 8$. In Sect. 2 required definitions and
procedure to obtain inner isothetic cover are explained in brief. While traversing
along the inner isothetic cover combinatorial rules are applied to obtain the

largest rectangle. The algorithm presented in this paper runs in $O(n \log n)$ time, where n is the number of pixels on the contour of the digital object. In Sect. 3, the procedure to obtain largest rectangle is stated in details including rules, algorithm, time complexity, and demonstration. Experimental results are given in Sect. 4 to verify the algorithm. Section 5 contains concluding remarks.

2   Definitions and Preliminaries

A digital object A is a 8-connected component [8]. The *background grid* is given by $\mathbb{G} = (\mathbb{H}, \mathbb{V})$, where $\mathbb{H}$ and $\mathbb{V}$ represent the respective sets of (equi-spaced) horizontal grid lines and vertical grid lines. The *grid size* g is defined as the distance between two consecutive horizontal/vertical grid lines. A *grid point* is the point of intersection of a horizontal and a vertical grid line. A *unit grid block* (UGB) is the smallest square having its four vertices as four grid points and edges as grid edges. An *isothetic polygon* P is a simple polygon (i.e., with non-intersecting sides) of finite size in $\mathbb{Z}^2$ whose alternate sides are subsets of the members of $\mathbb{H}$ and $\mathbb{V}$. The polygon P, hence given by a finite set of UGBs, is represented by the (ordered) sequence of its vertices, which are grid points. The border B_P of P is the set of points belonging to its sides. The interior of P is the set of points in the union of its constituting UGBs excluding the border of P. An isothetic cover has two type of vertices 90^0 (type **1**) and 270^0 (type **3**). The *inner (isothetic) cover* (IIC), denoted by $P(A)$, is a set of inner polygons and (inner) hole polygons, such that the region, given by the union of the inner polygons minus the union of the interiors of the hole polygons, contains a UGB if and only if it is a subset of A. An edge of P defined by two consecutive vertices of type **1** is termed as a *convex edge*, as it gives rise to a *convexity*. Similarly, an edge defined by two consecutive type **3** vertices gives rise to a *concavity*, and hence termed as a *concave edge*.

Using the algorithm in [2,3], we obtain (the ordered set of vertices of) P for A, which is, therefore, the maximum-area isothetic polygon inscribing A. During the construction of P, the vertices and grid points lying on the edges of P are dynamically inserted in a circular doubly-linked list, L, and simultaneously in two lexicographically sorted (in increasing order) lists, L_x and L_y, with respective primary and secondary keys as x- and y-coordinates for L_x, and opposite for L_y. Each node of the list L has two level pointers, the lower level pointers are used to link both edge and corner points of IIC and the top level pointers are used to link only the corner points of IIC.

3   Procedure to Determine Largest Rectangle

To find a largest rectangle, the inner isothetic cover P (constructed using algorithm in [2,3]) is traversed in anti-clockwise direction from its top left corner. During this traversal, whenever a convex edge is encountered, corresponding histogram polygon (i.e., a portion of the main polygon) is constructed, as explained in Sect. 3.1. Largest rectangle inscribed in the histogram polygon

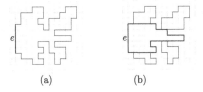

Fig. 2. Histogram polygon w.r.t. a convex edge.

is determined (Sect. 3.2). The convexity encountered in P is reduced using the appropriate reduction rule explained in Sect. 3.3. After reduction, it may give rise to a convex edge, corresponding to which the histogram polygon, thereof the largest rectangle inscribed in it, is determined. Thus, the traversal continues till it reaches the start point of P, the largest of all such largest rectangles, inscribed in the histogram polygons of convex edges, is the resulting largest rectangle of P.

3.1   Finding Histogram Polygon

During traversal, whenever a convex edge (say e) is detected, histogram polygon is considered, whose base is e. The base may lie horizontally (at top or bottom) or vertically (at left or right). The base is extended to extract the histogram polygon which does not contain any concavity horizontally (vertically) if the base is vertical (horizontal). Figure 2(a) shows a vertical convex edge e at left side, which is extended upto the right side of P in such a way that there is no horizontal concavities. Corresponding histogram polygon is shown in Fig. 2(b) in blue outline.

3.2   Finding Largest Rectangle in a Histogram Polygon

To find the largest rectangle in histogram polygon, the opposite side of the base, e, is traversed. Whenever a convex edge is encountered, the area of the corresponding rectangle is determined, which is compared with the stored largest rectangle (or global largest rectangle). Larger rectangle is stored as global largest rectangle. After considering the rectangle in histogram polygon, P is reduced based on the reduction rules stated in Sect. 3.3. This process continues until it traverses all the convex edges opposite to e. For example, in Fig. 3(a), the area corresponding to the first convexity, shown as the gray rectangle, is computed as $l_1 \times l_2$ and the convexity is reduced as shown in Fig. 3(b). In Fig. 3(b), the area corresponding second convexity is shown. Figure 3(c) shows the third convexity and its area, Fig. 3(d) shows the area corresponding to fourth convexity and finally Fig. 3(e) shows area corresponding to last convexity. It may be noted that the convexities in Fig. 3(c,d,e) are derived convexities.

3.3 Reduction Rules

The reduction rules are applied only when two consecutive type 1 vertices, i.e., convex edge is detected. Let v_1, v_2, v_3, and v_4 be the four most recent vertices in order and type of v_2 and v_3 be 1. Let v_0 be the vertex (if any) that precedes v_1. If v_0 exists then reduction rule is applied on the sequence $< v_0, v_1, v_2, v_3, v_4 >$ of vertices otherwise it is applied on the sequence $< v_1, v_2, v_3, v_4 >$ of vertices. Depending on the types of vertices v_1 and v_4, there will be four possibilities— (i) 3113, (ii) 3111, (iii)1111 and finally (iv) 1113. Two rules with their sub-rules are formulated, Rule 1 takes care of the pattern in (i) and (iv) and Rule 2 deals

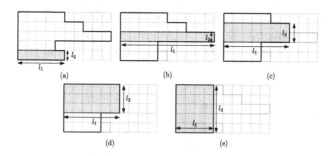

(a)   (b)   (c)

(d)   (e)

Fig. 3. Illustration of finding largest rectangle in a histogram polygon.

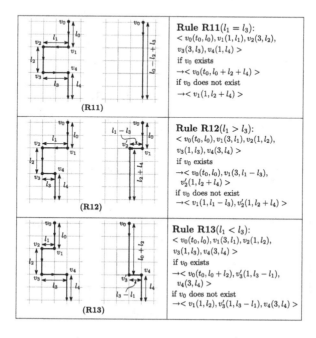

(R11)

Rule R11($l_1 = l_3$):
$< v_0(t_0, l_0), v_1(1, l_1), v_2(3, l_2),$
$v_3(3, l_3), v_4(1, l_4) >$
if v_0 exists
$\to < v_0(t_0, l_0 + l_2 + l_4) >$
if v_0 does not exist
$\to < v_1(1, l_2 + l_4) >$

(R12)

Rule R12($l_1 > l_3$):
$< v_0(t_0, l_0), v_1(3, l_1), v_2(1, l_2),$
$v_3(1, l_3), v_4(3, l_4) >$
if v_0 exists
$\to < v_0(t_0, l_0), v_1(3, l_1 - l_3),$
$v_2'(1, l_2 + l_4) >$
if v_0 does not exist
$\to < v_1(1, l_1 - l_3), v_2'(1, l_2 + l_4) >$

(R13)

Rule R13($l_1 < l_3$):
$< v_0(t_0, l_0), v_1(3, l_1), v_2(1, l_2),$
$v_3(1, l_3), v_4(3, l_4) >$
if v_0 exists
$\to < v_0(t_0, l_0 + l_2), v_3'(1, l_3 - l_1),$
$v_4(3, l_4) >$
if v_0 does not exist
$\to < v_1(1, l_2), v_3'(1, l_3 - l_1), v_4(3, l_4) >$

Fig. 4. Illustration of reduction Rule 1

with the pattern in (ii) and (iii). The proposed algorithm applies reduction rules only when patterns 3113 or 3111 are encountered. The reduction processes are explained below.

Pattern 3113: This pattern implies that two convex (Type 1) vertices preceded concave (Type 3) and followed by another concave vertex. Two consecutive Type 1 vertices in the middle of the pattern creates a convex edge which is to be removed. The reduction process is explained with the help of Fig. 4. Let l_i denote the length of outgoing edge from vertex v_i. Depending on the length l_1 and l_3 there are three sub-rules.

Rule 11 ($l_1 = l_3$)
To remove the convexity, there are two cases to be considered depending on the existence of the vertex v_0. If v_0 exists, vertices v_1, v_2, v_3, and v_4 are removed and the length l_0 is updated to $l_0 + l_3 + l_4$. On the other hand, if v_0 does not exist, vertices v_2, v_3, and v_4 are removed and length l_1 is updated to $l_3 + l_4$.

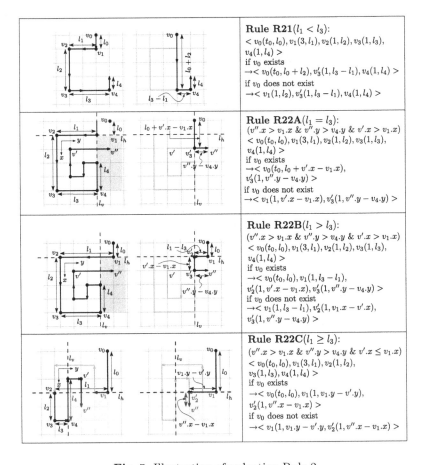

Rule R21($l_1 < l_3$):
$< v_0(t_0, l_0), v_1(3, l_1), v_2(1, l_2), v_3(1, l_3), v_4(1, l_4) >$
if v_0 exists
$\rightarrow < v_0(t_0, l_0 + l_2), v_3'(1, l_3 - l_1), v_4(1, l_4) >$
if v_0 does not exist
$\rightarrow < v_1(1, l_2), v_3'(1, l_3 - l_1), v_4(1, l_4) >$

Rule R22A($l_1 = l_3$):
$(v''.x > v_1.x$ & $v''.y > v_4.y$ & $v'.x > v_1.x)$
$< v_0(t_0, l_0), v_1(3, l_1), v_2(1, l_2), v_3(1, l_3), v_4(1, l_4) >$
if v_0 exists
$\rightarrow < v_0(t_0, l_0 + v'.x - v_1.x),$
$v_3'(1, v''.y - v_4.y) >$
if v_0 does not exist
$\rightarrow < v_1(1, v'.x - v_1.x), v_3'(1, v''.y - v_4.y) >$

Rule R22B($l_1 > l_3$):
$(v''.x > v_1.x$ & $v''.y > v_4.y$ & $v'.x > v_1.x)$
$< v_0(t_0, l_0), v_1(3, l_1), v_2(1, l_2), v_3(1, l_3), v_4(1, l_4) >$
if v_0 exists
$\rightarrow < v_0(t_0, l_0), v_1(1, l_3 - l_1),$
$v_2'(1, v'.x - v_1.x), v_3'(1, v''.y - v_4.y) >$
if v_0 does not exist
$\rightarrow < v_1(1, l_3 - l_1), v_2'(1, v_1.x - v'.x),$
$v_3'(1, v''.y - v_4.y) >$

Rule R22C($l_1 \geq l_3$):
$(v''.x > v_1.x$ & $v''.y > v_4.y$ & $v'.x \leq v_1.x)$
$< v_0(t_0, l_0), v_1(3, l_1), v_2(1, l_2),$
$v_3(1, l_3), v_4(1, l_4) >$
if v_0 exists
$\rightarrow < v_0(t_0, l_0), v_1(1, v_1.y - v'.y),$
$v_2'(1, v''.x - v_1.x) >$
if v_0 does not exist
$\rightarrow < v_1(1, v_1.y - v'.y, v_2'(1, v''.x - v_1.x) >$

Fig. 5. Illustration of reduction Rule 2

Rule 12 $(l_1 > l_3)$
To remove the convexity, vertex v_2 is modified to v_2' and its length is set to l_2+l_4. The length l_1 is modified as $l_1 - l_3$ and the vertices v_3 and v_4 are removed. This reduction is independent of presence of the vertex v_0.

Rule 13 $(l_1 < l_3)$
This rule depends on the presence of vertex v_0. If v_0 is present, its length is updated to $l_0 + l_2$, vertices v_1 and v_2 are removed, and vertex v_3 is modified to v_3' with its length set to $l_3 - l_1$. On the other hand, if v_0 is not present, length l_1 is modified as l_2, vertex v_2 is removed, and vertex v_3 is modified to v_3' with its length set to $l_3 - l_1$.

Pattern 3111: This pattern indicates that three consecutive convex (Type 1) vertices is preceded by a concave (Type 3) vertex and may create a convoluted sequence of vertices on the boundary of the object. To remove the convexity of this pattern, the traversal is continued until it comes out of the convoluted region (i.e., the shaded region) bounded by the horizontal line l_h through v_1 and the vertical line l_v through v_4 as shown in Fig. 5.

Rule 21 $(l_1 < l_3)$
This rule is same as **R13**. If v_0 exists, its length is updated to $l_0 + l_3$, vertices v_1 and v_2 are removed, and vertex v_3 is modified to v_3' with its length modified to $l_3 - l_1$. If v_0 does not exist, length l_1 is modified as l_2, vertex v_2 is removed, and finally vertex v_3 is modified as v_3' with its length modified to $l_3 - l_1$.

Rule 22 $(l_1 \geq l_3)$
To remove this type of convexity, the traversal is continued from vertex v_4 until it comes out of the convoluted region. This can be checked by comparing the coordinates of the vertices during the traversal. The middle and bottom rows of Fig. 5 illustrate this situation for one direction (downward or direction 3) of vertex v_2. In this case the traversal comes out of the convoluted region when it reaches the first vertex whose x-coordinate value is less than the x-coordinate value of v_4 and the y value is greater than the y value of vertex v_1. As shown in the figure, when the traversal reaches vertex v'', the above condition is satisfied and reduction is applied as follows. Let v' is the immediate previous vertex of v''. If $l_1 = l_3$, length of v_0 if exists, is updated to $l_0 + v'.x - v_1.x$, v_3 is modified to v_3' with its length modified to $v''.y - v_4.y$, and finally vertices v_1, v_2 and all the the vertices from v_4(including it) to v' are deleted. If v_0 does not exist, length of v_1 is updated to $v'.x - v_1.x$, vertex v_3 is modified to v_3' with its length modified to $v''.y - v_4.y$, and finally vertices v_2 and all the the vertices from v_4(including it) to v' are removed. This is explained by rule **R22A** in Fig. 5. If $l_1 > l_3$, length of v_1 is updated to $l_1 - l_3$, v_2 is modified to v_2' with its length modified to $v'.x - v_1.x$, v_3 is modified to v_3' with its length modified to $v''.y - v_4.y$, and finally all the vertices from v_4(including it) to v' are removed. In this case the reduction process is independent of existence of v_0. This is explained by rule **R22B** in Fig. 5. It is to be noted that after reduction rule is applied there still exists a convexity formed by vertices v_1, v_2', v_3' and v'' which will be removed subsequently by the application of one of the available rules.

3.4   Algorithm

The Algorithm FIND-RECT (Algorithm 1) is used to determine a largest rectangle in a digital object, which takes as input the digital object, A, and the largest rectangle is reported as the output. Inner isothetic cover of A is generated by calling the procedure IIC (Step 1) and L, L_x, and L_y are created (procedure IIC is based on Sect. 2 [2,3]). L_{curr} determines the current position of the vertex L whereas L_{end} is the last vertex in the vertex list L. Initially, L_{curr} is set to the start of L, i.e., L_{start} and the area of global largest rectangle, $rect.area$, is set to '0' (Step 2). The list L is traversed until it reaches the end L_{end} (Steps 3-6). The procedure CHECK-CONVEX checks whether there is a convex edge (Step 4) and if it is so, procedure FIND-HIST is called (Step 5). L_{curr} advances one step in L (Step 6). In the Procedure FIND-HIST (Procedure 1), L_H stores all the vertices of histogram polygon in anticlockwise manner. Initially L_H is set to NULL (Step 1). SEARCH-NEXT procedure finds out the next vertex, v, of the histogram polygon (explained in Sect. 3.1) (Step 2). The vertices v_1, v_2, and v are appended in L_H (Step 3). All the vertices in histogram polygon has to be found out till it is in range (Steps 4-6) which is determined by the procedure CHK-RANGE (Step 4). In Step 5, the next vertex, v' in histogram polygon is found out by the procedure SEARCH-NEXT (Step 5) and v' is appended in L_H (Step 6). The FIND-RECT (Procedure 2) procedure is called to determine the rectangles in histogram polygon (Step 7). Reduction rules (discussed in Sect. 3.3) are applied on convex edge from which histogram polygon has been generated, by calling the procedure APPLY-RULE (Step 8). The L_{curr} will be updated accordingly (Step 9) and will be returned to FIND-LR (Step 10).

In the Procedure 2, FIND-RECT, L_{H_v} is initialized to the next vertex of the convex edge in anticlockwise manner in the histogram polygon (Step 1). Steps 3–9 are repeated until the condition in Step 2 is false, i.e., all the vertices of the histogram polygon will be traversed except its base. If a convex edge is detected by calling the procedure CHECK-CONVEX in Step 3, corresponding rectangle, $rect'$, is determined by calling the procedure CAL-RECT (Step 4). Corresponding area of the rectangle is determined by the procedure CAL-AREA (Step 5). The area of $rect'$ is compared with $rect$, if the area of $rect'$ is larger, then the global largest rectangle is updated (Steps 6-7). Reduction rules (discussed in Sect. 3.3) are applied on the convex edge (Step 8). L_{H_v} advances one step in L (Step 9).

Algorithm 1. FIND-LR
Input: A
Output: $rect$
1 $L, L_x, L_y \leftarrow$ IIC(A) ;
2 $L_{curr} \leftarrow L_{start}, rect.area \leftarrow 0$;
3 **while** $L_{curr} \neq L_{end}$ **do**
4   **if** CHECK-CONVEX$(L_{curr} \rightarrow type, (L_{curr} \rightarrow next) \rightarrow type)$ **then**
5   FIND-HIST$(L_{curr}, L_{curr} \rightarrow next)$
6      $L_{curr} \leftarrow L_{curr} \rightarrow next$;
7 **return** $rect$;

Procedure 1. Find-Hist(v_1, v_2)
1 $L_H \leftarrow \{\phi\}$;
2 $v \leftarrow$ SEARCH-NEXT(v_2, L_x, L_y);
3 $L_H \leftarrow L_H \cup \{v_1, v_2, v\}$;
4 **while** CHK-RANGE(v_1, v_2) **do**
5      $v' \leftarrow$ SEARCH-NEXT(v, L_x, L_y);
6      $L_H \leftarrow L_H \cup \{v'\}$;
7 FIND-RECT(L_H);
8 APPLY-RULE(v_1, v_2);

Procedure 2. Find-Rect(L_H)

---

1 $L_{H_v} \leftarrow L_{H_{start}} \rightarrow next \rightarrow next$;
2 **while** $L_{H_v} \neq L_{H_{end}}$ **do**
3   **if** CHECK-CONVEX($L_{H_v} \rightarrow type$,
      $(L_{H_v} \rightarrow next) \rightarrow type$) **then**
4          $rect' \leftarrow$
          CAL-RECT($L_{H_v}, L_{H_v} \rightarrow next$);
5          $rect'.area \leftarrow$ CAL-AREA($rect'$) ;
6   **if** $rect'.area > rect.area$ **then**
7              $rect \leftarrow rect'$;
8   APPLY-RULE($L_v, L_v \rightarrow next$);
9      $L_{H_v} \leftarrow L_{H_v} \rightarrow next$;

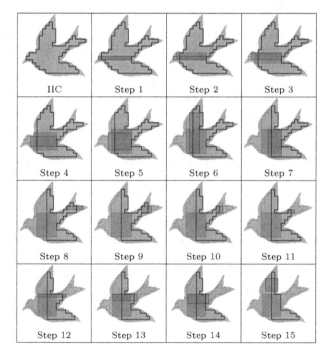

Fig. 6. Demonstration of the proposed algorithm on an object (Bird)

3.5 Demonstration

An illustration of obtaining largest rectangle is shown in Fig. 6. The top-left figure shows the IIC of the object, the green polygons in each step indicates the reduced polygon, the yellow rectangle indicates the largest rectangle found so far and the pink polygon indicates the largest rectangle for immediate previous convexity or the current convexity. Step 1 shows the result of application of reduction rule **R13** for the first convexity it encounters and the largest rectangle (in yellow) found corresponding to this convexity. Step 2 shows the removal of second convexity with rule **R12**. Since the largest area rectangle obtained for this convexity is greater than the largest rectangle obtained so far (before this), the global largest rectangle is updated with this rectangle. Continuing this way the convexity at Step 5, gives largest rectangle for this object since the rectangles corresponding to all subsequent convexities are smaller. It is to be noted that reduction rules are applied for patterns 3113 and 3111, so the IIC will be reduced to a histogram polygon whose base will be the bottom edge of the reduced polygon. At the last step the largest rectangle for this histogram polygon is to be computed for potential largest rectangle. In this case the largest rectangle for this reduced polygon is smaller than the one already computed.

3.6   Time Complexity

To estimate the running time of the algorithm, let us look at the steps involved and their individual running time. Also, let n be the number of pixels on the contour of the digital object, A and g be the grid size. To construct inner isothetic cover along with L, L_x, and L_y $O((n/g)\log(n/g))$ time is required. To determine the largest rectangle, the inner isothetic cover is linearly traversed once and whenever an unexplored convex edge is encountered the procedure FIND-HIST and thereby procedure FIND-RECT is called to find the largest rectangle corresponding to this convex edge. To find the histogram polygon with respect to a convex edge, $O(n/g + \log(n/g))$ time is required, as it includes searching in L_x and L_y which takes $O(\log n/g)$ time. Procedure CHECK-CONVEX, CHK-RANGE, APPLY-RULE, UPDATE, CAL-AREA, and FIND-RECT take $O(1)$ time. Procedure FIND-RECT traverses linearly only the vertices lying opposite to the base of the histogram polygon found out by the procedure FIND-HIST. So, FIND-RECT takes $O(n/g)$ time to calculate the largest rectangle inside the histogram polygon. It is to be noted that each convex edge is traversed only once and the other convex edges which are in opposite direction but totally contained within the convex edge currently being considered need not be checked as largest rectangle corresponding to these convex has already been taken care of. If there are in total k number of convex regions where $k \ll n$, then total time complexity will be $O(k.n/g + (n/g)\log(n/g))$.

3.7   Proof of Correctness

The algorithm identifies a convex edge and finds out the corresponding histogram polygon, then the largest rectangle inside the histogram polygon is computed. After a convex edge is considered, it is reduced following a certain combinatorial reduction rules. After the reduction if it gives rise to secondary convex, it is treated in the similar manner stated above to find the corresponding largest rectangle. The largest of all these rectangles corresponding to each convex edge is reported as largest rectangle. To prove that the algorithm indeed finds out largest rectangle we have to show that the algorithm considers all convex edges.

The traversal procedure ensures that all convex edges (and secondary convex edges) are considered as it starts from the top left vertex of the IIC and continues till the traversal reaches the start point. Sides of a largest rectangle, as it tries to maximise the area, will either be a convex edge or a concave edge may constitute a part of its side. The procedure for finding out largest rectangle inside a histogram polygon of a convex edge, ensures that it finds a largest rectangle as it walks along the sky-line of the histogram polygon (reduces whenever required) and determines the rectangle it produces with the base (the convex edge). To prove that the algorithm also terminates, it is to be noted that during the traversal, whether or not a convex edge is encountered, the traversal advances to the next vertex of IIC. So, eventually the traversal reaches the start point and terminates.

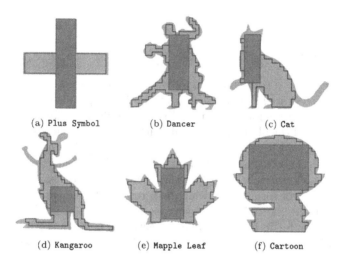

(a) Plus Symbol   (b) Dancer   (c) Cat

(d) Kangaroo   (e) Mapple Leaf   (f) Cartoon

Fig. 7. Largest area rectangle (shaded in yellow) inside six different objects for $g = 8$.

Table 1. Different data for various digital objects at grid size, $g = 8$.

Object	Object perimeter	Object area	Perimeter of IIC	Area of IIC	Perimeter of LR	Area of LR
Bird	625	27768	688	7168	208	2688
Kangaroo	1151	57828	1088	11008	224	3136
Dancer	1057	50398	1320	14400	352	6144
Cartoon	918	65536	1152	35456	528	17024
Cat	730	42920	616	8896	256	2816
Hand	1300	72675	1632	28224	464	13312
Mapple leaf	861	50882	992	17472	352	6720
Plus Symbol	978	70488	960	23744	608	13888

4   Experimental Results

The proposed algorithm is implemented in C in Ubuntu 12.04, 64-bit, kernel version 3.5.0-43-generic, the processor being Intel i5-3570, 3.4 GHz FSB. Experimental results of six different digital objects are shown in Fig. 7. In Table 1, different data, e.g., area and perimeter of digital object, IIC, and largest rectangle, are given. It is seen that largest rectangle occupies approximately one-third area of inner isothetic polygon. The results show that the algorithm can be useful for shape analysis of digital object, as most of the time the largest rectangle determined is at the central position of the digital object. A digital object my be characterized by positioning the successive largest rectangle inside it. These type of information may be useful to determine some topological information from digital objects when the largest rectangles are placed recursively in the rest of the IIC.

5   Conclusion

This paper describes a combinatorial algorithm to find largest rectangle inside a digital object in $O(k.n/g + (n/g)\log(n/g))$ time. The paper also presents the rules for the algorithm, a demonstration, and time complexity. Experimental results show the efficacy of the algorithm. The algorithm can be applied for shape analysis of digital object. LR problem has some industrial application also which has been stated in Sect. 1. One largest rectangle can divide the object in several parts. We can generate iteratively largest rectangle in each part upto a certain limit. LR in each level will form a tree which will represents the connectivity among several parts inside the object. In future, some topological information for the digital objects can be derived from above mentioned technique. All these implies the practical importance of the problem in various shape related applications.

References

1. Aggarwal, A., Suri, S.: Fast algorithms for computing the largest empty rectangle. In: Proceedings of the 3rd Annual Symposium on Computational Geometry, pp. 278–290 (1987)
2. Biswas, A., Bhowmick, P., Bhattacharya, B.B.: TIPS: on finding a tight isothetic polygonal shape covering a 2D object. In: Kalviainen, H., Parkkinen, J., Kaarna, A. (eds.) SCIA 2005. LNCS, vol. 3540, pp. 930–939. Springer, Heidelberg (2005)
3. Biswas, A., Bhowmick, P., Bhattacharya, B.B.: Construction of isothetic covers of a digital object: A combinatorial approach. J. Vis. Comun. Image Represent. **21**(4), 295–310 (2010)
4. Chang, J., Yap, C.: A polynomial solution for the potato-peeling problem. Discrete Comput. Geom. **1**, 155–182 (1986)
5. Chazelle, B., Drysdale, R.L., Lee, D.T.: Computing the largest empty rectangle. In: STACS-1984, pp. 43–54. Springer, Heidelberg (1984)
6. Chazelle, B., III, R.D., Lee, D.: Computing the largest empty rectangle. SIAM J. Comput. **15**, 300–315 (1986)

 7. Daniels, K., Milenkovic, V., Roth, D.: Finding the largest area axis-parallel rectangle in a polygon. Comput. Geom.: Theor. Appl. **7**, 125–148 (1997)
 8. Klette, R., Rosenfeld, A.: Digital Geometry: Geometric Methods for Digital Picture Analysis. Morgan Kaufmann, San Francisco (2004)
 9. McKenna, M., O'Rourke, J., Suri, S.: Finding the largest rectangle in an orthogonal polygon. In: Proceedings of the 23rd Allerton Conference on Communication, Control and Computing, pp. 486–495 (1985)
10. Nandy, S.C., Bhattacharya, B.B., Ray, S.: Efficient algorithms for identifying all maximal isothetic empty rectangles in VLSI layout design. In: Nori, K.V., Veni Madhavan, C.E. (eds.) Foundations of Software Technology and Theoretical Computer Science. LNCS, vol. 472, pp. 255–269. Springer, Heidelberg (1990)
11. Ullman, J.: Chap. 9: Algorithms for VLSI Design Tools. Computational Aspects of VLSI. Computer Science Press, Rockville (1984)

Shape Matching of 3D Topologically Segmented Objects

Nilanjana Karmakar[⊠] and Arindam Biswas

Department of Information Technology,
Indian Institute of Engineering Science and Technology, Shibpur, Howrah, India
nilanjana.nk@gmail.com, barindam@gmail.com

Abstract. Shape matching of 3D digital objects is an important domain of study from topological as well as geometric point of view. Shape matching of two or more digital objects by an efficient segmentation-based method is reported in this paper. The method receives input objects after segmentation of their articulated components and exploits the topological relation between the articulated components and the central section of the objects for shape matching. The method involves simple calculations and is primarily based on the extent of articulations in the objects. The accuracy of shape matching is dependent on the object size and segmentation of the object and is invariant to rotation. Experimental results are provided to demonstrate the structural similarity in various digital objects.

1 Introduction

Shape matching of 3D digital objects is a well-explored area of research that often leads to feature extraction and 3D shape retrieval mechanisms. With the increase in the tools available for efficient extraction and storage of three-dimensional data, 3D shape matching has been useful for a wide variety of disciplines including computer vision, mechanical engineering, artifact searching, molecular biology, chemistry, CAD, virtual reality, medicine, entertainment, etc. Shape matching has been attempted from various perspectives including topological, graph-based, feature-based, etc. In [3], a shape descriptor called shape context is attached to each point on two comparable shapes and the correspondences between such points are used to estimate an aligning transform that aligns the two shapes. Content based 3D shape retrieval by different 3D shape descriptors have been proposed [9,13]. Similarity between polyhedral models has been accurately identified by a topology matching approach using multi-resolution Reeb graphs [5]. This method has been extended [4] for accurate and fast determination of search key. The shape retrieval method in [2] uses extended Reeb graphs to encode different shape characteristics and compare the graphs by a method of multiple kernel learning. Another approach [7] is based on converting 3D models into skeletal graphs and matching the graphs thereby preserving topological and geometric information. The topology-varying shape matching method in [1] establishes correspondences between shapes with large topological

© Springer International Publishing Switzerland 2016
A. Bac and J.-L. Mari (Eds.): CTIC 2016, LNCS 9667, pp. 170–179, 2016.
DOI: 10.1007/978-3-319-39441-1_16

discrepancies by topological operations like part split, duplication, and merging. Other 3D shape matching techniques involve continuous geodesic eccentricity transform [6], representing the signature of an object as a shape distribution [10], using multivariate Gaussian distribution of real valued shape descriptors [12], etc. Shape matching by object segmentation has been proposed in [11] where curve skeleton of an object has been used for segmentation.

The shape matching method proposed in this paper uses 3D segmentation as a tool for shape matching. The areas of the segment of the objects are used to identify geometric and topological properties to be used for shape matching. The rest of the paper is organized as follows. In Sect. 2, the segmented object is represented as topological space and the surface areas of the segments are processed to find similar objects. The paper is concluded with a few shape matching results in Sect. 3.

2   Proposed Work

Given a digital object A and a set of other digital objects $\{A_1, A_2, ..., A_n\}$, a shape matching method comparing A with each of A_i, $1 \leqslant i \leqslant n$, is presented in this paper. The digital objects, represented as topological spaces, are segmented and the separated articulated components are used to find the extent of similarity between two or more objects.

A digital object A represented as a triangulated surface is closed and orientable (2-manifold). The triangulation is such that exactly two triangles are incident on each edge and the interiors of no two triangles intersect. Let A be imposed on a 3D digital grid $\mathbb{G}$ represented as a set of unit grid cubes (UGCs) each of grid size g. Each grid line of $\mathbb{G}$ represents a grid level g_i, where $0 \leqslant i < l$, l representing the length of the object along a given coordinate plane.

2.1   Representation of Digital Object as Topological Space

Let $\mathcal{W}$ be a topological space defined by the set of triangles representing A and endowed with the topology $\Gamma_{\mathcal{W}}$. Let $\beta_{\mathcal{W}}$ be a basis for $\mathcal{W}$ defined as a collection of basis elements such that

(i.) a basis element $\mathcal{P}_i$ consists of triangles (elements) corresponding to the object-occupied UGCs intercepted between grid values g_i and g_{i+1}, where $0 \leqslant i < l$.

(ii.) if $\exists \mathcal{P}_i, \mathcal{P}_j \in \beta_{\mathcal{W}}$ such that $\mathcal{P}_i \cap \mathcal{P}_j \neq \emptyset$, then $\exists \mathcal{P}_k \in \beta_{\mathcal{W}}$ such that $\mathcal{P}_k \subseteq \mathcal{P}_i \cap \mathcal{P}_j$.

For instance, in Fig. 1, the UGCs intercepted between g_0 and g_1 are intersected by the triangles t_1, t_2, t_3, t_5, t_6, t_7, t_8, and t_9. Hence, the basis element $\mathcal{P}_0 = \{t_1, t_2, t_3, t_5, t_6, t_7, t_8, t_9\}$. Similarly, $\mathcal{P}_1 = \{t_1, t_2, t_3, t_4, t_5, t_6, t_7, t_8, t_9, t_{10}\}$, $\mathcal{P}_2 = \{t_4, t_5, t_6, t_{10}\}$, and $\mathcal{P}_3 = \mathcal{P}_0 \cap \mathcal{P}_2 = \{t_5, t_6\}$. Thus, $\beta_{\mathcal{W}} = \{\mathcal{P}_0, \mathcal{P}_1, \mathcal{P}_2, \mathcal{P}_3\}$.

A digital object represented as the topological space $\mathcal{W}$ is segmented by an efficient algorithm [8]. As $\mathcal{W}$ is a closed and orientable 2-manifold, it is homeomorphic to the topological space representing the 3D isothetic cover of the

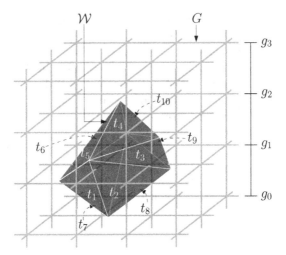

Fig. 1. Digital object A is represented as the topological space $\mathcal{W}$ defined by the set of triangles $\{t_1, t_2, t_3, t_4, t_5, t_6, t_7, t_8, t_9, t_{10}\}$. The basis elements of the basis $\beta_{\mathcal{W}}$ are defined according to the grid ranges in $\mathbb{G}$.

object for sufficiently small grid size. Quotient topology is imposed on $\mathcal{W}$ such that the basis elements representing the topologically invariant sections of the object are mapped to the elements of the corresponding quotient space. The quotient spaces are represented by weighted Reeb graphs along the yz-, zx-, and xy-planes which are segmented by an efficient algorithm. Natural segmentation is ensured by using dynamic threshold of segmentation decided by exponential averaging of the node weights belonging to the same segment. The segmented quotient spaces corresponding to the three coordinate planes are related to each other and transformed topologically to yield the segmented object in a topological space $\mathcal{W}'$ which is used here for further analysis.

Let triangles t_1 and t_2 be two elements of $\mathcal{W}$ such that they are incident on the same edge. Hence, there exists elements in $\mathcal{W}$ that do not, in general, possess disjoint neighborhoods. That is, $\mathcal{W}$ is not a Hausdorff space. The segmented topological space $\mathcal{W}'$ having topology $\Gamma_{\mathcal{W}'}$ is defined as a collection of disjoint subsets of $\mathcal{W}$. Hence, each element $t_i \in \mathcal{W}'$ belongs to a single open set in $\mathcal{W}'$ so that the elements (triangles) defining the segmentation contour virtually have disjoint neighborhoods.

2.2 Shape Matching Through Segmentation of Topological Spaces

Let $\mathcal{W}'_1$ and $\mathcal{W}'_2$ be the topological spaces representing the two objects A_1 and A_2 after segmentation. Let r be the number of disjoint open sets in $\mathcal{W}'_1$ that represent the segments of A_1. Let each open set S_i in $\mathcal{W}'_1$, $1 \leqslant i \leqslant r$, be represented by k number of triangles, where $k > 0$. The surface area covered by each open set (segment) is calculated as the total surface area of k number of triangles.

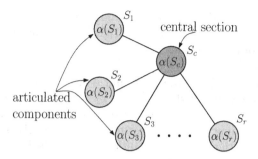

Fig. 2. Graph-theoretic representation of a segmented object.

The segmentation procedure separates the articulated sections from the rest of the object. The 'rest of the object' is henceforth referred to as the 'central section'. The topological relation between the articulated components and the central section of the object is explored by studying certain properties explained next. Shape matching of A_1 and A_2 is carried out using these properties.

Let $\mathcal{R} = \{S_1, S_2, ..., S_c, ..., S_r\}$ denotes the set of segments of an object. Let S_c represents the central component such that $\mathcal{R}' = \mathcal{R} \setminus \{S_c\}$. Let μ be the mean surface area of the segments in $\mathcal{R}'$. Let $\sigma_{\mathcal{R}}$ represents the standard deviation of the surface areas of the r segments in $\mathcal{R}$ and let $\sigma_{\mathcal{R}'}$ represents the same for the $r-1$ segments in $\mathcal{R}'$. The topological relation between S_c and the other articulated components is determined by the terms,

(i.) $\frac{|\alpha(S_c) - \mu|}{\alpha(S_c)}$

(ii.) $\frac{|\sigma_{\mathcal{R}} - \sigma_{\mathcal{R}'}|}{\sigma_{\mathcal{R}}}$

where $\alpha(S_c)$ denotes the surface area of S_c.

Figure 2 gives a graph-theoretic representation of the segmented object where each node denotes a segment. As the segmentation procedure separates the articulated components from the rest of the object, each of the nodes in $\mathcal{R}'$ representing the articulated components are adjacent to the node S_c representing the body of the object. Thus, the general structure of the graph, excluding degenerate cases, contains several pendant vertices adjacent to a node of higher degree. The surface area covered by each segment is mentioned in the corresponding node. The variation between the areas of the articulated segments and the central segment provides an idea about the variation of object topology. The term $\frac{|\alpha(S_c) - \mu|}{\alpha(S_c)}$ gives a measure of the variation considering the mean μ of the articulated segment areas. For instance, if the central segment of an object is of much larger area than its articulated segments, then the value of $\frac{|\alpha(S_c) - \mu|}{\alpha(S_c)}$ is large. Since different articulated components may have different structures and surface areas, use of the mean surface area in the above expression is justified. Again, the variation of the structures of the articulated segments from their mean μ is captured in the standard deviation $\sigma_{\mathcal{R}'}$. Similarly, variation of the structures of the articulated and the central segment from their mean is measured by $\sigma_{\mathcal{R}}$.

The term $\frac{|\sigma_{\mathcal{R}}-\sigma_{\mathcal{R}'}|}{\sigma_{\mathcal{R}}}$ provides information about the contribution of the central segment to the total area of the object surface. For instance, if the articulated components of an object are much smaller in area than the central section, then the value of $\frac{|\sigma_{\mathcal{R}}-\sigma_{\mathcal{R}'}|}{\sigma_{\mathcal{R}}}$ is large.

Let us consider two digital objects A_1 and A_2 with mean surface areas μ_1 and μ_2 such that they are segmented into r_1 and r_2 number of segments respectively. Let $\mathcal{R}1$ ($\mathcal{R}2$) be the set of r_1 (r_2) number of segments and let $\mathcal{R}1'$ ($\mathcal{R}2'$) be the set of segments excluding the central segment S_{c_1} (S_{c_2}). A_1 and A_2 are said to be similar in shape if the following conditions are satisfied.

(i.) $\Delta r = |r_1 - r_2| \leqslant \xi_1$

(ii.) $\Delta s = |(\frac{|\alpha(S_{c_1})-\mu_1|}{\alpha(S_{c_1})}) - (\frac{|\alpha(S_{c_2})-\mu_2|}{\alpha(S_{c_2})})| \leqslant \xi_2$

(iii.) $\Delta t = |(\frac{|\sigma_{\mathcal{R}1}-\sigma_{\mathcal{R}1'}|}{\sigma_{\mathcal{R}1}}) - (\frac{|\sigma_{\mathcal{R}2}-\sigma_{\mathcal{R}2'}|}{\sigma_{\mathcal{R}2}})| \leqslant \xi_3$

The condition (i.) is a preliminary condition for checking the similarity of two objects. Here, the threshold ξ_1 ranges from 0 to 1 so that A_1 (A_2) has at most one segment more than A_2 (A_1). If condition (i.) is satisfied then only we proceed further with the other conditions. For instance, in Table 1, the object Horse is checked for similarity with Hand, Leopard, and Dinosaur ($\Delta r \leqslant 1$) and not with Spider and Ant ($\Delta r > 1$).

For a reasonable range of areas of articulated components, the value of $\frac{|\alpha(S_{c_1})-\mu_1|}{\alpha(S_{c_1})}$ in condition (ii.) determines the extent of articulation in the topology of A_1. Similarly, the value of $\frac{|\alpha(S_{c_2})-\mu_2|}{\alpha(S_{c_2})}$ determines the extent of articulation in the topology of A_2. Similarity of A_1 and A_2 w.r.t. their articulations is measured by the value of Δs. Two objects are more similar if the value of Δs is lower. As a convention, the threshold ξ_2 varies between 0 and 0.05 for the objects considered in this paper. That is, we allow a variation of 5 % in the shapes of A_1 and A_2. For instance, in Table 2, the objects Table and Human are not comparable because $\Delta s = 0.39$ or $\Delta s = 0.31$. It is evident that the variation of the central section area from the mean of the articulated segment areas is much larger (larger than the threshold) in case of Table than that in case of Human. On the other hand, Horse and Leopard are comparable as $\Delta s = 0.01$. The value of ξ_2 depends on the object size, nature of triangulation, and density of triangulation.

The contribution of articulated components and the central component in the total surface area of A_1 is determined by the value of $\frac{|\sigma_{\mathcal{R}1}-\sigma_{\mathcal{R}1'}|}{\sigma_{\mathcal{R}1}}$ in condition (iii.). The contribution of articulated components and the central component in the total surface area of A_2 is determined by the value of $\frac{|\sigma_{\mathcal{R}2}-\sigma_{\mathcal{R}2'}|}{\sigma_{\mathcal{R}2}}$. Similarity of A_1 and A_2 w.r.t. the relation between the articulated and non-articulated sections of the objects is determined by the value of Δt. Two objects are more similar if the value of Δt is lower. For instance, in Table 2, the central section in the object Horse has a much greater contribution in its total surface area than the articulated segments. Similar is the case in Dinosaur; hence Dinosaur may be compared to Horse ($\Delta t = 0.14$). In Human, the contribution of the central section and the articulated components to the total surface area are comparable; hence Human is not comparable to Horse ($\Delta t = 0.68$). The value of threshold ξ_3 also depends on the object size, nature of triangulation, and density of triangulation. If A_1 and A_2 satisfy all the three conditions, then they are considered as similar.

3   Results and Conclusion

Shape matching results for some topologically segmented objects are shown in Figs. 3 and 4 and Tables 1 and 2 where the articulated components have been

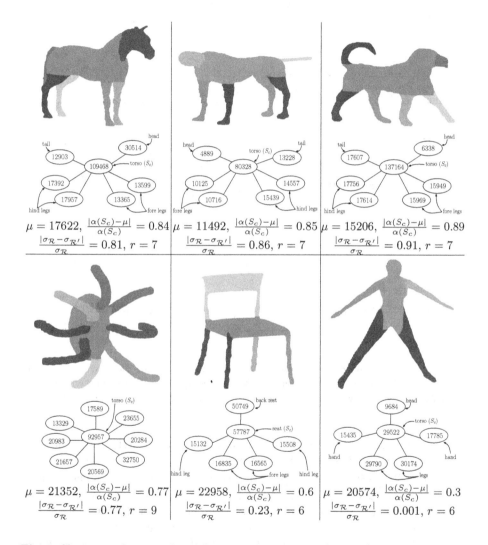

Fig. 3. Shape matching results with search object **Horse** (top left). According to the values of $\frac{|\alpha(S_c)-\mu|}{\alpha(S_c)}$ and $\frac{|\sigma_\mathcal{R}-\sigma_{\mathcal{R}'}|}{\sigma_\mathcal{R}}$ terms the objects **Leopard** (top middle) and **Dog** (top right) are more similar to **Horse** than **Octopus**, **Chair**, and **Human** (bottom left, middle, and right).

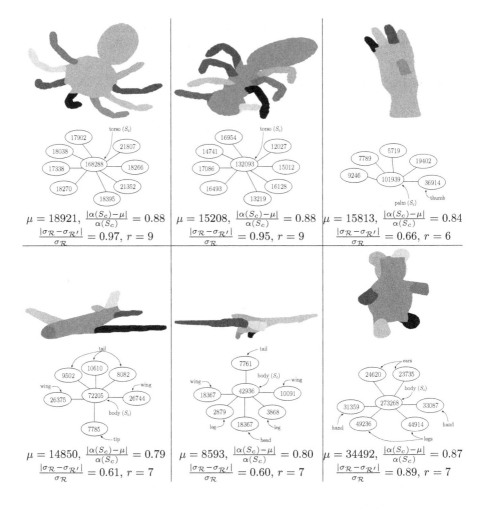

Fig. 4. Top: Shape matching results with search object `Spider` (left) which is more similar to `Ant` (middle) than `Hand` (right). Bottom: Shape matching results with search object `Airplane` (left) which is more similar to `Bird` (middle) than `Teddy` (right).

separated from the central section. Figure 3 shows graph-theoretic representation of the objects `Horse`, `Leopard`, `Dog`, `Octopus`, `Chair`, and `Human`. Comparing the values of r, $\frac{|\alpha(S_c)-\mu|}{\alpha(S_c)}$, and $\frac{|\sigma_\mathcal{R}-\sigma_{\mathcal{R}'}|}{\sigma_\mathcal{R}}$ shows that the search object `Horse` is more similar with `Leopard` and `Dog`, than with `Octopus`, `Chair`, and `Human`. Similarly, in Fig. 4, the search object `Spider` is more similar to `Ant` than `Hand`, and the search object `Airplane` is more similar to `Bird` than `Teddy`. Databases for shape matching of different segmented objects and at various postures are presented in Tables 1 and 2. The values of Δr, Δs, and Δt show that the same object at different postures are the most similar (diagonal entries in the table). The similarity of `Horse` with `Leopard` and `Dinosaur`, or `Ant` with `Spider`, and the dissimilar-

Table 1. Shape matching results for the objects **Hand**, **Spider**, **Ant**, **Animal** (**Horse** and **Leopard**), and **Dinosaur** at different postures. Each object in the database is matched against all other objects in the database. Δr denotes the difference in the number of segments of the two objects. Δs gives the difference when the term $\frac{|\alpha(S_c) - \mu|}{\alpha(S_c)}$ is evaluated for the two objects. Δt denotes the difference when the term $\frac{|\sigma_{\mathcal{R}} - \sigma_{\mathcal{R}'}|}{\sigma_{\mathcal{R}}}$ is evaluated for the two objects. Same objects at different postures are the most similar.

	$\Delta r = 0$ $\Delta s = 0.05$ $\Delta t = 0.02$	$\Delta r = 3$ $\Delta s = 0$ $\Delta t = 0.02$	$\Delta r = 3$ $\Delta s = 0.02$ $\Delta t = 0.05$	$\Delta r = 1$ $\Delta s = 0.02$ $\Delta t = 0.10$	$\Delta r = 1$ $\Delta s = 0.06$ $\Delta t = 0.29$
	$\Delta r = 3$ $\Delta s = 0.04$ $\Delta t = 0.04$	$\Delta r = 0$ $\Delta s = 0.01$ $\Delta t = 0$	$\Delta r = 0$ $\Delta s = 0.01$ $\Delta t = 0.07$	$\Delta r = 2$ $\Delta s = 0.03$ $\Delta t = 0.12$	$\Delta r = 2$ $\Delta s = 0.07$ $\Delta t = 0.31$
	$\Delta r = 3$ $\Delta s = 0.02$ $\Delta t = 0.11$	$\Delta r = 0$ $\Delta s = 0.02$ $\Delta t = 0.13$	$\Delta r = 0$ $\Delta s = 0.04$ $\Delta t = 0.06$	$\Delta r = 2$ $\Delta s = 0$ $\Delta t = 0.01$	$\Delta r = 2$ $\Delta s = 0.04$ $\Delta t = 0.18$
	$\Delta r = 1$ $\Delta s = 0.08$ $\Delta t = 0.13$	$\Delta r = 2$ $\Delta s = 0.03$ $\Delta t = 0.17$	$\Delta r = 2$ $\Delta s = 0.05$ $\Delta t = 0.10$	$\Delta r = 0$ $\Delta s = 0.01$ $\Delta t = 0.05$	$\Delta r = 0$ $\Delta s = 0.03$ $\Delta t = 0.14$
	$\Delta r = 1$ $\Delta s = 0.08$ $\Delta t = 0.25$	$\Delta r = 2$ $\Delta s = 0.03$ $\Delta t = 0.29$	$\Delta r = 2$ $\Delta s = 0.05$ $\Delta t = 0.32$	$\Delta r = 0$ $\Delta s = 0.01$ $\Delta t = 0.17$	$\Delta r = 0$ $\Delta s = 0.03$ $\Delta t = 0.02$

ity of **Human** with **Table** and **Horse** are also substantiated by the values of Δr, Δs, and Δt in Tables 1 and 2. The segmentation method is rotation-invariant preserving the areas of the segments during rotation. Hence, shape matching of two objects is preserved under rotation. Shape matching w.r.t. scaling can be taken care of by normalizing the sizes of the two objects before segmentation. The objects are segmented by topological means and the topological variations on the object surface are incorporated in the graph-theoretic representations. The scheme of shape matching in this paper, however, is independent of the

Table 2. Shape matching results for the objects Animal (Horse and Leopard), Dinosaur, Table, Chair, and Human at different postures. Each object in the database is matched against all other objects in the database. Δr denotes the difference in the number of segments of the two objects. Δs gives the difference when the term $\frac{|\alpha(S_c)-\mu|}{\alpha(S_c)}$ is evaluated for the two objects. Δt denotes the difference when the term $\frac{|\sigma_R - \sigma_{R'}|}{\sigma_R}$ is evaluated for the two objects. Same objects at different postures are the most similar.

	$\Delta r = 0$ $\Delta s = 0.01$ $\Delta t = 0.05$	$\Delta r = 0$ $\Delta s = 0.03$ $\Delta t = 0.14$	$\Delta r = 2$ $\Delta s = 0.06$ $\Delta t = 0.18$	$\Delta r = 1$ $\Delta s = 0.53$ $\Delta t = 0.74$	$\Delta r = 1$ $\Delta s = 0.42$ $\Delta t = 0.68$
	$\Delta r = 0$ $\Delta s = 0.01$ $\Delta t = 0.17$	$\Delta r = 0$ $\Delta s = 0.03$ $\Delta t = 0.02$	$\Delta r = 2$ $\Delta s = 0.06$ $\Delta t = 0.30$	$\Delta r = 1$ $\Delta s = 0.53$ $\Delta t = 0.62$	$\Delta r = 1$ $\Delta s = 0.42$ $\Delta t = 0.56$
	$\Delta r = 2$ $\Delta s = 0.04$ $\Delta t = 0.10$	$\Delta r = 2$ $\Delta s = 0$ $\Delta t = 0.29$	$\Delta r = 0$ $\Delta s = 0.03$ $\Delta t = 0.03$	$\Delta r = 1$ $\Delta s = 0.50$ $\Delta t = 0.89$	$\Delta r = 1$ $\Delta s = 0.39$ $\Delta t = 0.83$
	$\Delta r = 1$ $\Delta s = 0.60$ $\Delta t = 0.78$	$\Delta r = 1$ $\Delta s = 0.56$ $\Delta t = 0.59$	$\Delta r = 1$ $\Delta s = 0.53$ $\Delta t = 0.91$	$\Delta r = 0$ $\Delta s = 0.06$ $\Delta t = 0.01$	$\Delta r = 0$ $\Delta s = 0.17$ $\Delta t = 0.05$
	$\Delta r = 1$ $\Delta s = 0.38$ $\Delta t = 0.53$	$\Delta r = 1$ $\Delta s = 0.34$ $\Delta t = 0.34$	$\Delta r = 1$ $\Delta s = 0.31$ $\Delta t = 0.66$	$\Delta r = 0$ $\Delta s = 0.16$ $\Delta t = 0.26$	$\Delta r = 0$ $\Delta s = 0.05$ $\Delta t = 0.20$

topological equivalence of objects. That is why we attempt to compare the shape of an object like Chair, which is homeomorphic to a torus, to that of a spherical object Human.

Acknowledgements. A part of this research is funded by Council of Scientific and Industrial Research (CSIR), Government of India.

References

1. Alhashim, I.: Topology-Varying Shape Matching and Modeling. Ph.D. thesis, Simon Fraser University (2015)
2. Barra, V., Biasotti, S.: 3D shape retrieval and classification using multiple kernel learning on extended reeb graphs. Vis. Comput. **30**(11), 1247–1259 (2014)
3. Belongie, S., Malik, J., Puzicha, J.: Shape matching and object recognition using shape contexts. IEEE Trans. Pattern Anal. Mach. Intell. **24**(4), 509–522 (2002)
4. Chen, D.Y., Ouhyoung, M.: A 3D object retrieval system based on multi-resolution reeb graph. In: Proceedings of Computer Graphics Workshop (CG) (2002)
5. Hilaga, M., Shinagawa, Y., Kohmura, T., Kunii, T.L.: Topology matching for fully automatic similarity estimation of 3d shapes. In: Proceedings of the 28th Annual Conference on Computer Graphics and Interactive Techniques, SIGGRAPH 2001, Los Angeles, USA, pp. 203–212. ACM, New York (2001)
6. Ion, A., Artner, N., Peyre, G., Marmol, S., Kropatsch, W., Cohen, L.: 3D shape matching by geodesic eccentricity. In: Proceedings of the IEEE Computer Society Conference on Computer Vision and Pattern Recognition Workshops, CVPRW 2008, pp. 1–8, June 2008
7. Iyer, N., Jayanti, S., Lou, K., Kalyanaraman, Y., Ramani, K.: A multi-scale hierarchical 3D shape representation for similar shape retrieval. In: Proceedings of Tools and Methods for Competitive Engineering Conference, Lausanne, Switzerland (2004)
8. Karmakar, N., Biswas, A., Bhowmick, P.: Reeb graph based segmentation of articulated components of 3D digital objects. Theoret. Comput. Sci. 624, 25–40 (2016). doi:10.1016/j.tcs.2015.11.013
9. Körtgen, M., Novotni, M., Klein, R.: 3D shape matching with 3D shape contexts. In: Proceedings of the 7th Central European Seminar on Computer Graphics, Vienna, Austria (2003)
10. Osada, R., Funkhouser, T., Chazelle, B., Dobkin, D.: Matching 3D models with shape distributions. In: Proceedings of the International Conference on Shape Modeling & Applications SMI 2001, Genoa, Italy, pp. 154–166. IEEE Computer Society, Washington, DC (2001)
11. Serino, L., Arcelli, C., di Baja, G.S.: From skeleton branches to object parts. Comput. Vis. Image Underst. **129**, 42–51 (2014)
12. Shilane, P., Funkhouser, T.: Selecting distinctive 3D shape descriptors for similarity retrieval. In: Proceedings of the IEEE International Conference on Shape Modeling and Applications SMI 2006, Matsushima, Japan, pp. 18–23. IEEE Computer Society, Washington, DC (2006)
13. Veltkamp, R., Tangelder, J.: Content based 3D shape retrieval. In: Furht, B. (ed.) Encyclopedia of Multimedia, pp. 88–96. Springer, Heidelberg (2006)

Construction of an Approximate 3D Orthogonal Convex Skull

Nilanjana Karmakar[✉] and Arindam Biswas

Department of Information Technology,
Indian Institute of Engineering Science and Technology, Shibpur, Howrah, India
nilanjana.nk@gmail.com, barindam@gmail.com

Abstract. Orthogonal convex skull of a 3D digital object is a maximal volume orthogonal convex polyhedron lying entirely inside the object. An efficient combinatorial algorithm to construct an approximate 3D orthogonal convex skull of a digital object is presented in this paper. The 3D orthogonal inner cover, an orthogonal polyhedron which tightly inscribes the digital object, is divided into slab polygons and 2D orthogonal skulls of these slab polygons are combined together using combinatorial techniques to obtain an approximate 3D orthogonal convex skull. The algorithm operates in integer domain and requires at most two passes. The current version of the algorithm deals with non-intersecting objects free from holes and cavities. Experimentation on a wide range of digital objects has provided expected results, some of which are presented here to demonstrate the efficacy of the algorithm.

Keywords: Approximate 3D orthogonal convex skull · Orthogonal slicing · 3D orthogonal inner cover · 3D concavity

1 Introduction

Shape description of digital objects is a prominent area of research in the realm of image analysis. Convex skull can be used as an effective tool for shape description of digital objects. The concept of convex skull was initially studied as the potato-peeling problem which dealt with finding the convex polygon of maximum area contained in a given simple (non-convex) polygon [7,9]. The solution for a planar n-gon, $n \leq 5$ [9] was followed by polynomial time algorithms of $O(n^9 \log n)$ and $O(n^7)$ [5,6]. The same problem has been addressed later under the name of the convex skull problem [12]. Variations of the potato-peeling problem using triangulated polygons with or without holes is addressed in [2] and a near-optimal near-linear time algorithm based on visibility graph is proposed in [4]. An orthogonal version of the problem is addressed in [13] with an improved complexity of $O(n^2)$, where the maximal-area orthoconvex polygon is determined by computing the maximal 'staircase' boundary of the polygon. The method of finding the orthogonal convex skull of a digital object used in the current work has been reported in [8]. The convex skull problem has been extended to the

© Springer International Publishing Switzerland 2016
A. Bac and J.-L. Mari (Eds.): CTIC 2016, LNCS 9667, pp. 180–192, 2016.
DOI: 10.1007/978-3-319-39441-1_17

(a) $\underline{P}_{\mathbb{G}}(A)$ (b) slab polygons (xy) (c) Approximate $3OCS(A)$

Fig. 1. 3D orthogonal inner cover of a digital object A, slab polygons due to orthogonal slicing along the xy-plane, and an approximate 3D orthogonal convex skull.

3D orthogonal domain where the maximal volume convex subset enclosed in the object is determined by using a constrained distance transform [3].

A novel and efficient algorithm for the construction of an approximate 3D orthogonal convex skull of a digital object is presented in this paper. A two-pass algorithm is used to determine the approximate orthogonal convex skull irrespective of the object size or grid resolution. The algorithm accounts for non-intersecting objects free from holes and cavities. Figure 1(a) shows a digital object, its slice polygons for xy-plane (Fig. 1(b)), and its approximate $3OCS(A)$ is shown in Fig. 1(c). The rest of the paper is organized as follows. The problem is defined in Sect. 2. The algorithm with its time complexity are explained in Sect. 3. The paper is concluded with experimental results in Sect. 4.

2   Definitions and Preliminaries

A *digital object* A is defined as a finite subset of $\mathbb{Z}^3$, with all its constituent points (i.e., voxels) having integer coordinates and connected in 26-neighborhood. Each voxel is equivalent to a *3-cell* [11] centered at the concerned integer point (Fig. 2(Left)). A *digital grid* $\mathbb{G}$ consists of three orthogonal sets of equi-spaced grid lines along the x-, y-, and z-axes. A larger (smaller) value of the grid size g implies a sparser (denser) grid. A *unit grid cube* (UGC) is a (closed) cube of length g. A UGC-face, f_k, has two adjacent UGCs, U_1 and U_2, such that $f_k = U_1 \cap U_2$ (Fig. 2(Left)). A UGC consists of $g \times g \times g$ voxels and each UGC-face consists of $g \times g$ voxels.

An *orthogonal polyhedron* is a 3D polytope with all its vertices as grid vertices, all its edges made of grid edges, and all its faces being simple isothetic polygons lying on face planes. An *orthogonal convex polyhedron* is an orthogonal polyhedron whose intersection with a face plane parallel to any coordinate plane is either empty or a collection of projection-disjoint orthogonal convex polygons[1]. The *3D orthogonal inner cover* of A, $\underline{P}_{\mathbb{G}}(A)$, is defined as the set of orthogonal polyhedrons that tightly inscribes A; i.e.,

i. $\mathbf{P}_{\mathbb{G}}(A) \subseteq A$
ii. for each $p \in \underline{P}_{\mathbb{G}}(A)$, $0 < d_{\top}(p, A') \leqslant g$

---

[1] Orthogonal convex polygons are also known as "hv-convex" polygon in literature..

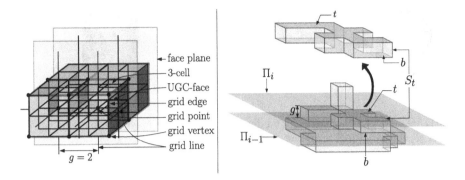

Fig. 2. Left: α-adjacent 3-cells for $g = 2$. **Right:** Slab S_t of height g, bounded on top by slab polygon t lying on Π_i and in the bottom by b, the projection of t on Π_{i-1}.

where $\mathbf{P}_{\mathbb{G}}(A)$ denotes the entire inner cover including its surface $\underline{P}_{\mathbb{G}}(A)$ and interior region. Here, $A' = \mathbb{Z}^3 \setminus A$ and $d_\top$ denotes isothetic distance[2]. In this work, we consider objects such that its $\underline{P}_{\mathbb{G}}(A)$ contains only one orthogonal polyhedron.

The *3D orthogonal convex skull* of a digital object A, denoted by $3OCS(A)$, is a maximal volume orthogonal polyhedron such that

i. no point $p \in \mathbb{Z}^3 \setminus A$ lies on or inside $3OCS(A)$ and
ii. $3OCS(A)$ is orthogonally convex.

3   Proposed Work

Given a digital object, A, its inner orthogonal cover, $\underline{P}_{\mathbb{G}}(A)$, is sliced into slab polygons [10] (Sect. 3.1) along one plane (say xy-). The concavities in these slab polygons are removed and the convex slab polygons are regrouped to form an orthogonal polyhedron. The resulting polyhedron is again sliced along another plane (say yz-) and the concavities are removed from the slab ploygons. Similarly, this procedure is repeated for zx-plane. This constitutes one pass of the algorithm. After the second pass, the resulting orthogonal polyhedron, which is devoid of any concavity along any plane, is an approximate orthogonal convex skull.

3.1   Slicing and Orthogonal Slabs

The 3D object A is provided as a set of voxels. A is imposed on a 3D digital grid $\mathbb{G}$ represented as a set of UGCs each of grid size g. A UGC U_l containing object

---

[2] *Isothetic distance* between two points $p(x_1, y_1, z_1)$ and $q(x_2, y_2, z_2)$ is defined as the Minkowski norm L_∞ given by $d_\top(p, q) = \max\{|x_1 - x_2|, |y_1 - y_2|, |z_1 - z_2|\}$. Isothetic distance of a point p from an object A is $d_\top(p, A) = \min\{d_\top(p, q) : q \in A\}$. It may be noted that isothetic distance may also be expressed in terms of Chebyshev distance [1] which is a special case of Minkowski norm..

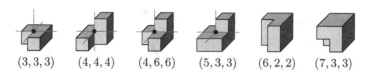

$(3,3,3)$ $(4,4,4)$ $(4,6,6)$ $(5,3,3)$ $(6,2,2)$ $(7,3,3)$

Fig. 3. Types of concave vertex (that do not belong to intersecting polyhedron).

voxels is defined as a *partially object-occupied* UGC. A UGC U_k lying completely inside A is defined as a *fully object-occupied* UGC.

Let $\Pi = \{\Pi_1, \Pi_2, ..., \Pi_r\}$ be a set of *slicing planes* separated by g and parallel to the zx-plane (or yz- or xy-plane) which intersects $\underline{P}_\mathbb{G}(A)$. A UGC-face f_k (f_l) is considered as *fully object-occupied* (*partially object-occupied*) if there exists a U_k (U_l) below (in case of zx-plane) f_k (f_l). The inner boundary of A intersected by Π_i is traversed orthogonally keeping fully occupied UGC-faces f_k to the left. Thus, a *slab polygon* on Π_i is obtained. Let t be a slab polygon on Π_i and b be the projection of t on Π_{i-1}. The section of $\underline{P}_\mathbb{G}(A)$ of height g intercepted between Π_i and Π_{i-1} and bounded horizontally on top and bottom by t and b respectively is defined as the slab S_t (Fig. 2(Right)). Since b is the projection of t, their shapes are identical, that is, t does not vary from Π_i to Π_{i-1}. Hence, S_t can be represented by t. It is evident that the UGCs contained in a slab are fully object-occupied.

3.2   Concavity in Three Dimensions

Depending on the fully object-occupied neighboring UGCs a grid vertex v may be classified into different types where each type is represented by a 3-tuple defined as (#incident UGCs, #incident edges, #incident faces) w.r.t. v. The grid vertices of types $(3,3,3)$, $(4,4,4)$, $(4,6,6)$, $(5,3,3)$, $(6,2,2)$, and $(7,3,3)$ are classified as concave vertices. In Fig. 3, some instances of the possible concave vertices (which do not form intersecting polyhedrons) are shown.

While traversing a slab polygon t (which represents S_t) a concavity is detected if we encounter at least two consecutive concave vertices. In Fig. 4(a), the concave vertices on t and b are shown in blue color. For a nested concavity the number of consecutive concave vertices is more than two, as shown in Fig. 4(c). The rectangular faces of S_t having width g and incident on the concave vertices are defined as *concavity faces*. If two concavity faces are parallel to each other, then they are referred to as parallel concavity faces. A concavity on a slab has at least three concavity faces. Two of them must be parallel concavity faces (see Fig. 4(a) and (c)).

3.3   Resolving the Concavities of a Slab

During the traversal along the boundary of a slab, whenever a concavity is detected, it is resolved as follows. A face plane Π_f perpendicular to a slab S_t and passing through the concave vertices of a concavity divides the slab into

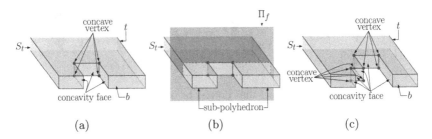

Fig. 4. (a) A concavity on slab S_t containing two consecutive concave vertices (blue) on each of the slab polygons t and b, and a pair of parallel concavity faces. (b) Sub-polyhedrons formed when a face plane Π_f passes through the concave vertices of S_t. (c) Nested concavity on S_t containing four consecutive concave vertices on each of t and b (Color figure online).

three different parts: two separate sub-polyhedrons lying on one side of Π_f, and the rest of the slab on the other side (Fig. 4(b)). To maintain convexity of the slab one of the sub-polyhedrons has to be dropped depending on whether the concavity is defined by two or more consecutive 270° grid vertices. While traversing a concavity, the sub-polyhedron appearing before the concavity has already been processed in the previous steps. Hence, that sub-polyhedron does not contain any concavity. The next sub-polyhedron is checked recursively for concavity. If deleting a sub-polyhedron disconnects the slab into two parts, then the sub-polyhedron is not deleted. Otherwise, the sub-polyhedron having the smaller volume is dropped. This ensures that the sub-polyhedron having the larger volume is included in $OCS(S_t)$, thereby striving to achieve a larger volume of $3OCS(A)$. As retaining the larger volume sub-polyhedron is a local decision, it does not ensure that $3OCS(A)$ will be of the largest possible volume.

Figure 5 shows a brief demonstration of resolving the concavities in a slab. Concavity C_1 (category $\mathbf{C}_{z,x}$) has two sub-polyhedrons $\mathcal{A}$ and $\mathcal{B}$ (Fig. 5(a)). The sub-polyhedron $\mathcal{A}$, occurring after C_1, is checked recursively for concavity. Concavity C_2 (category $\mathbf{C}_{z,y}$) is detected on $\mathcal{A}$ (Fig. 5(b)). C_2 has sub-polyhedrons $\mathcal{A}_1$ and $\mathcal{A}_2$. $\mathcal{A}_2$ is devoid of concavity. Volume of $\mathcal{A}_2$ is smaller than $\mathcal{A}_1$ and deleting $\mathcal{A}_2$ does not disconnect the slab. Hence, $\mathcal{A}_2$ is deleted. In case of concavity C_3 (category $\mathbf{C}_{z,x}$), sub-polyhedrons $\mathcal{A}_3$ and $\mathcal{A}_4$ are of equal volume (Fig. 5(c)). As deleting $\mathcal{A}_4$ disconnects the slab, $\mathcal{A}_4$ is not deleted. Hence, $\mathcal{A}_3$ is deleted. Now, sub-polyhedron $\mathcal{A}$ of concavity C_1 contains no further concavity (Fig. 5(d)). As volume of $\mathcal{B}$ is less than $\mathcal{A}$, sub-polyhedron $\mathcal{B}$ is deleted to resolve C_1 (Fig. 5(e)). Similarly, concavity C_4 (category $\mathbf{C}_{z,y}$) is resolved by deleting sub-polyhedron $\mathcal{D}$. Thus, all the concavities on the given slab are resolved (Fig. 5(f)). The process is repeated for all the slabs parallel to a given coordinate plane. After each deletion of a sub-polyhedron, $\underline{P}_{\mathbb{G}}(A)$ is modified accordingly.

3.4   Finding Approximate 3D Orthogonal Convex Skull

Construction of an approximate 3D orthogonal convex skull of A involves resolving the concavities of $\underline{P}_{\mathbb{G}}(A)$ so that $3OCS(A)$ is a convex orthogonal polyhedron. Along each coordinate plane the following steps are carried out:

i. A is sliced orthogonally to form a set of orthogonal slabs that represent $\underline{P}_{\mathbb{G}}(A)$ (Sect. 3.1).
ii. The concavities on each slab are detected and resolved, thereby modifying $\underline{P}_{\mathbb{G}}(A)$ (Sect. 3.3).
iii. If $\underline{P}_{\mathbb{G}}(A)$ has been disconnected into more than one polyhedrons, then all the polyhedrons except the one having the largest volume are discarded.

The above steps are repeated along another coordinate plane considering the modified $\underline{P}_{\mathbb{G}}(A)$ as input.

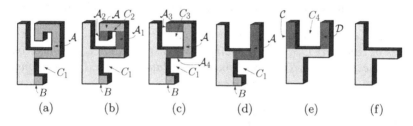

Fig. 5. Demonstration of resolving the concavities on a slab. Concavity C_1 is of category $\mathbf{C}_{z,x}$, C_2 of category $\mathbf{C}_{z,y}$, C_3 of category $\mathbf{C}_{z,x}$, and C_4 of category $\mathbf{C}_{z,y}$.

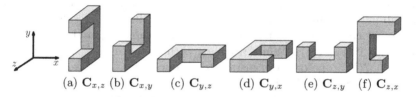

(a) $\mathbf{C}_{x,z}$ (b) $\mathbf{C}_{x,y}$ (c) $\mathbf{C}_{y,z}$ (d) $\mathbf{C}_{y,x}$ (e) $\mathbf{C}_{z,y}$ (f) $\mathbf{C}_{z,x}$

Plane	Concavity	Induced concavity		
		yz-plane	zx-plane	xy-plane
yz	$\mathbf{C}_{x,z}$	$\mathbf{C}_{y,z}$	-	-
	$\mathbf{C}_{x,y}$	$\mathbf{C}_{z,y}$		
zx	$\mathbf{C}_{y,z}$	-	$\mathbf{C}_{x,z}$	-
	$\mathbf{C}_{y,x}$		$\mathbf{C}_{z,x}$	
xy	$\mathbf{C}_{z,y}$	-	-	$\mathbf{C}_{x,y}$
	$\mathbf{C}_{z,x}$			$\mathbf{C}_{y,x}$

Fig. 6. Two categories of possible concavities w.r.t. a slab along each coordinate plane. (a) Resolving concavity of category $\mathbf{C}_{x,z}$ induces concavity of category $\mathbf{C}_{y,z}$ along yz-plane, (b) resolving $\mathbf{C}_{x,y}$ induces $\mathbf{C}_{z,y}$ along yz-plane, (c) resolving $\mathbf{C}_{y,z}$ induces $\mathbf{C}_{x,z}$ along zx-plane, (d) resolving $\mathbf{C}_{y,x}$ induces $\mathbf{C}_{z,x}$ along zx-plane, (e) resolving $\mathbf{C}_{z,y}$ induces $\mathbf{C}_{x,y}$ along xy-plane, and (f) resolving $\mathbf{C}_{z,x}$ induces $\mathbf{C}_{y,x}$ along xy-plane.

Along each coordinate plane there exists exactly two categories of concavity on a slab S_t, as shown in the table in Fig. 6. A concavity can be resolved only along the coordinate plane to which it is parallel, i.e., $C_{x,z}$ (belonging to category $\mathbf{C}_{x,z}$) and $C_{x,y}$ (belonging to category $\mathbf{C}_{x,y}$) are resolved only along the yz-plane, etc. Resolving a concavity along a coordinate plane may induce another concavity along a different coordinate plane which leads to the following observation.

Observation 1. *W.l.o.g., resolving an instance of concavity of category $\mathbf{C}_{x,z}$, while traversing the slab along the yz-plane, may induce one (or more) instance(s) of concavity of category $\mathbf{C}_{y,z}$.*

It is observed that resolving a concavity of category $\mathbf{C}_{x,z}$ may induce a concavity of category $\mathbf{C}_{y,z}$, and resolving a concavity of category $\mathbf{C}_{y,z}$ may induce a concavity of category $\mathbf{C}_{x,z}$; resolving $\mathbf{C}_{x,y}$ may induce $\mathbf{C}_{z,y}$ and vice versa; resolving $\mathbf{C}_{y,x}$ may induce $\mathbf{C}_{z,x}$ and vice versa. W.l.o.g. let us consider concavity $C_{y,z}$ (belongs to $\mathbf{C}_{y,z}$). $C_{y,z}$ and $C_{x,z}$ (belongs to $\mathbf{C}_{x,z}$) may be induced from each other for a finite number of times, which leads to the following lemma.

Lemma 1. *Let resolving an instance of concavity $C_{y,z}$ induces one or more instances of concavity $C_{x,z}$ and resolving those instances of $C_{x,z}$ induces one or more instances of $C_{y,z}$. Then resolving the instances of the induced concavity $C_{y,z}$ does not induce any further concavity.*

Proof. Let the concavities of slab polygons of P along the zx-plane be resolved first, followed by the yz-plane and the xy-plane. Since resolving a concavity may induce another concavity, more than one pass may be required to resolve the induced concavities, as will be elaborated later in Theorem 1. In the first pass, let the concavity $C_{y,z}$ on slab Sy_1 of P (Fig. 7(a)) be resolved by deleting one of its sub-polyhedrons $\mathcal{A}$ along the zx-plane (Fig. 7(b)). As a result one or more instances of concavity $C'_{x,z}$ may be induced on slabs Sx_1 and Sx_2 of P', by Observation 1 (Fig. 7(c)). While resolving each instance of $C'_{x,z}$ along the yz-plane (Fig. 7(d)), one or more instances of concavity $C'_{y,z}$ may be induced on slabs Sy_2 and Sy_3 of P'' (Fig. 7(e)). In this case, no concavity is detected along the xy-plane in the first pass. Hence, it is not shown in Fig. 7. In the second pass, one or more instances of concavity $C'_{y,z}$ on P'' (Fig. 7(f)) are resolved along the zx-plane. Let us assume, by contradiction, that this induces one or more instances of $C''_{x,z}$ on slab Sx_3 of P''' (Fig. 7(h)).

A concavity on a slab is characterized by two sub-polyhedrons and is resolved by deleting any one of them according to certain rules (Sect. 3.3). During the first pass along the zx-plane, $C_{y,z}$ is the only concavity detected on slab Sy_1 of P (Fig. 7(a)). During the second pass along the zx-plane (Fig. 7(e)), two instances of concavity $C'_{y,z}$ are detected on slabs Sy_2 and Sy_3. They are resolved by deleting sub-polyhedrons $\mathcal{D}$ and $\mathcal{E}$ respectively from P''. Since $C'_{y,z}$ is not detected on P in the first pass along the zx-plane, it is concluded that the sub-polyhedrons $\mathcal{D}$ and $\mathcal{E}$ existed as a part of P (Fig. 7(a)). Hence, it is justified that if resolving an instance of concavity $C_{y,z}$ induces one or more instances of concavity $C_{x,z}$, then resolving those instances of $C_{x,z}$ induces one or more instances of $C_{y,z}$ (Fig. 7 (a–e)).

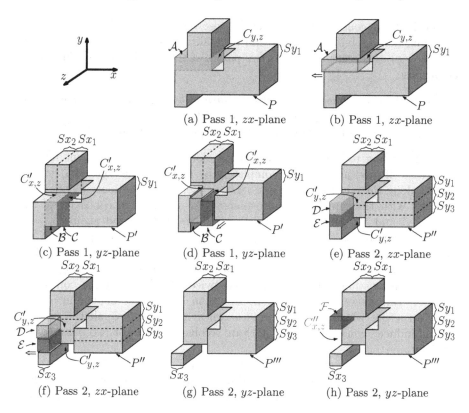

(a) Pass 1, zx-plane

(b) Pass 1, zx-plane

(c) Pass 1, yz-plane

(d) Pass 1, yz-plane

(e) Pass 2, zx-plane

(f) Pass 2, zx-plane

(g) Pass 2, yz-plane

(h) Pass 2, yz-plane

Fig. 7. (a) When an instance of concavity $C_{y,z}$ is resolved along the zx-plane (b) by deleting a sub-polyhedron $\mathcal{A}$, (c) two instances of concavity $C'_{x,z}$ are induced, which are resolved along the yz-plane (d) by deleting sub-polyhedrons $\mathcal{B}$ and $\mathcal{C}$. As a result (e) two instances of $C'_{y,z}$ are induced which are resolved along the zx-plane (f) by deleting sub-polyhedrons $\mathcal{D}$ and $\mathcal{E}$. (g) Polyhedron P''' is obtained after resolving all the concavities of P. (h) It is assumed that an instance of concavity $C''_{x,z}$ exists on P'''.

During the first pass along the yz-plane (Fig. 7(c)), two instances of concavity $C'_{x,z}$ are detected on the slabs Sx_1 and Sx_2 of P'. If it is assumed that concavity $C''_{x,z}$ exists on slab Sx_3 (Fig. 7(h)), then the sub-polyhedron $\mathcal{F}$ should be present on the slab Sx_3 of P''' during the second pass along the yz-plane. But the sub-polyhedron $\mathcal{F}$ did not exist as a part of P' during the first pass along the yz-plane (Fig. 7(c)). It implies that $\mathcal{F}$ has been deleted before the first pass along the yz-plane, i.e., $\mathcal{F}$ has been deleted while resolving $C_{y,z}$ along the zx-plane (Fig. 7(b)). Hence, $\mathcal{F}$ cannot exist on Sx_3 during the second pass along the yz-plane. In absence of the sub-polyhedron $\mathcal{F}$, concavity $C''_{x,z}$ cannot exist on Sx_3 of P''' (Fig. 7(h)). This contradicts our assumption. Hence, no concavity is induced while resolving an instance of $C'_{y,z}$ (Fig. 7(g)). Hence proved.   □

The induced concavities may not be resolved within a single pass of the algorithm, i.e., applying the algorithm once along the yz-, zx-, and xy-planes. The following theorem proves that two passes of the algorithm are sufficient for the purpose.

Theorem 1. *The orthogonal polyhedron formed by applying the proposed algorithm at most twice on $\underline{P}_G(A)$ for the set of three coordinate planes (yz-, zx-, and xy-planes) gives an approximate 3D orthogonal convex skull $3OCS(A)$.*

Proof. According to Lemma 1, a concavity along with its subsequent induced concavities (if any) are resolved completely by applying the concavity removal method along three coordinate planes. The sequence of coordinate planes starts and ends with the same coordinate plane. For example, a concavity of category $\mathbf{C}_{y,z}$ and the induced concavities of category $\mathbf{C}_{x,z}$ and again $\mathbf{C}_{y,z}$ are resolved along the zx-, yz-, and zx-planes (Fig. 7). This is possible only if two passes of the algorithm are used. Since there exists categories of concavity along each of the three coordinate planes, application of the algorithm along each of yz-, zx-, and xy-planes may be required twice. Resolving all the concavities of $\underline{P}_G(A)$ can, however, conclude in less than two passes depending on the object structure which does not induce concavity. Therefore, all the concavities and induced concavities on $\underline{P}_G(A)$ are resolved in at most two passes of the algorithm.

A slab refers to a section of $\underline{P}_G(A)$ at a given slicing plane. Since $\underline{P}_G(A)$ is of maximum volume that can be inscribed in the object, at the given slicing plane a slab is also of maximum volume. While resolving a concavity on a slab, the sub-polyhedron with the larger volume is included in the 3D orthogonal convex skull of the slab, thereby trying to achieve a larger volume of $3OCS(A)$. It may be noted that the 3D orthogonal convex skull of the slab may not be unique due to the varying starting point and initial direction of traversal. Also, the volume of $3OCS(A)$ may vary with different order of the coordinate planes along which the algorithm is applied. Due to the variation in volume, the result given by the algorithm is an approximate 3D orthogonal convex skull. Hence proved.   □

3.5   Algorithm

Given an object A as a set of voxels, its approximate 3D orthogonal convex skull is constructed by the two-pass algorithm presented in Fig. 8. The three coordinate planes are considered in sequence (**for** loop Steps 3–15), for each slicing plane along a coordinate plane the concavities in a slab are removed (Steps 6–11). On a slicing plane, each slab $S[i]$ is subjected to a method   explained in Sect. 3.3 to construct the 2D orthogonal convex skull $OCS(S[i])$ (Step 10). Consequently, the slab corresponding to the 2D orthogonal convex skull $K[i]$ along each slicing plane is accumulated in $\underline{P}'_G(A)$ (Step 11).

While finding the 2D orthogonal convex skull of the slab polygon w.r.t. a slab, the volume of the slab is determined by computing the area of the slab polygon. The total volume of a set of consecutive slabs in the direction perpendicular to

```
Algorithm 3OCS(A, G)
01.  count ← 0                        ▷ # passes
02.  do
03.       for each coordinate plane j      ▷j ∈ {yz, zx, xy}
04.            set P'_G(A) ← 0             ▷ empty set of polyhedrons
05.            slice P_G(A) with Π        ▷Π = {Π₁, Π₂, ..., Π_r}
06.            for each slicing plane Π_k   ▷1 ≤ k ≤ r
07.                 set s ← # slabs on Π_k
08.                 set S ← set of slabs on Π_k
09.                 for i = 1 to s
10.                      set K[i] ← 2D orthogonal convex skull OCS(S[i])
11.                      P'_G(A) ← P'_G(A) ∪ K[i]
12.                 set m ← # polyhedrons in P'_G(A)
13.                 set M ← set of polyhedrons in P'_G(A)
14.                 P_G(A) ← largest(vol(M[1]), vol(M[2]), ..., vol(M[m]))
15.            count ← count + 1
16.  while(count < 2)
17.  return P_G(A)
```

Fig. 8. Brief outline of the proposed algorithm.

the given coordinate plane gives the volume of the polyhedron composed of the set of slabs. Thus, the volumes of all the orthogonal polyhedrons in $\underline{P}'_G(A)$ are determined. If $\underline{P}'_G(A)$ contains a single polyhedron, then $\underline{P}'_G(A)$ represents the modified 3D orthogonal inner cover $\underline{P}_G(A)$. If $\underline{P}'_G(A)$ contains more than one polyhedron, then the polyhedron having the largest volume is assigned to $\underline{P}_G(A)$ (Step 14) and the rest of the polyhedrons are discarded.

The process is repeated with the modified $\underline{P}_G(A)$ along the other coordinate planes (Steps 3–15) and the algorithm is repeated exactly once along all the three coordinate planes (Steps 2–16). Finally, the modified 3D orthogonal inner cover $\underline{P}_G(A)$ is reported as the approximate 3D orthogonal convex skull (approximate $3OCS(A)$) (Step 17). The algorithm deals with non-intersecting objects free from holes and cavities.

3.6   Time Complexity

Let n be the number of voxels on the object surface connected in 26-neighborhood. A UGC is a cube of length g which contributes a maximum of five faces to the cover. Therefore, the number of UGCs on the object surface containing object voxels is $O(n/g)$ in the worst case, which implies that the number of UGC-faces on the object surface is given by $O(n/g)$. The full object occupancy of a UGC is determined by checking six UGC-faces completely in $O(g^2)$ time.

W.r.t. each slicing plane, orthogonal slicing involves traversal of the grid vertices on the slicing plane exactly once. Therefore, considering all the slicing planes, the UGCs on the object surface are traversed exactly once. This traversal requires $O(n/g)$ time. $O(g^2)$ time is required to check whether a UGC-face is fully object-occupied. Hence, the direction of traversal at each grid vertex is determined in $O(g^2)$ time. Therefore, the orthogonal slicing along a coordinate plane is completed in $O(n/g) \times O(g^2) = O(ng)$ time.

The grid vertices on the object surface are sorted lexicographically in $O((n/g) \log(n/g))$ time. Once a concavity is detected on a slab, the terminal vertices of the next sub-polyhedron are found from the lexicographically sorted lists. For all the slabs this operation is completed in $O(\log n)$ time. In case of nested concavity, connectivity of a sub-polyhedron is checked in $O(1)$ time. Volumes of the sub-polyhedrons are computed in $O(n/g)$ time to remove the sub-polyhedron of smaller volume in every case. Hence, the overall time complexity for resolving the concavities on all the slabs is bounded by $O(n \log n)$. Volume of the approximate 3D orthogonal convex skull is determined by computing the volume of all the slabs parallel to a given coordinate plane in $O(n/g)$ time. Therefore, the total time complexity for finding an approximate $3OCS(A)$ is given by $O(ng) + O(n \log n) = O(n \log n)$.

4   Experimental Results and Conclusions

The proposed algorithm has been implemented in C in Linux Fedora Release 13, Dual Intel Xeon Processor 2.8 GHz, 800 MHz FSB. The experimental results in

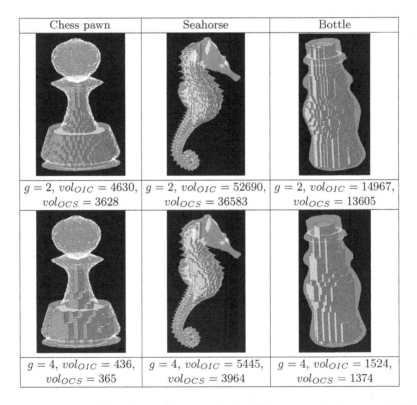

Chess pawn	Seahorse	Bottle
$g = 2$, $vol_{OIC} = 4630$, $vol_{OCS} = 3628$	$g = 2$, $vol_{OIC} = 52690$, $vol_{OCS} = 36583$	$g = 2$, $vol_{OIC} = 14967$, $vol_{OCS} = 13605$
$g = 4$, $vol_{OIC} = 436$, $vol_{OCS} = 365$	$g = 4$, $vol_{OIC} = 5445$, $vol_{OCS} = 3964$	$g = 4$, $vol_{OIC} = 1524$, $vol_{OCS} = 1374$

Fig. 9. Approximate 3D orthogonal convex skull of Chess pawn, Seahorse, and Bottle.

Fig. 9 display an approximate 3D orthogonal convex skull of each of the objects Chess pawn, Seahorse, and Bottle for different grid sizes. The volumes of both the 3D orthogonal inner cover (vol_{OIC}) and the approximate 3D orthogonal convex skull (vol_{OCS}) decrease exponentially with increasing grid size. It may be noted that the number of concave vertices of all the types except $(7, 3, 3)$ is less in the approximate $3OCS(A)$ than in the 3D orthogonal inner cover.

The approximate 3D orthogonal convex skull of an object may vary with the order of the coordinate planes along which the algorithm is applied. If the two sub-polyhedrons due to a concavity on a slab are of equal volume, then more than one result may exist. The approximate 3D orthogonal convex skull may also vary depending on the starting point and initial direction of traversal (anti-clockwise or clockwise) while resolving the concavities on a slab. The result will be unique only if none of the concavities on a slab have more than two consecutive 270° vertices or the sub-polyhedron with the larger volume is not deleted to maintain connectivity of the slab. The volume of $3OCS(A)$ may vary with its structure, reporting an approximate 3D orthogonal convex skull in every case. Figure 10 illustrates the variation of the approximate 3D orthogonal convex skull when the proposed algorithm is applied on the object along the coordinate planes in different orders, like, along the yz-plane first, followed by the zx-plane and the xy-plane, etc. The current version of the algorithm is limited to non-intersecting digital objects and objects free from holes and cavities. Extension of the algorithm to account for those objects may be attempted in future.

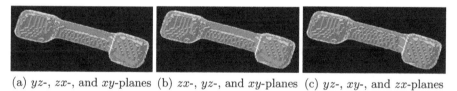

(a) yz-, zx-, and xy-planes (b) zx-, yz-, and xy-planes (c) yz-, xy-, and zx-planes

Fig. 10. Variation of the approximate 3D orthogonal convex skull of Phone due to application of the algorithm along the coordinate planes in different orders.

Acknowledgement. A part of this research is funded by CSIR, Govt. of India.

References

1. Abello, J.M., Pardalos, P.M., Resende, M.G.C. (eds.): Handbook of Massive Data Sets, vol. I. Springer, Heidelberg (2002)
2. Aronov, B., van Kreveld, M., Löffler, M., Silveira, R.I.: Peeling meshed potatoes. Algorithmica **60**(2), 349–367 (2011)
3. Borgefors, G., Strand, R.: An approximation of the maximal inscribed convex set of a digital object. In: Roli, F., Vitulano, S. (eds.) ICIAP 2005. LNCS, vol. 3617, pp. 438–445. Springer, Heidelberg (2005)

4. Cabello, S., Cibulka, J., Kynčl, J., Saumell, M., Valtr, P.: Peeling potatoes near-optimally in near-linear time. In: Proc. 13th Annual Symposium on Computational Geometry, SOCG 2014, Kyoto, Japan. pp. 224–231. ACM, New York (2014)
5. Chang, J., Yap, C.: A polynomial solution for potato-peeling and other polygon inclusion and enclosure problems. In: Proceedings of the 25th Annual Symposium on Foundations of Computer Science, SFCS 1984, Singer Island, Florida, pp. 408–416. IEEE Computer Society, Washington, DC (1984)
6. Chang, J., Yap, C.: A polynomial solution for the potato-peeling problem. Discrete Comput. Geom. **1**(2), 155–182 (1986)
7. Chassery, J.M., Coeurjolly, D.: Optimal shape and inclusion. In: Ronse, C., Najman, L., Decenciére, E. (eds.) Mathematical Morphology: 40 Years On. Computational Imaging and Vision, vol. 30, pp. 229–248. Springer, Netherlands (2005)
8. Dutt, M., Biswas, A., Bhowmick, P., Bhattacharya, B.B.: On finding an orthogonal convex skull of a digital object. Int. J. Imaging Syst. Technol. **21**, 14–27 (2011)
9. Goodman, J.E.: On the largest convex polygon contained in a non-convex n-gon, or how to peel a potato. Geometriae Dedicata **11**(1), 99–106 (1981)
10. Karmakar, N., Biswas, A., Bhowmick, P.: Fast slicing of orthogonal covers using DCEL. In: Barneva, R.P., Brimkov, V.E., Aggarwal, J.K. (eds.) IWCIA 2012. LNCS, vol. 7655, pp. 16–30. Springer, Heidelberg (2012)
11. Klette, R., Rosenfeld, A.: Digital Geometry: Geometric Methods for Digital Picture Analysis. Morgan Kaufmann, San Francisco (2004)
12. Woo, T.: The convex skull problem. Technical report, Department of Industrial and Operations Engineering, University of Michigan, Ann Arbor, MI (1986)
13. Wood, D., Yap, C.K.: The orthogonal convex skull problem. Discrete Comput. Geom. **3**(4), 349–365 (1988)

Designing a Topological Algorithm
for 3D Activity Recognition

Maria-Jose Jimenez[1(✉)], Belen Medrano[1], David Monaghan[2],
and Noel E. O'Connor[2]

[1] Applied Math Department, School of Computer Engineering, University of Seville,
Campus Reina Mercedes, 41012 Sevilla, Spain
{majiro,belenmg}@us.es
[2] INSIGHT Centre for Data Analytics, Dublin City University, Dublin, Ireland
{david.monaghan,noel.oconnor}@dcu.ie

Abstract. Voxel carving is a non-invasive and low-cost technique that
is used for the reconstruction of a 3D volume from images captured from
a set of cameras placed around the object of interest. In this paper we
propose a method to topologically analyze a video sequence of 3D reconstructions representing a tennis player performing different forehand and
backhand strokes with the aim of providing an approach that could be
useful in other sport activities.

Keywords: 3D video sequence · Voxel carving · Volume reconstruction ·
Persistent homology · Activity recognition

1 Introduction

The combinatorial nature of a 3D digital image is a suitable material to homology
computation by taking as input the (algebraic) cubical complex associated to the
image (whose building blocks are vertices, edges, squares and cubes). Homology
is a topological invariant that characterizes "holes" in any dimension (in the case
of a 3D space, connected components, tunnels and cavities). *Persistent homology*
[5,20] studies homology classes and their life-times (persistence) in an increasing
nested sequence of subcomplexes (a filtration on the cubical complex).

Space or voxel carving [2,4,12,18] is a technique for creating a three-dimensional reconstruction of an object from a series of two-dimensional images
captured from cameras placed around the object at different viewing angles.
The technique involves capturing a series of synchronised images of an object,

M.-J. Jimenez and B. Medrano—Author partially supported by IMUS, Junta de
Andalucia under grant FQM-369, Spanish Ministry under grant MTM2012-32706
and MTM2015-67072-P and ESF ACAT programme.
D. Monaghan and N.E. O'Connor—Part of this work was supported by Science Foun-
dation Ireland through the Insight Centre for Data Analytics under grant number
SFI/12/RC/2289.

A. Bac and J.-L. Mari (Eds.): CTIC 2016, LNCS 9667, pp. 193–203, 2016.
DOI: 10.1007/978-3-319-39441-1_18

and, by analysis of these images and with prior knowledge of the exact three-dimensional location of the cameras, deriving an approximation of the shape of the object.

There are numerous research papers dealing with the problem of human activity recognition from 3D data (see [1] for a recent review). An important subgroup of these works provide algorithms for activity recognition from a set of silhouettes of the subject, such as [19] or [11]. In [19] Fourier Transform in cylindrical coordinates is performed to compare *motion history volumes* representing different actions. In [11], the so-called *action volume* is produced from a set of human body silhouettes from the same view angle. They combine multiview angles to obtain a set of representative action volumes that are used to classify the action.

There have been some papers (see [13–15]) dealing with the application of persistent homology to the problem of gait recognition. Different silhouettes are extracted from a whole gait cycle (from only one viewpoint) and stacked together to form some kind of action volume to be topologically analyzed by using persistent homology.

In this paper we focus on sequences of 3D reconstructions of volumes that are captured from a small set of cameras with different viewpoints in a tennis court. From that input, we construct another 3D object containing the motion history information and that we analyze it from a persistent homology perspective.

In the following section, we describe the specific context in which we develop our work. Section 3 describes the design of our method to apply persistent homology to such specific context with the aim of recognizing the activity in a video sequence of voxel carving reconstructions. Reports on the computations performed as well as some conclusions are collected in Sect. 4. We draw some ideas for future work in the last section.

2   Voxel Carving Video Sequences

Voxel carving techniques are very useful for 3D reconstruction since they are non-invasive and they can cover a very large environment. They can be implemented with an array of low-cost cameras to produce a synchronised set of images. In each image, the subject of interest is identified and then segmented from the background of the image (this is commonly known by silhouette extraction). The subject silhouette is segmented from the background and a 3D bounding box is then drawn around the subject's approximate position in 3D space. This bounding box defines a volume that has a corresponding real world three-dimensional coordinate system. The different silhouettes are used to "carve" the defined volume accordingly. A sequence of reconstructed volumes can be seen in Fig. 1.

In the real world coordinate system the approximate subject volume is populated with voxels, that are set at a particular distance apart or spatial resolution, i.e. if the distance between voxels decreases then the spatial resolution increases. From experimental observation, authors in [16] found that a three dimensional

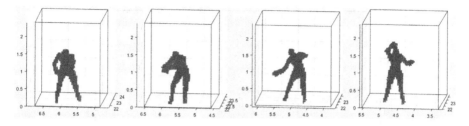

Fig. 1. A sequence of 3D reconstructions by voxel carving. Each frame is a 3D point cloud.

spatial frequency of 4 cm, i.e. 15,625 samples per cubic metre, was sufficiently adequate for their purposes and in [7] they concluded that higher resolutions did not contribute to a better topological model in the reconstruction process. That means that the spacing considered between each voxel is 4 cm in the OX, OY and OZ directions. This way, it is satisfied that the final reconstructions are qualitatively detailed enough to be used as a 3D visualisation tool, and, at the same time, based on the computational performance of a single PC, this resolution allows to run the algorithm at near to real-time. Regarding the quality of space carving results, persistent homology was proposed first in [8] as a tool for a topological analysis of the carving process along the sequence of 3D reconstructions with increasing number of cameras.

The general voxel technique proposed in [12] was modified and adapted to a specific task, as fully detailed in [16,17]. And it is, in fact, that specific voxel carving technique that we are using in this paper, fixing the number of cameras to five, since this is the usual constraint we can find in practise.

Once we get the sequence of voxel carving results, the first step is to segment the frames involving each action accomplished by the subject. This can be done by a visual inspection of the video, but an attempt to automatically recognize the beginning and end of each movement (either forehand or backhand strokes) led us to compute the variation of the mass center of each 3D frame with respect to previous and next ones with a kind of second derivative. That is, for each frame F_i, consider the mass center $(c_{i,1}, c_{i,2}, c_{i,3})$ and compute the list of values $|2c_{i,1} - c_{i-1,1} - c_{i+1,1}| + |2c_{i,2} - c_{i-1,2} - c_{i+1,2}| + |2c_{i,3} - c_{i-1,3} - c_{i+1,3}|$ whose graphic representation can be seen in Fig. 2. One can observe that peaks are mainly grouped around five points corresponding to five movements of the player (three forehand and two backhand).

3 Persistent Homology for 3D Activity Recognition

Persistent homology has been proved to be a useful tool in the study of 3D shape comparison. For example, in the paper [3] the authors provide an algorithm to approximate the matching distance (which is computationally costly) when comparing 3D shapes represented by triangle meshes.

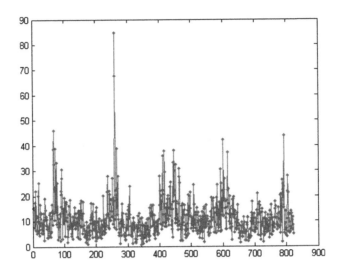

Fig. 2. Graphic representation of variation of mass center of each frame in the sequence with respect to previous and next ones.

However, as far as we know there is no work on activity recognition using persistent homology, except for the related topic of gait recognition which has already been explored from the persistent homology viewpoint in [13–15].

We are concerned with the application of persistent homology computation to provide topological analysis of a time sequence of 3D reconstructions by the voxel carving technique. We consider a sequence of voxel carving results under a fixed number of cameras, so it is convenient to have in mind that each frame is referring to a 3D reconstruction, that is, a set of 3D points in space.

The input data is a sequence $\{F_t\}_t$ of 3D binary digital images or subsets of points F_t of $\mathbb{Z}^3$ considered under the $(26, 6)$–adjacency relation for the foreground (F_t) and background $(\mathbb{Z}^3 \setminus F_t)$, respectively. Due to the nature of our input data, we focus on a special type of cell complex: *cubical complex*. A cubical complex Q in $\mathbb{R}^3$, is given by a finite collection of p-cubes such that a 0-cube is a vertex, a 1-cube is an edge, a 2-cube is a square and a 3-cube is a solid cube (or simply a cube); together with all their faces and such that the intersection between two of them is either empty or a face of each of them. The cubical complex $Q(F_t)$ associated to F_t is given by identification of each 3D point of F_t with the unit cube centered at that point and then considering all those 3-cubes together with all their faces (square faces, edges and vertices), such that shared faces are considered only once. Sometimes we will refer to p-cubes with the more general term of *cells* (corresponding to the more general concept of cell complex, see [10]).

Given a cell complex, homology groups can be computed using a variety of methods. Incremental Algorithm for computing AT-model (Algebraic Topological Model) [9], computes homology information of the cell complex by an incremental technique, considering the addition of a cell each time following a full order on the set of cells of the complex. In [6], the authors revisited this algorithm with the aim of setting its equivalence with persistent homology computation algorithms [5,20] working over $\mathbb{Z}/2\mathbb{Z}$ as ground ring. We make use of algorithm in [6] for the persistent homology computation, though any other algorithm for computing persistent homology, adapted to cubical complexes, could have been applied. We will use the generated *persistence barcode* as a source to create a feature vector characterising the movement. Recall that a persistence barcode encodes "times" (indexes in the ordering) of birth and death of each homology class (see [5,20]).

The method described in this paper consists in the following steps starting from a segmented sequence of 3D frames reconstructed by voxel carving: (1) from each reconstructed volume, take the projection on a plane parallel to the net in the tennis court; (2) produce a stack with the 2D images from the previous step; (3) topologically analyze the volume by considering different directions; (4) create several topological feature vectors associated to the volume; (5) compare vectors by using a similarity measure.

Step 1. In this specific context, a particular viewpoint that can be useful for recognizing the action is a front view from the net in the tennis court. Having a 3D reconstruction obtained from different viewpoints allows to reproduce the result from a viewpoint of interest even though there is no camera in that viewing angle. For each 3D reconstructed volume, hence, we project the points onto a plane parallel to the net (see Fig. 3). If necessary, this projection could be done onto other planes of interest depending on the action to be recognized. Even more, one could combine the information obtained from different projections, that is the advantage of having a 3D reconstruction of the subject.

Step 2. Form a stack with all the 2D images from the previous step, by aligning the mass centre of every 2D projection. This way, a volume is constructed that can be considered a motion history volume since contains information of the whole movement. In this volume, we will convene that OX is the axis that is perpendicular to the net (in the tennis court), OY is parallel to the net and OZ means the hight of the points in the volume (see Fig. 4).

Step 3. Consider the cubical complex Q associated to the 3D digital image from previous step. We must consider a full ordering of its cubes $\{c^1, \ldots, c^n\}$ such that if c^i is a face of c^j, then $i < j$. Such ordering will be determined by different filter functions given by the distance to certain planes in the 3D space. Then we will have a nested sequence of subcomplexes $\emptyset = Q^0 \subseteq Q^1 \cdots \subseteq Q^m$ (a filtration over Q determined by the value of the filter function induced on the cells of the complex) for which persistent homology can be computed.

Set the minimum and maximum coordinates of the points in the considered volume, $\{x_{min}, x_{max}, y_{min}, y_{max}, z_{min}, z_{max}\}$, and consider the following "directions" to provide the filters:

- direction given by OX axis, x^+: the filter function x^+ is provided then by the distance to the plane $x = x_{min}$;
- directions given by OY axis, y^+ and y^-: the filter function y^+ (resp. y^-) is provided then by the distance to the plane $y = y_{min}$ (resp. minus the distance);
- directions given by OZ axis, z^+ and z^-: the filter function z^+ (resp. z^-) is provided then by the distance to the plane $z = z_{min}$ (resp. minus the distance);

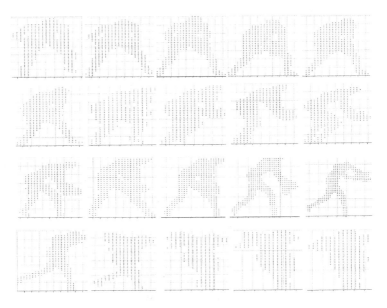

Fig. 3. Set of silhouettes obtained, from a sequence of 3D reconstructions, by projection on a plane parallel to the net in the tennis court

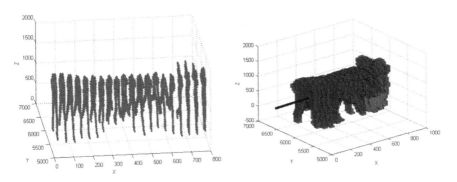

Fig. 4. Stack of silhouettes obtained, from a sequence of 3D reconstructions representing a backhand movement

- 45° direction on the OYZ plane, oyz^+ and oyz^-: the filter function oyz^+ (resp. oyz^-) is provided then by the distance to the plane $y + z = y_{min} + z_{min}$ (resp. minus the distance);
- $(-45)°$ direction on the OYZ plane, ozy^+ and ozy^-: the filter function ozy^+ (resp. ozy^-) is provided then by the distance to the plane $y - z = y_{max} - z_{min}$ (resp. minus the distance);

These directions are represented in Fig. 5. However, direction given by OX axis would provide poor information when applied to the whole complex, since in normal conditions, it will produce a unique connected component. That is why we propose a subdivision of the initial complex into 9 volumes (see Fig. 6) in order to compute persistent homology of each of these volumes separately along x^+ direction. This way, each silhouette is divided into a 3 by 3 array that may separate the evolution of movement of extremities from the central part of the body. More specifically, the volumes are given by $V_{ij} = \{(x, y, z), y_i \leq y \leq y_{i+1}, z_j \leq z \leq z_{j+1}\}$ for $i, j = 0, 1, 2$, with $y_0 = y_{min}; z_0 = z_{min}; y_i = y_{min} + \frac{i}{3}(y_{max} - y_{min})$ and $z_i = z_{min} + \frac{i}{3}(z_{max} - z_{min})$ for $i = 1, 2; y_3 = y_{max}; z_3 = z_{max}$.

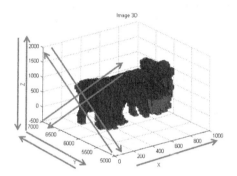

Fig. 5. Each of the 9 possible directions described to provide a filter function to order the cells in the complex.

Step 4. The filter function considered in the previous step set an ordering of all que cells in the cubical complex. Next step is to compute the persistence barcode for the cubical complex representing the motion volume. We make use of the concept of *simplified barcode* stated in [7] by which bars shorter than the distance between two consecutive subcomplexes in the considered filtration are discarded. In the case of the subdivision in the nine volumes, the computation is performed for each one of them. Hence, out of each computed barcode, a vector is produced in the style of Lamar et al. [13–15]. That is, consider the ordered set of cells in the whole volume $\{c_1, \ldots, c_N\}$ and a partition of this ordered set into n equal parts. Then, for each of the n intervals $(c_i^j, c_k^{j+1}]$, $j = 1, \ldots n$, compute

1. a_j = the number of homology classes living along the interval;
2. b_j = the number of homology classes that are born in the interval;

and compose the vector $[a_1, b_1, a_2, b_2, \ldots, a_n, b_n]$.

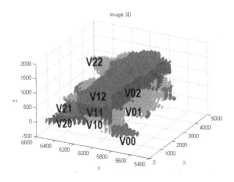

Fig. 6. Color representation of the 9 volumes segmented from the motion history volume

Step 5. Finally, a similarity measure has to be considered for comparison of the feature vectors. We adopt the cosine of the angle between two vectors to measure how similar the corresponding barcodes are, that is, for two vectors V_1 and V_2 computed on the same direction, compute

$$S_{1,2} = \frac{V_1 \cdot V_2}{|V_1| \cdot |V_2|} .$$

Notice that each barcode produces a feature vector so the final similarity measure would be computed as the total sum of all the partial comparison measures between the corresponding vectors.

4   Experiments

We have considered 8 video sequences for forehand stroke and other 8 for backhand strokes. Such video sequences correspond to synthetic 3D reconstructions by voxel carving with coordinates on $0.4\mathbb{Z}^3$. Due to the fact that the result of voxel carving process may carry eventual numerical errors that produce some missing points, and after taking some experiments, we discarded the 1-homology classes and considered only dimension 0, that is, connected components.

By an initial evaluation on the computed barcodes (see Fig. 7, last column), we have confirmed the intuition that the direction z^- (that is, from top to bottom), is not very informative, so we have skipped it to compute the similarity measure. We have implemented the partition for $n = 5$ and $n = 10$ and realized that the latter provides much better results. This was also quite intuitive from observing Figs. 7 and 8 since $n = 5$ is too low to distinguish the numerous small bars from the few more significant bars that appear in the barcode.

We have also come up with the conclusion that the division into the 9 volumes to follow up the movement direction x^+ does not provide good results, what was also clear by watching the corresponding barcodes. The problem is that the connection of the whole object is lost and the division can be very different depending on the inclination of the subject yielding to different results. In the first column of results of Fig. 9, the normalized similarity measure has been computed from the sum of similarity measures of each pair of vectors in directions y^+, y^-, z^+, oyz^+, oyz^-, ozy^+ and ozy^-, as well as those of volumes V_{00}, V_{01}, V_{02}, V_{10}, V_{12}, V_{20}, V_{21}, V_{22}, for $n = 5$ in direction x^+.

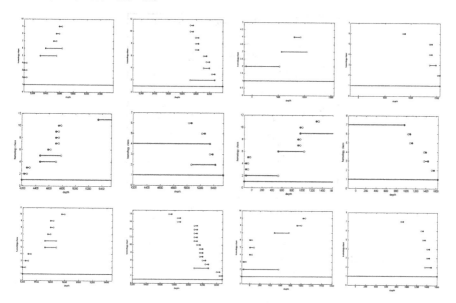

Fig. 7. Persistence 0-barcodes of three forehand movements (three rows) on each of the directions y^+, y^-, z^+ and z^-

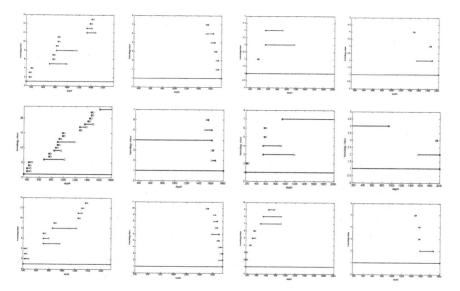

Fig. 8. Persistence 0-barcodes of three forehand movements (three rows) on each of the directions oyz^+, oyz^-, ozy^+ and ozy^-

Second and third columns of results in Fig. 9 have been computed without considering volumes V_{ij}, for partitions $n = 5$ and $n = 10$ respectively. It is clear that only for $n = 10$ does the method provide good results.

	All the filter functions for n=5	Filter functions OY+, OY-, OZ+, OYZ+, OYZ+, OZY-, for n=5	Filter functions OY+, OY-, OZ+, OYZ+, OYZ+, OZY- for n=10
F1-F2	0,90	0,87	**0,83**
F2-F3	0,88	0,83	**0,79**
F3-B1	0,83	0,75	0,54
B1-B2	0,86	0,81	**0,78**
B2-B3	0,76	0,70	**0,67**
B1-B3	0,41	0,40	**0,84**
F2-B1	0,38	0,33	0,59
F3-B2	0,38	0,37	0,57
B1-B3	0,39	0,42	**0,84**
F1-B1	0,84	0,77	0,59
F2-B2	0,85	0,79	0,58
F3-B3	0,74	0,64	0,55
F1-B2	0,86	0,81	0,60
F2-B3	0,78	0,72	0,58
F1-B3	0,77	0,73	0,58

Fig. 9. Results of normalized similarity measures between three forehand and three backhand strokes with different partitions and filter functions.

5   Conclusions and Future Work

Fixing a certain number of cameras and considering a video sequence of 3D reconstructions (by voxel carving), we propose a method for activity recognition of a tennis player stroke based on persistent homology. This work could set the ground for extension to other activities recognition. Depending on the context, different projections could be used to form the stack of silhouettes to be analyzed and different directions of interest could be selected.

References

1. Aggarwal, J.K., Xia, L.: Human activity recognition from 3D data: a review. Pattern Recogn. Lett. **48**, 70–80 (2014)
2. Broadhurst, A., Drummond, T., Cipolla, R.: A probabilistic framework for space carving. In: Conference on Computer Vision, vol. 1, p. 388 (2001)
3. Cerri, A., Di Fabio, B., Jablonski, J., Medri, F.: Comparing shapes through multi-scale approximations of the matching distance. Comput. Vis. Image Underst. **121**, 43–56 (2014)
4. Culbertson, W.B., Malzbender, T., Slabaugh, G.G.: Generalized voxel coloring. In: Triggs, B., Zisserman, A., Szeliski, R. (eds.) ICCV-WS 1999. LNCS, vol. 1883, pp. 100–115. Springer, Heidelberg (2000)
5. Edelsbrunner, H., Letscher, D., Zomorodian, A.: Topological persistence and simplification. In: FOCS 2000, pp. 454–463. IEEE Computer Society (2000)

6. Gonzalez-Diaz, R., Ion, A., Jimenez, M.J., Poyatos, R.: Incremental-decremental algorithm for computing at-models and persistent homology. In: Real, P., Diaz-Pernil, D., Molina-Abril, H., Berciano, A., Kropatsch, W. (eds.) CAIP 2011, Part I. LNCS, vol. 6854, pp. 286–293. Springer, Heidelberg (2011)

7. Gutierrez, A., Jimenez, M.J., Monaghan, D., O'Connor, N.E.: Topological evaluation of volume reconstructions by voxel carving. Comput. Vis. Image Underst. **121**, 27–35 (2014)

8. Gutierrez, A., Monaghan, D., Jiménez, M.J., O'Connor, N.E.: Persistent homology for 3D reconstruction evaluation. In: Ferri, M., Frosini, P., Landi, C., Cerri, A., Di Fabio, B. (eds.) CTIC 2012. LNCS, vol. 7309, pp. 30–38. Springer, Heidelberg (2012)

9. Gonzalez-Diaz, R., Real, P.: On the cohomology of 3D digital images. Discrete Appl. Math. **147**(2–3), 245–263 (2005)

10. Hatcher, A.: Algebraic Topology. Cambridge University Press, Cambridge (2002)

11. Iosifidis, A., Tefas, A., Pitas, I.: Multi-view action recognition based on action volumes, fuzzy distances and cluster discriminant analysis. Sig. Process. **93**, 1445–1457 (2013)

12. Kutulakos, K.N., Seitz, S.M.: A theory of shape by space carving. Intern. J. Comput. Vision. **38**, 199–218 (2000)

13. Lamar-León, J., García-Reyes, E.B., Gonzalez-Diaz, R.: Human gait identification using persistent homology. In: Alvarez, L., Mejail, M., Gomez, L., Jacobo, J. (eds.) CIARP 2012. LNCS, vol. 7441, pp. 244–251. Springer, Heidelberg (2012)

14. Leon, J.L., Cerri, A., Reyes, E.G., Diaz, R.G.: Gait-based gender classification using persistent homology. In: Ruiz-Shulcloper, J., Sanniti di Baja, G. (eds.) CIARP 2013, Part II. LNCS, vol. 8259, pp. 366–373. Springer, Heidelberg (2013)

15. Lamar-Leon, J., Baryolo, R.A., Garcia-Reyes, E., Gonzalez-Diaz, R.: Gait-based carried object detection using persistent homology. In: Bayro-Corrochano, E., Hancock, E. (eds.) CIARP 2014. LNCS, vol. 8827, pp. 836–843. Springer, Heidelberg (2014)

16. Monaghan, D., Kelly, P., O'Connor, N.E.: Quantifying human reconstruction accuracy for voxel carving in a sporting environment. In: ACM MM, Scottsdale, AZ, 28 November–1 December 2011

17. Monaghan, D., Kelly, P., O'Connor, N.E.: Dynamic voxel carving in tennis based on player localisation using a low cost camera network. In: 2011 IEEE International Conference on Image Processing (ICIP 2011), Brussels, Belgium, 11–14 September 2011

18. Seitz, S.M., Curless, B., Diebel, J., Scharstein, D., Szeliski, R.: A comparison and evaluation of multi-view stereo reconstruction algorithms. In: IEEE Conference on Computer Vision and Pattern Recognition, vol. 1, pp. 519–528 (2006)

19. Weinland, D., Ronfard, R., Boyer, E.: Free viewpoint action recognition using motion history volumes. Comput. Vis. Image Underst. **104**, 249–257 (2006)

20. Zomorodian, A., Carlsson, G.: Computing persistent homology. Discrete Comput. Geom. **33**(2), 249–274 (2005)

Robust Computations of Reeb Graphs in 2-D Binary Images

Antoine Vacavant[1]([⊠]) and Aurélie Leborgne[1,2]

[1] ISIT, Université d'Auvergne, UMR/CNRS/6284, BP10448,
63000 Clermont-Ferrand, France
antoine.vacavant@udamail.fr
[2] Université de Lyon, INSA-Lyon, LIRIS, UMR/CNRS/5205,
69621 Villeurbanne, France
aurelie.leborgne@liris.cnrs.fr
http://isit.u-clermont1.fr/~anvacava/index.html
http://liris.cnrs.fr/aurelie.leborgne

Abstract. In this article, we present a novel approach devoted to robustly compute the Reeb graph of a digital binary image, possibly altered by noise. We first employ a skeletonization algorithm, named DECS (Discrete Euclidean Connected Skeleton), to calculate a discrete structure centered within the object. By means of an iterative process, valid with respect to Morse theory, we finally obtain the Reeb graph of the input object. Our various experiments show that our methodology is capable of computing the Reeb graph of images with a high impact of noise, and is applicable in concrete contexts related to medical image analysis.

Keywords: Skeletonization · Reeb graph · Topology

1 Introduction

The Reeb graph [15] is a compact discrete structure representing the topology of a graphical object by associating edges to its branches and vertices to their junctions. This graph is calculated on any compact manifold in two or in three dimensions (2-D or 3-D respectively) thanks to the definition of a given function h, in the sense of Morse theory [8,14]. The critical points of this h function (extrema and saddle points) are related to the vertices of the graph. As a natural consequence, Reeb graphs have been extensively explored by optimizing algorithms for its construction, and especially for 3-D meshes [4,9,19]. For this construction, one of the keys is the definition of h. A classic viewpoint is to consider a one-directional function as Morse induced (also named *height function, e.g.* along one space axis X, Y or Z), but it could also be defined as a geodesic distance in 2-D or in 3-D [10,19]. The first option, illustrated in Fig. 1(a), can be justified since the topology of many objects may be represented along an axis (statues, animals, persons, *etc.*), but it is generally not sufficient

© Springer International Publishing Switzerland 2016
A. Bac and J.-L. Mari (Eds.): CTIC 2016, LNCS 9667, pp. 204–215, 2016.
DOI: 10.1007/978-3-319-39441-1_19

to model the topology of every kinds of complex objects, requiring geodesic-like functions (see Fig. 1(b)). More generally, in pattern recognition, the Reeb graph has been employed to model 2-D and 3-D shapes for many applications, *e.g.* object retrieval [2,20], character recognition in license plates [18], noisy contour vectorization [22], implicit curve tracing [21] and image segmentation [12,23].

Skeletons, medial axes and their extensions [1,3,5,13] are other digital structures capable of capturing some topological features of the processed shapes. As a consequence, a natural strategy is to compute a skeleton or a medial axis of an object to obtain its Reeb graph [10,16] and *vice-versa* [7]. However, in practice, these structures cannot be linked directly to Reeb graphs since they are very sensible to image noise. Generally, they produce extra branches or other unwanted data that do not belong to the Reeb graph, and the specific treatment of these artefacts is a difficult task. The closest and most recent related work (to the best of our knowlegde) following this strategy is presented in [10], wherein the authors use a classic skeletonization scheme and local binary patterns to obtain the Reeb graph of the input binary image.

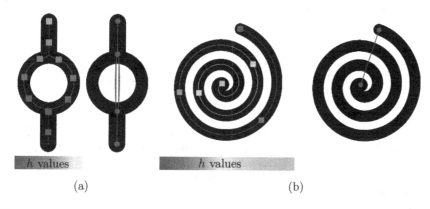

Fig. 1. From a skeleton computed in 2-D binary shapes, we can obtain a valid Reeb graph, by adopting h as a height function, along Y axis for instance, in (a), or by respecting shape's geometry with a geodesic distance (b). Skeleton edge pixels are colored depending on their h function values (see palette below), and colored squares represent the set of these pixels having the same value

Our paper focuses on the computation of the Reeb graph of 2-D binary shapes, by employing a robust skeletonization scheme [13] (Fig. 1). Thanks to an iterative process, we can build the Reeb graph of complex and possibly very noisy objects by respecting a given height function, but also by considering other functions (by following a centrifugal force for example). The reminder of our article is the following: in Sect. 2, we recall the robust skeletonization algorithm introduced in a previous work, so that we can obtain a valid Reeb graph in Sect. 3. We propose in Sect. 4 experiments showing the robustness of our approach, and its application in our context of medical image analysis.

2   Skeleton Extraction

Througout our article, we will use the following notations. From any image I, we denote a pixel by $\mathbf{p} \in \mathbb{Z}^2$ belonging to I with its X- and Y-coordinates x and y respectively. To access randomly a pixel in I, we also use the notation $I(x, y)$.

In this section, we remind the robust skeletonization algorithm introduced in [13] named Discrete Euclidean Connected Skeleton (DECS for short). Algorithm 1 summarizes the workflow of DECS, first calculating the sparse reduced discrete medial axis (RDMA) from [5] of the foreground object E in the input binary image I, as a part of the union of maximal balls:

$$E = \bigcup_{1 \leq k \leq K} B\big(\mathbf{p}_k, \delta(\mathbf{p}_k)\big), \text{where } B(\mathbf{p}, r) = \big\{\mathbf{q} \in \mathbb{Z}^2 \: : \: d_E(\mathbf{p}, \mathbf{q}) < r\big\}. \quad (1)$$

E represents the union of K balls $(\mathbf{p}_k, \delta(\mathbf{p}_k)) \in \mathbb{Z}^2 \times \mathbb{N}$, $\delta(\mathbf{p}_k)$ is the radius of the maximal ball centered in $\mathbf{p}_k$, and d_E is the classic Euclidean distance. These radii are obtained thanks to the computation of the Euclidean distance transform of I (EDT_I) by any algorithm of the literature [6]. The RDMA removes the balls which are not maximal in E, and is generally illustrated by the set of balls' centers $\{\mathbf{p}_k\}_{k=1,K}$.

Algorithm 1. DECS Algorithm [13].

    input   : A binary image I.
    output : The DECS of the foreground object in I.
1 **begin**
2   compute the Euclidean distance map EDT_I of I ;   {See [6]}
3   compute $RDMA_I$ the reduced discrete medial axis from EDT_I ; {From [5]}
4   compute Laplacian-of-Gaussian filtering of D_I as RDG_I ;
5   combine RDG_I and $RDMA_I$ to calculate a coarse skeleton S_I ;
6   thin and prune S_I to obtain S_I^* ;
7   **return** S_I^* ;
8 **end**

With the map EDT_I, we also define the ridgeness map RDG_I of I as

$$RDG_I(x, y) = \frac{1}{\pi \sigma^4} \left(1 - \frac{x^2 + y^2}{2\sigma^2}\right) \exp\left(-\frac{x^2 + y^2}{2\sigma^2}\right) \times EDT_I(x, y). \quad (2)$$

In [13], the authors suggest to set $\sigma = 1$. This map is indeed obtained by applying the Laplacian of Gaussian operator on EDT_I, and represents the ridges wherein main branches of the input object are located. A simple thresholding operation on RDG_I is not sufficient to obtain a valid skeleton, and a more relevant process has to be designed for this purpose.

In this way, by combining both $RDMA_I$ and RDG_I, the DECS algorithm then leads to a coarse and thick skeleton S_I of the image I, which is then pruned and thinned to obtain the final skeleton S_I^*. It should be noted that (1) the complexity of this algorithm has been proved to be linear with respect to the size of I in [13]; (2) the few parameters of this method can be set once for a wide range of images, always leading to a robust skeletonization. Every phases of DECS are illustrated in Fig. 2.

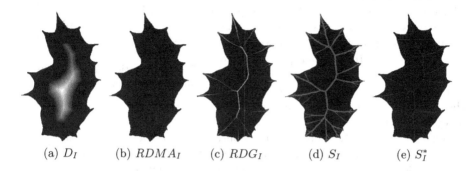

(a) D_I (b) $RDMA_I$ (c) RDG_I (d) S_I (e) S_I^*

Fig. 2. Example of the application of DECS algorithm on a sample binary image I (for notations, refer to Algorithm 1)

3   Reeb Graph Computation

3.1   Reeb Graph Definition

The goal of this section is to show that the calculation of a valid Reeb graph of any binary image I can be carried out using the robust skeleton obtained by DECS. We first need definitions related to topological spaces (Definition 1 below). Suppose we have an *equivalence relation* $\sim$ defined on a topological space M. Let $M_\sim$ be the set of equivalence classes and let $\psi : M \to M_\sim$ map each point $\mathbf{p}$ to its equivalence class (also called the *quotient map*).

Definition 1 (Quotient topology and space). *The* quotient topology *of* $M_\sim$ *consists of all subsets* $U \subseteq M_\sim$ *whose preimages,* $\psi^{-1}(U)$ *are open in* M*. The set* $M_\sim$ *together with the quotient topology is the* quotient space *defined by* $\sim$*.*

Then, we can present the definition of the Reeb graph in the continuous case as:

Definition 2 (Reeb graph). *Let* h *be a continuous function defined on a compact variety* M*,* $h : M \to \mathbb{R}$*. The* Reeb graph *of* M*, denoted by* $G(h)$*, is the quotient space defined by the equivalence relation* $\mathbf{p} \sim \mathbf{q} \Leftrightarrow (\mathbf{p}, h(\mathbf{p})) \sim (\mathbf{q}, h(\mathbf{q}))$ *such that:*

$$\begin{cases} h(\mathbf{p}) = h(\mathbf{q}), \\ \mathbf{p} \text{ and } \mathbf{q} \text{ belongs to the same connected component of } h^{-1}(h(\mathbf{p})). \end{cases} \quad (3)$$

From this definition, we can extract the following properties of Reeb graphs.

By construction, $G(h)$ is defined by considering the *level-sets* of the function h,
and by associating points belonging to the
same connected component (equivalence relation $\sim$), for every level-sets of h. If we consider
h as a height function along an axis, as in
Fig. 3, $G(h)$ is built by considering increasing
values of h along this axis. This shows that
Reeb graphs bring together topology (as it
is equivalent to a topological quotient space)
and geometry (by the expression of the function h over the shape of M). As a consequence,
the definition of h upon the geometry of M is
a key for the Reeb graph computation.

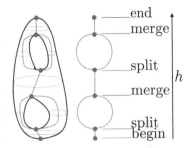

Fig. 3. Notations of $G(h)$ for an illustrative continuous shape

Once a h function is decided, the construction of the Reeb graph $G(h) = (V, E)$ is composed of edges in E associated to the shape's branches, *i.e.* the points belonging to the same connected component for any h value. In $G(h)$, vertices of V represent the critical points of the h function (see Fig. 3) as: *begin* for the minimal h values and *end* for maximal h values, both having a degree of one; *merge* and *split* for saddle values, with higher degrees. These points are defined according to the construction of edges by *merging* or *splitting* them respectively.

All previous notions hold in the discrete case. The consequence of our observations is that algorithms designed in the construction of $G(h)$ employ an iterative propagation process, througout a finite number of level-sets of h, calculating
vertices and edges of the graph of the input digital object. To obtain those elements, our strategy is to use the robust skeletonization process presented in the
previous section so that we compute a valid Reeb graph even for altered binary
images.

3.2 Robust Reeb Graph Computation with DECS

The construction of the Reeb graph based on the DECS is described in
Algorithm 2. Once a starting point p_S is selected, a breadth-first search for
vertices and edges of the Reeb graph is launched. During this process, the function h is calculated in a discrete way by attributing increasing values to scanned
points of the graph (line 14). Vertices are added in V by associating the correct
label, with respect to the h function's critical points, in line 10 (see also Fig. 3).

Proposition 1. *For any binary image I, Algorithm 2 computes the Reeb graph
of the foreground object $G_I(h)$.*

We propose to justify this proposition with two axes: (1) the DECS skeleton
is able to represent the medial topological branches of the input foreground
object; (2) the rest of Algorithm 2 actually calculates its Reeb graph with its
edges, vertices and the associated h function.

Algorithm 2. Reeb graph computation algorithm.

---

   input   : A binary image I.
   output : The Reeb graph $G_I(h)$ of the foreground object in I.
 1 **begin**
 2   compute the DECS S_I^* with Alg. 1 ;
 3     $G_I(h) = (V, E) \leftarrow \emptyset$;
 4   select a starting point $\mathbf{p}_S$ in S_I^* ;
 5     $h(\mathbf{p}_S) \leftarrow 0$;
 6     $Q \leftarrow \{\mathbf{p}_S\}$;
 7   **while** $Q \neq \emptyset$ **do**
 8        $\mathbf{p} \leftarrow top(Q)$;   {Breadth-first construction of $G_I(h)$}
 9   **if** $\mathbf{p}$ *is associated to a critical point of h* **then**
10   add $\mathbf{p}$ in V with correct label amongst $\{begin, end, merge, split\}$;
11   add in E edges connected to $\mathbf{p}$;
12   **foreach** *point $\mathbf{p}'$ adjacent to $\mathbf{p}$ in S_I^** **do**
13   **if** $\mathbf{p}'$ *is not treated* **then**
14              $h(\mathbf{p}') \leftarrow h(\mathbf{p}) + 1$;   {Increasing level-set}
15              $Q \leftarrow Q \cup \{\mathbf{p}'\}$;

16   **return** $G_I(h)$;
17 **end**

The medial axis $RDMA_I$, used in the DECS algorithm, is able to capture the topology of the foreground object in I [17]. As illustrated in Fig. 2, this is a sparse and disconnected representation, but it is also able to reconstruct the whole geometry of the input shape by calculating the union of maximal balls expressed in Eq. 1. This minimal set of the maximal balls included in the object is a relevant basis for the DECS algorithm, since this is an incomplete but sufficient representation of its branches, from a topological point of view. Conversely, the ridgeness map RDG_I (see Fig. 2 again) of the input image is a dense and complete model which smoothly locates main branches by ridges for each image pixel.

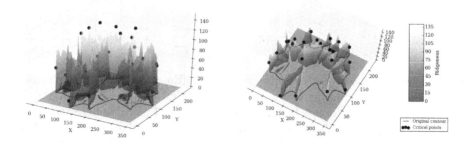

Fig. 4. Illustration of the location of Reeb graph vertices points with respect to RDG_I map

Thanks to the combination of $RDMA_I$ and RDG_I, the DECS skeleton leads to the complete branches, stored in the edges of the Reeb graph $G_I(h)$ and, at their junctions, the vertices of $G_I(h)$. Definition 2 shows that this graph co-exists with a function defined on the input compact variety. In our case, the values of this function are calculated during the breadth-first scanning of the object; the starting point of our process is associated to zero (and is a *begin* critical point), then scanned points are assigned with an increasing value. They can be saddle critical points of h (*merge, split*) or maximal points (*end*).

Figure 4 shows the location of the Reeb graph vertices (black circles) with respect to the RDG_I map (see Eq. 2 for its formulation) plotted as an elevation map, for each input image pixel. This figure illustrates that the Reeb graph vertices correspond to the highest RDG_I values, and so located on the most relevant branches of the input shape.

As a validation of our contribution, the next section is devoted to the experimental evaluation of our robust Reeb graph computation algorithm thanks to synthetic and real images.

4   Experimental Evaluation

We first propose to build the DECS skeleton and the Reeb graph of noisy images, generated from the synthetic examples given in Fig. 1. For this purpose, we use a noise generation model close to the one proposed by Kanungo *et al.* [11], iterated several times to increase its impact. This alters the contour of the input object by switching the values of pixels belonging to the foreground object border. Figure 5 groups the results of our algorithm with noise generated once, then 5, 10, 50 and 100 times on the same image. We can observe that the S_I^* skeleton obtained by the DECS algorithm is very robust, and enables to extract the Reeb graph of the foreground object, which stays stable despite of the increasing noise. For the simple one-hole object, we select the lowest skeleton point as the starting point of the Reeb graph construction; for the spiral object, this is the most external one. We can consider an h function along an axis (first case) or defined in a geodesic way to respect the geometry of the input shape (spiral object).

To numerically represent the topology of the processed objects, we can calculate the Euler number χ from the Reeb graph $G_I(h) = (V, E)$, thanks to this formulation [15]:

$$\chi = \sum_{n \in V, n=begin \lor n=end} deg(n) - \sum_{n \in V, n=split \lor n=merge} deg(n) - 2. \qquad (4)$$

For the first one-hole object, we always obtain an Euler number $\chi = 0$, meaning that it is homeomorpheous to a torus; for the spiral shape, we obtain $\chi = 2$, the same number as a point. We also remind that the Euler number can be also calculated as $\chi = 2 - 2 \times \#holes$, which confirms the values we obtain. This numerical analysis will be further useful for the comparison of our contribution with other computations of Reeb graphs based on skeletonization schemes.

In Fig. 6, we propose to test our algorithm in a concrete application about analysis of histological image of liver cells. In (a), we first enhance specific colors of the complex input image of size 1280×1024 pixels[1], to highlight then hepatocytes (cell centers) by applying a double thresolding operation. Then, we calculate the DECS of the image by considering the hepatocytes as background objects (and the rest obviously as foreground). By using this structure, we can then select different starting points to construct the Reeb graph (b). For example, we can select the most upper left point of the skeleton, meaning that the graph is built along a diagonal axis (from top-left to bottom-right). We can also choose the central DECS point, leading to a breadth-first scanning following a centrifugal direction. In this experiment, we have selected one starting point as in Algorithm 2, *i.e.* $G_I(h)$ contains only one *begin* node, and we could

(a) $n_{it} = 1$ \hspace{4cm} (b) $n_{it} = 5$

(c) $n_{it} = 10$ \hspace{3.5cm} (d) $n_{it} = 50$

(e) $n_{it} = 100$

Fig. 5. Extraction of DECS and Reeb graph for more and more noisy synthetic images, with an increasing number of iterations n_{it}. The value of h function is depicted with the palette used in Fig. 1

---

[1] Available at https://embryology.med.unsw.edu.au/embryology/index.php/ Histology.

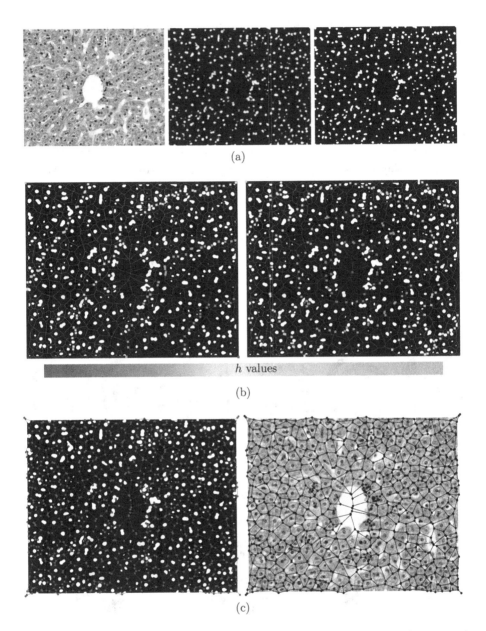

Fig. 6. Extraction of Reeb graph in a liver histological image, enhanced and segmented to extract hepatocytes (a). In (b), we show that the graph can be computed by considering a one-directional height function h, from top-left to bottom-right (left), or a centrifugal function (right). Graph edge pixels are colored depending on their h function values (see palette below), and colored squares represent the set of these pixels having the same value. (c): Obtained Reeb graph superimposed on the segmented and original images (color figure online)

select multiple starting points as an extension. In this case, we would have to handle several breadth-first constructions of separated graphs, and merge them into a single graph. We show in Fig. 6(c) the Reeb graph we obtain, whatever the starting point chosen, superimposed on the segmented image and on the original image.

By using Eq. 4 expressed earlier, we can determine the number of holes within the processed binary object thanks to the Euler number. In our case, we have obtained $\chi = -812$, leading to 407 holes in the image (and so 407 cell centers); this number can be verified with the original segmented image, depicted in Fig. 6(a), by counting the number of connected components. Besides this numerical evaluation, the Reeb graph brings obviously more information about the shape and organization of hepatocytes in the histological image. Indeed, the graph calculated as we propose would be of high interest to analyse further the structure of the liver, since homogeneous and regular cells arranged around the central vein imply that the liver is healthy, contrary to a cirrhotic one wherein cells have irregular shapes and are disrupted around the vein.

We finally propose in Fig. 7 to compute the Reeb graph of the liver vessels represented within a sample angiogram. The input image[2] is first segmented by considering a simple thresholding process based on angiogram intensities, similar to the one we used in the previous experiment. This binary object is then treated by our algorithm, to obtain the vascular structure with the Reeb graph in Fig. 7(b). The starting point of our process has been selected at the entrance of the vein (most right-bottom point in DECS), which permits to have the complete path of blood inside the liver, from this entrance to the finest veins.

Fig. 7. Reeb graph computation in an angiogram of liver vessels

5   Conclusion and Future Works

In this article, we have proposed a novel approach able to compute robustly the Reeb graph of a 2-D shape contained in a binary image. Our algorithm is capable

---

[2] From   http://health.siemens.com/ct_applications/somatomsessions/index.php/ minimally-invasive-treatment-of-hepatocellular-carcinoma-using-a-siemens-miyabi-system/.

of constructing this graph, by considering a relevant h function depending on the shape of the input object. We have shown the performance of our approach throughout two main experiments involving synthetic and real images. In the first case, we have confirmed that our contribution can compute the Reeb graph despite of a very strong contour-based noise model. We have then illustrated the application of our algorithm in a concrete context of medical image analysis.

A first future direction of our research concerns, still in the 2-D case, the validation of our approach in a medical context as shown in Sect. 4. We would like to test its performance on a large database of vascular images, requiring that we extract finely the vessel structures and their bifurcations. Moreover, we would like to compare our pipeline with other skeletonization algorithms, to ensure that DECS is the most robust way to obtain a valid Reeb graph. To do so, we can use a numerical evaluation by using the Euler number together with a structural comparison by using graph matching techniques as graph edit distance for example. Another important future work is to extend our approach to 3-D. As explained in Sect. 2, the DECS computation is based on several stages that can be adapted rather easily to higher dimensions. Then, the Reeb graph construction employs a breadth-first search-like strategy, which could be adapted to n-D. This work could also be highlighted in a medical context, to analyze the vessels in 3-D volumes acquired from CT-scans or MRI for example. Finally, we aim at designing matching algorithms for (2-D and then in higher dimensions) patterns or objects based on Reeb graphs.

References

1. Arcelli, C., di Baja, G.S., Serino, L.: Distance-driven skeletonization in voxel images. IEEE Trans. Pattern Anal. Mach. Intell. **33**(4), 709–720 (2010)
2. Barra, V., Biasotti, S.: 3D shape retrieval and classification using multiple kernel learning on extended Reeb graphs. Vis. Comput. Int. J. Comput. Graph. **30**(11), 1247–1259 (2014)
3. Bertrand, G., Couprie, M.: Powerful parallel and symmetric 3D thinning schemes based on critical kernels. J. Math. Imaging Vis. **48**(1), 134–148 (2014)
4. Biasotti, S., Giorgi, D., Spagnuolo, M., Falcidieno, B.: Reeb graphs for shape analysis and applications. Theor. Comput. Sci. **392**(1–3), 5–22 (2008)
5. Coeurjolly, D., Montanvert, A.: Optimal separable algorithms to compute the reverse Euclidean distance transformation and discrete medial axis in arbitrary dimension. IEEE Trans. Pattern Anal. Mach. Intell. **29**(3), 437–448 (2007)
6. Coeurjolly, D., Vacavant, A.: Separable distance transformation and its applications. In: Brimkov, V., Barneva, R. (eds.) Theoretical Foundations and Applications to Computational Imaging Digital Geometry Algorithms, vol. 2, pp. 189–214. Springer, Heidelberg (2012)
7. Ge, X., Safa, I.I., Belkin, M., Wang, Y.: Data skeletonization via reeb graphs. In: ShaweTaylor, J., Zemel, R.S., Bartlett, P., Pereira, F.C.N., Weinberger, K.Q., (eds.) Advances in Neural Information Processing Systems, vol. 24, pp. 837–845 (2011)
8. Gramain, A.: Topologie des surfaces. Presses Universitaires Françaises, Paris (1971)

9. Harvey, W., Wang, Y., Wenger, R.: A randomized O(mlogm) time algorithm for computing Reeb graphs of arbitrary simplicial complexes. In: Proceedings of Symposium on Computational Geometry (SCG), pp. 267–276 (2010)
10. Janusch, I., Kropatsch, W.G.: Reeb graphs through local binary patterns. In: Liu, C.-L., Luo, B., Kropatsch, W.G., Cheng, J. (eds.) GbRPR 2015. LNCS, vol. 9069, pp. 54–63. Springer, Heidelberg (2015)
11. Kanungo, T., Haralick, R., Baird, H., Stuezle, W., Madigan, D.: A statistical, nonparametric methodology for document degradation model validation. IEEE Trans. Pattern Anal. Mach. Intell. **22**(11), 1209–1223 (2000)
12. Karmakar, N., Biswas, A., Bhowmick, P.: Reeb graph based segmentation of articulated components of 3D digital objects. Theoret. Comput. Sci. **624**, 25–40 (2016)
13. Leborgne, A., Mille, J., Tougne, L.: Noise-resistant digital euclidean connected skeleton for graph-based shape matching. J. Vis. Commun. Image Represent. **31**, 165–176 (2015)
14. Morse, M.: The Calculus of Variations in the Large, vol. 18. American Mathematical Society Colloquium Publication, New York (1934)
15. Reeb, G.: Sur les points singuliers d'une forme de Pfaff complétement intégrable ou d'une fonction numérique. Comptes Rendus de l'Académie des Sciences, Paris **222**, 847–849 (1946)
16. Pascucci, V., Scorzelli, G., Bremer, P.T., Mascarenhas, A.: Robust on-line computation of reeb graphs: Simplicity and speed. ACM Trans. Graph. **26**(3), 1–9 (2007). Article number 58
17. Sherbrooke, E., Patrikalakis, N.M., Wolter, F.E.: Differential and topological properties of medial axis transforms. Graph. Models Image Process. **58**, 574–592 (1996)
18. Thome, N., Vacavant, A., Robinault, L., Miguet, S.: A cognitive and video-based approach for multinational license plate recognition. Mach. Vis. Appl. **22**(2), 389–407 (2011)
19. Tierny, J., Vandeborre, J.P., Daoudi, M.: Invariant high level reeb graphs of 3D polygonal meshes. In: Proceedings of IEEE International Symposium on 3D Data Processing, Visualization and Transmission (3DPVT 2006), pp. 105–112 (2006)
20. Tierny, J., Vandeborre, J.P., Daoudi, M.: Partial 3D shape retrieval by reeb pattern unfolding. Comput. Graph. Forum **28**(1), 41–55 (2009). Wiley
21. Vacavant, A., Coeurjolly, D., Tougne, L.: A framework for dynamic implicit curve approximation by an irregular discrete approach. Graph. Models **71**(3), 113–124 (2009)
22. Vacavant, A., Roussillon, T., Kerautret, B., Lachaud, J.O.: A combined multiscale/irregular algorithm for the vectorization of noisy digital contours. Comput. Vis. Image Underst. **117**(4), 438–450 (2013)
23. Werghi, N., Xiao, Y., Siebert, J.: A functional-based segmentation of human body scans in arbitrary postures. IEEE Trans. Syst. Man. Cybern. Part B Cybern. **36**(1), 153–165 (2006)

The Coherent Matching Distance
in 2D Persistent Homology

Andrea Cerri[1]([⊠]), Marc Ethier[2,3], and Patrizio Frosini[4]

[1] IMATI – CNR, Genova, Italy
andrea.cerri@ge.imati.cnr.it
[2] Faculté des Sciences, Université de Saint-Boniface, Winnipeg, MB, Canada
methier@ustboniface.ca
[3] Institute of Computer Science and Computational Mathematics,
Jagiellonian University, Kraków, Poland
[4] Dipartimento di Matematica, Università di Bologna, Bologna, Italy
patrizio.frosini@unibo.it

Abstract. Comparison between multidimensional persistent Betti numbers is often based on the multidimensional matching distance. While this metric is rather simple to define and compute by considering a suitable family of filtering functions associated with lines having a positive slope, it has two main drawbacks. First, it forgets the natural link between the homological properties of filtrations associated with lines that are close to each other. As a consequence, part of the interesting homological information is lost. Second, its intrinsically discontinuous definition makes it difficult to study its properties. In this paper we introduce a new matching distance for 2D persistent Betti numbers, called *coherent matching distance* and based on matchings that change coherently with the filtrations we take into account. Its definition is not trivial, as it must face the presence of monodromy in multidimensional persistence, i.e. the fact that different paths in the space parameterizing the above filtrations can induce different matchings between the associated persistent diagrams. In our paper we prove that the coherent 2D matching distance is well-defined and stable.

Keywords: Multidimensional matching distance · Multidimensional persistent betti numbers · Monodromy

1 Introduction

In the last twenty-five years the concept of *topological persistence* has become of common use in computational geometry and topological data analysis. It is

Work carried out under the auspices of INdAM-GNSAGA. M.E. has been partially supported by the Toposys project FP7-ICT-318493-STREP, as well as an ESF Short Visit grant under the Applied and Computational Algebraic Topology networking programme. A.C. is partially supported by the FP7 Integrated Project IQmulus, FP7-ICT-2011–318787, and the H2020 Project Gravitate, H2020 - REFLECTIVE - 7 - 2014 - 665155.

© Springer International Publishing Switzerland 2016
A. Bac and J.-L. Mari (Eds.): CTIC 2016, LNCS 9667, pp. 216–227, 2016.
DOI: 10.1007/978-3-319-39441-1_20

based on the idea that the most important properties of a filtered topological space are the ones that persist under large changes of the parameters defining the sublevel sets in the filtration. The concept of persistence revealed quite useful in extracting information from data that can be described by functions taking values in $\mathbb{R}^h$ and defined on a topological space (e.g., images or point clouds representing 3D-models, via a distance function).

The theory of topological persistence was initially developed for the case $h = 1$, but in recent years the interest in the case $h > 1$ has rapidly increased, leading to new theoretical developments and computational methods (cf., e.g., [1,4,5,12]). One of this methods is based on a reduction of the h-dimensional case to the 1-dimensional setting by using a suitable family of derived real-valued functions [2,3,6]. If $h = 2$, it consists of changing the 2D filtration given by a filtering function $f = (f_1, f_2) : M \to \mathbb{R}^2$ into the 1D filtrations associated with the real-valued functions $f^*_{a,b} : M \to \mathbb{R}$ defined as $f^*_{a,b}(x) := \min\{a, 1 - a\} \cdot \max\left\{ \frac{f_1(x)-b}{a}, \frac{f_2(x)+b}{1-a} \right\}$, for $a \in]0,1[$ and $b \in \mathbb{R}$. This approach allows for introducing a distance $D_{match}(\beta_f, \beta_g)$ between the persistent Betti numbers associated with f and g, which is defined as the supremum of the classical bottleneck distance between the persistent diagrams of $f^*_{a,b}$ and $g^*_{a,b}$, varying a and b.

While this method brings back the problem to the 1D case, it opens the way to new issues of interest. First of all, the distance D_{match} forgets the natural link between the homological properties of filtrations associated with pairs (a, b) that are close to each other. This fact implies that part of the homological information is lost. Second, its intrinsically discontinuous definition makes it difficult to study its properties.

As a possible answer to these observations, we introduce in this paper a new matching distance for 2D persistent Betti numbers, called *coherent matching distance* and based on the use of matchings that change continuously with respect to the filtrations we take into account. In order to state its definition, we have to manage the problem of monodromy, consisting of the fact that a loop in the space parameterizing the above filtrations can induce a transformation that changes a matching σ between the associated persistent diagrams into a matching $\tau \neq \sigma$ [7].

The paper is organized as follows. In Sect. 2, we recall the definitions of multidimensional persistent Betti number and multidimensional matching distance, together with the monodromy phenomenon in 2D persistent homology. In Sect. 3, we introduce the coherent 2D matching distance and prove that it is well-defined and stable.

2   Mathematical Setting

Let $f = (f_1, f_2)$ be a continuous map from a finitely triangulable topological space M to the real plane $\mathbb{R}^2$.

2.1   Persistent Betti Numbers

As a reference for multidimensional persistent Betti numbers we use [6]. According to the main topic of this paper, we will also stick to the notations and working assumptions adopted in [7]. In particular, we build on the strategy adopted in the latter to study certain instances of monodromy for multidimensional persistent Betti numbers. Roughly, the idea is to reduce the problem to the analysis of a collection of persistent Betti numbers associated with a real-valued function, and their compact representation in terms of *persistence diagrams*.

We use the following notations: Δ^+ is the open set $\{(u, v) \in \mathbb{R} \times \mathbb{R} : u < v\}$. Δ represents the diagonal $\{(u, v) \in \mathbb{R} \times \mathbb{R} : u = v\}$. We can extend Δ^+ with points at infinity of the kind (u, ∞), where $|u| < \infty$. Denote this set Δ^*. For a continuous function $\varphi : M \to \mathbb{R}$, and for any $n \in \mathbb{N}$, if $u < v$, the inclusion map of the sublevel set $M_u = \{x \in M : \varphi(x) \leq u\}$ into the sublevel set $M_v = \{x \in M : \varphi(x) \leq v\}$ induces a homomorphism from the nth homology group of M_u into the nth homology group of M_v. The image of this homomorphism is called the *nth persistent homology group of (M, φ) at (u,v)*, and is denoted by $H_n^{(u,v)}(M, \varphi)$. In other words, the group $H_n^{(u,v)}(M, \varphi)$ contains all and only the homology classes of n-cycles born before or at u and still alive at v. By assuming to work with coefficients in a field $\mathbb{K}$, we get that homology groups are vector spaces. Therefore, they can be completely described by their dimension, leading to the following definition [10].

Definition 1 (Persistent Betti Numbers). *The* persistent Betti numbers function *of φ, briefly PBN, is the function $\beta_\varphi : \Delta^+ \to \mathbb{N} \cup \{\infty\}$ defined by*

$$\beta_\varphi(u, v) = \dim H_n^{(u,v)}(M, \varphi).$$

Under the above requirements for M, it is possible to show that β_φ is finite for all $(u, v) \in \Delta^+$ [6]. Obviously, for each $n \in \mathbb{Z}$, we have different PBNs of φ (which might be denoted by $\beta_{\varphi,n}$, say), but for the sake of notational simplicity we omit adding any reference to n.

Following [6], we assume the use of Čech homology, and refer the reader to that paper for a detailed explanation about preferring this homology theory to others. For the present work, it is sufficient to recall that, with the use of Čech homology, the PBNs of a real-valued function can be completely described by the corresponding *persistence diagrams*. Formally, a persistence diagram can be defined via the notion of *multiplicity* [8,11]. Following the convention used for PBNs, any reference to n will be dropped in the sequel.

Definition 2 (Multiplicity). *The multiplicity $\mu_\varphi(u, v)$ of $(u, v) \in \Delta^+$ is the finite, non-negative number given by*

$$\min_{\substack{\varepsilon > 0 \\ u+\varepsilon < v-\varepsilon}} \beta_\varphi(u + \varepsilon, v - \varepsilon) - \beta_\varphi(u - \varepsilon, v - \varepsilon) - \beta_\varphi(u + \varepsilon, v + \varepsilon) + \beta_\varphi(u - \varepsilon, v + \varepsilon).$$

The multiplicity $\mu_\varphi(u, \infty)$ of (u, ∞) is the finite, non-negative number given by

$$\min_{\varepsilon > 0,\, u+\varepsilon < v} \beta_\varphi(u + \varepsilon, v) - \beta_\varphi(u - \varepsilon, v).$$

Definition 3 (Persistence Diagram). *The persistence diagram $Dgm(\varphi)$ is the multiset of all points $(u, v) \in \Delta^*$ such that $\mu_\varphi(u, v) > 0$, counted with their multiplicity, union the points of Δ, counted with infinite multiplicity.*

Each point $(u, v) \in \Delta^*$ with positive multiplicity will be called a *cornerpoint*. A cornerpoint (u, v) will be said to be a *proper cornerpoint* if $(u, v) \in \Delta^+$, and a *cornerpoint at infinity* if $(u, v) \in \Delta^* \setminus \Delta^+$.

Persistence diagrams are stable under the *bottleneck distance* (a.k.a. *matching distance*). Roughly, small changes in the considered function induce small changes in the position of the cornerpoints which are far from the diagonal in the associated persistence diagram, and possibly produce variations close to the diagonal [8,9]. A visual intuition of this fact is given in Fig. 1. Formally, we have the following definition:

Definition 4 (Bottleneck distance). *Let Dgm_1, Dgm_2 be two persistence diagrams. The bottleneck distance $d_B(Dgm_1, Dgm_2)$ is defined as*

$$d_B(Dgm_1, Dgm_2) = \min_\sigma \max_{X \in Dgm_1} d(X, \sigma(X)),$$

where σ varies among all the bijections between Dgm_1 and Dgm_2 and

$$d\left((u, v), (u', v')\right) = \min\left\{\max\left\{|u - u'|, |v - v'|\right\}, \max\left\{\frac{v - u}{2}, \frac{v' - u'}{2}\right\}\right\} \quad (1)$$

for every $(u, v), (u', v') \in \Delta^* \cup \Delta$.

In practice, the distance d defined in (1) compares the cost of moving a point X to a point Y with that of annihilating them by moving both X and Y onto Δ, and takes the most convenient. Therefore, $d(X, Y)$ can be considered a measure of the minimum cost of moving X to Y along two different paths. These remarks easily yield that d is actually a pseudo-distance, that is, a distance without the property $d(X, Y) = 0 \Rightarrow X = Y$.

The stability of persistence diagrams can then be formalized as follows [8,9]:

Theorem 1 (Stability Theorem). *Let $\varphi, \psi : M \to \mathbb{R}$ be two continuous functions. Then $d_B(Dgm(\varphi), Dgm(\psi)) \le \|\varphi - \psi\|_\infty$.*

2.2   2-Dimensional Setting

The definition of persistent Betti numbers can be easily extended to functions taking values in $\mathbb{R}^h$ [6]. It has been proved that, in this case, the information enclosed in the persistent Betti numbers is equivalent to that represented by the set of persistent Betti numbers associated with a certain family of real-valued functions. We discuss this for the specific case of the above function $f : M \to \mathbb{R}^2$, referring the reader to Fig. 2 for a pictorial representation.

Consider the pairs $(a, b) \in \,]0, 1[\, \times \mathbb{R}$. Any such pair identifies an oriented line $r_{a,b} \in \mathbb{R}^2$ of positive slope, parameterized by t with equation

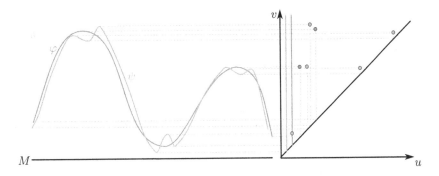

Fig. 1. Changing the function φ to ψ induces a change in the persistence diagram. In this example, the graphs on the left represent the real-valued functions φ and ψ, defined on the space M (a segment). The corresponding persistence diagrams (restricted to the 0th homology) are displayed on the right.

$(u, v) = t \cdot (a, 1 - a) + (b, -b)$. The space Λ of lines obtained according to this procedure is referred to as the *set of admissible lines*, whereas $P(\Lambda)$ denotes the set of pairs (a, b) parameterizing Λ. The generic point $(u, v) = t \cdot (a, 1 - a) + (b, -b)$ of $r_{a,b}$ can be associated with the sublevel set of M defined as $\{x \in M : f_1(x) \le u, f_2(x) \le v\}$, which is equivalent to that given by $\{x \in M : f_{a,b}(x) \le t\}$ induced by the real-valued function $f_{a,b} : M \to \mathbb{R}$ with $f_{a,b}(x) := \max\left\{\frac{f_1(x) - b}{a}, \frac{f_2(x) + b}{1 - a}\right\}$.

In this setting, the Reduction Theorem proved in [6] states that the persistent Betti numbers β_f can be completely recovered by considering all and only the persistent Betti numbers $\beta_{f_{a,b}}$ associated with the admissible lines $r_{a,b}$, which are in turn encoded in the corresponding persistence diagrams $\mathrm{Dgm}(f_{a,b})$.

2-Dimensional Matching Distance. Assume now that we have two continuous functions $f, g : M \to \mathbb{R}^2$. We consider the persistence diagrams $\mathrm{Dgm}(f_{a,b})$, $\mathrm{Dgm}(g_{a,b})$ associated with the admissible line $r_{a,b}$, and normalize them by multiplying their points by $\min\{a, 1 - a\}$. This is equivalent to consider the *normalized persistence* diagrams $\mathrm{Dgm}(f^*_{a,b})$, $\mathrm{Dgm}(g^*_{a,b})$, with $f^*_{a,b} = \min\{a, 1 - a\} \cdot f_{a,b}$ and $g^*_{a,b} = \min\{a, 1 - a\} \cdot g_{a,b}$, respectively. The 2-dimensional matching distance $D_{match}(\beta_f, \beta_g)$ [2] is then defined as

$$D_{match}(\beta_f, \beta_g) = \sup_{P(\Lambda)} d_B(\mathrm{Dgm}(f^*_{a,b}), \mathrm{Dgm}(g^*_{a,b})),$$

with $d_B(\mathrm{Dgm}(f^*_{a,b}), \mathrm{Dgm}(g^*_{a,b}))$ denoting the bottleneck distance between the normalized persistence diagrams $\mathrm{Dgm}(f^*_{a,b})$ and $\mathrm{Dgm}(g^*_{a,b})$.

Remark 1. The introduction of normalized persistence diagrams is crucial here. Indeed, the bottleneck distance $d_B(\mathrm{Dgm}(f^*_{a,b}), \mathrm{Dgm}(g^*_{a,b}))$ is stable against functions' perturbations when measured by the sup-norm, while this is not true for the distance $d_B(\mathrm{Dgm}(f_{a,b}), \mathrm{Dgm}(g_{a,b}))$, see [6, Theorem 4.2] for details.

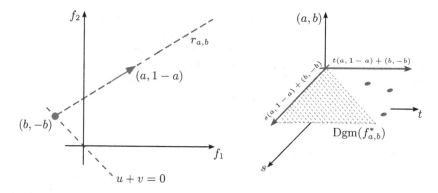

Fig. 2. Correspondence between an admissible line $r_{a,b}$ and the persistence diagram $\mathrm{Dgm}(f^*_{a,b})$. Left: a 1D filtration is constructed by sweeping the line $r_{a,b}$. The vector $(a, 1-a)$ and the point $(b, -b)$ are used to parameterize this line as $r_{a,b} = t \cdot (a, 1-a) + (b, -b)$. Right: the persistence diagram of the 1D filtration can be found on a planar section of the domain of the 2D persistent Betti numbers.

Monodromy in 2-Dimensional Persistent Homology. We know that normalized persistence diagrams are stable with respect to changes of the underlying functions, when the sup-norm is considered (Remark 1). Since each function $f^*_{a,b}$ depends continuously on the parameters a and b with respect to the sup-norm, it follows that the set of points in $\mathrm{Dgm}(f^*_{a,b})$ depends continuously on the parameters a and b. Analogously, the set of points in $\mathrm{Dgm}(g^*_{a,b})$ depends continuously on the parameters a and b. Suppose that $\sigma_{a,b} : \mathrm{Dgm}(f^*_{a,b}) \to \mathrm{Dgm}(g^*_{a,b})$ is an *optimal matching*, i.e. one of the matchings achieving the bottleneck distance $d_B(\mathrm{Dgm}(f^*_{a,b}), \mathrm{Dgm}(g^*_{a,b}))$. Given the above arguments, a natural question arises, whether $\sigma_{a,b}$ changes continuously varying a and b. In other words, we wonder if it is possible to straightforwardly introduce a notion of *coherence* for optimal matchings with respect to the elements of $P(\Lambda)$.

Perhaps surprisingly, the answer is no. A first obstruction is given by the fact that, trying to continuously extend a matching $\sigma_{a,b}$, the identity of points in the (normalized) persistent diagrams is not preserved when considering an admissible pair $(\bar{a}, \bar{b})$ for which either $\mathrm{Dgm}(f^*_{\bar{a},\bar{b}})$ or $\mathrm{Dgm}(g^*_{\bar{a},\bar{b}})$ has points with multiplicity greater than 1. In other words, we cannot follow the path of a cornerpoint when it collides with another cornerpoint. On the one hand, this problem can be solved by replacing $P(\Lambda)$ with its subset $\mathrm{Reg}(f) \cap \mathrm{Reg}(g)$, with

$$\mathrm{Reg}(f) = \{(a,b) \in P(\Lambda) | \mathrm{Dgm}(f^*_{a,b}) \text{ does not contain multiple points}\},$$
$$\mathrm{Reg}(g) = \{(a,b) \in P(\Lambda) | \mathrm{Dgm}(g^*_{a,b}) \text{ does not contain multiple points}\}.$$

Throughout the rest of the paper, we will talk about *singular pairs for f* to denote the pairs $(a,b) \in P(\Lambda) \setminus \mathrm{Reg}(f)$, and about *regular pairs for f* to denote the pairs $(a,b) \in \mathrm{Reg}(f)$. An analogous convention holds referring to the singular and regular pairs for g.

On the other hand, however, continuously extending a matching $\sigma_{a,b}$ presents some problems even in this setting. Roughly, the process of extending $\sigma_{a,b}$ along a path $c : [0,1] \to \mathrm{Reg}(f) \cap \mathrm{Reg}(g)$ depends on the homotopy class of c relative to its endpoints. This phenomenon is referred to as *monodromy in 2-dimensional persistent homology*, and has been studied for the first time in [7]. In what follows we will show how to overcome this issue in order to define a *coherent* modification of the standard 2-dimensional matching distance D_{match}.

2.3 Working Assumptions

To simplify the exposition, in what follows we state our results by assuming that M is homeomorphic to the m-sphere S^m, with $m \geq 2$. In particular, this implies that all normalized persistence diagrams $\mathrm{Dgm}(f_{a,b}^*)$, $\mathrm{Dgm}(g_{a,b}^*)$ contain a single cornerpoint at infinity in degree 0 and n, and no cornerpoint at infinity in the other homology degrees. In this way, the problem of continuously extending a matching can be restricted to considering only proper cornerpoints, as there are no ambiguities in following the evolution of cornerpoints at infinity. Also, we assume that

1. the functions $f, g : M \to \mathbb{R}^2$ are *normal*, i.e. the sets of singular pairs for f and g are discrete [7];
2. a constant real value $k > 0$ exists such that if two proper cornerpoints X_1, X_2 of $\mathrm{Dgm}^*(f_{a,b})$ have Euclidean distance less than k from Δ, then the Euclidean distance between X_1 and X_2 is not smaller than k, for all $(a,b) \in P(\Lambda)$. The same property holds for $\mathrm{Dgm}^*(g_{a,b})$.

3 The Coherent 2-Dimensional Matching Distance

The existence of monodromy implies that each loop in $\mathrm{Reg}(f)$ induces a permutation on $\mathrm{Dgm}(f_{a,b}^*)$. In other words, it is not possible to establish which point in $\mathrm{Dgm}(f_{a,b}^*)$ corresponds to which point in $\mathrm{Dgm}(f_{a',b'}^*)$ for $(a,b) \neq (a',b')$, since the answer depends on the path that is considered from (a,b) to (a',b') in the parameter space $\mathrm{Reg}(f)$. As a consequence, different paths going from (a,b) to (a',b') might produce different results while extending a matching $\sigma_{a,b}$. However, it is still possible to define a notion of coherent 2-dimensional matching distance.

3.1 Transporting a Matching Along a Path

First, we need to specify the concept of transporting a proper cornerpoint $X \in \mathrm{Dgm}(\varphi)$ along a homotopy $h(\tau, x) := (1-\tau) \cdot \varphi(x) + \tau \cdot \psi(x)$, with $\varphi, \psi : M \to \mathbb{R}$.

Definition 5 (Admissible path). *Let $p \in [0,1]$. A continuous path $P : [0,p] \to \Delta^+ \cup \Delta$ is said to be* admissible *for h at $\bar{p} \in [0,p]$ if the following hold:*

1. *$P(\tau) \in Dgm(h(\tau, \cdot))$ for $\tau \in [0,p]$;*
2. *$P([0,p]) \cap \Delta$ is finite;*

3. *if $P(\bar{p}) \in \Delta$ then there is no $p' \in]\bar{p}, p]$ such that $P([\bar{p}, p']) = \{P(\bar{p})\}$ and a continuous path $Q : [\bar{p}, p'] \to \Delta^+ \cup \Delta$ exists for which $Q(\tau) \in Dgm(h(\tau, \cdot))$ for $\tau \in [\bar{p}, p']$, $Q(\bar{p}) = P(\bar{p})$ and $Q([\bar{p}, p']) \neq \{P(\bar{p})\}$.*

P is called admissible for h if it is admissible for h at every point of its domain.

In other words, P is not admissible for a homotopy h if it "stops" at a point $P(\bar{p}) \in \Delta$ while it could "move on" in Δ^+. The set of all paths $P : [0, p] \to \Delta^+ \cup \Delta$ admissible for h is endowed with a partial order. For two paths $P_1 : [0, p_1] \to \Delta^+ \cup \Delta$, $P_2 : [0, p_2] \to \Delta^+ \cup \Delta$ admissible for h, we say that $P_1 \preceq P_2$ if $p_1 \leq p_2$ and $P_1(\tau) = P_2(\tau)$ for every $\tau \in [0, p_1]$.

In what follows, we focus on paths that are admissible for the homotopy induced on the function $f^*_{c(\tau)}$ by a continuous path $c : [0, 1] \to Reg(f)$. With a slight abuse of notation, we talk about paths admissible for c.

Proposition 1. *Let $c : [0, 1] \to Reg(f)$ be a continuous path with $c(0) = (a, b)$. For every proper cornerpoint $X \in Dgm(f^*_{a,b})$, a unique path $P : [0, 1] \to \Delta^+ \cup \Delta$ admissible for c exists, such that $P(0) = X$.*

Proof. For every real number $\alpha \geq 0$, consider the property

$$(*) \text{ a path } P_\alpha : [0, \alpha] \to \Delta^+ \cup \Delta \text{ admissible for c exists, with } P_\alpha(0) = X.$$

Define the set $A = \{\alpha \in [0, 1] : \text{ property } (*) \text{ holds}\}$. A is non-empty, since $0 \in A$. Set $\bar{\alpha} = \sup A$. We need to show that $\bar{\alpha} \in A$. First, let (α_n) be a non-decreasing sequence of numbers of A converging to $\bar{\alpha}$. Since $\alpha_n \in A$, for each n there is a path $P_n : [0, \alpha_n] \to \Delta^+ \cup \Delta$ admissible for c with $P_n(0) = X$. By the Hausdorff maximal principle, we can consider a maximal chain of paths P_n and define a function $P'_{\bar{\alpha}} : [0, \bar{\alpha}) \to \Delta^+ \cup \Delta$ by setting $P'_{\bar{\alpha}}(\tau) = P_n(\tau)$ for any P_n in the maximal chain whose domain contains τ.

In particular, the function $P'_{\bar{\alpha}}$ is such that $P'_{\bar{\alpha}}(0) = X$ and $P'_{\bar{\alpha}}(\tau) \in Dgm(f^*_{c(\tau)})$ for all $\tau \in [0, \bar{\alpha})$. However, to prove that $\bar{\alpha} \in A$ we still need to show that $P'_{\bar{\alpha}}$ can be continuously extended to the point $\bar{\alpha}$. The localization of cornerpoints [6, Proposition 3.8] implies that, possibly by extracting a convergent subsequence, we can assume that the limit $\lim_n P_n(\alpha_n) = \lim_n P'_{\bar{\alpha}}(\alpha_n)$ exists. By the 1-dimensional Stability Theorem 1, we have that $\lim_n P'_{\bar{\alpha}}(\alpha_n) \in Dgm(f^*_{c(\bar{\alpha})})$. Now, the function $P'_{\bar{\alpha}}$ can be extended to a path $P_{\bar{\alpha}} : [0, \bar{\alpha}] \to \Delta^+ \cup \Delta$ by setting $P_{\bar{\alpha}}(\bar{\alpha}) = \lim_n P'_{\bar{\alpha}}(\alpha_n)$. It is easy to check that $P_{\bar{\alpha}}$ is admissible for c, and hence $\bar{\alpha} \in A$.

Last, we prove by contradiction that $\bar{\alpha} = 1$. Suppose that $\bar{\alpha} < 1$. If $P_{\bar{\alpha}}(\bar{\alpha}) \notin \Delta$, again by the 1-dimensional Stability Theorem and the fact that $c(\bar{\alpha}) \in Reg(f)$, for any sufficiently small $\varepsilon > 0$ we could take a real number $\eta > 0$ such that there is exactly one proper cornerpoint $X'(\tau) \in Dgm(f^*_{c(\tau)})$ with $d(X'(\tau), P_{\bar{\alpha}}(\bar{\alpha})) \leq \varepsilon$ for every τ with $\bar{\alpha} \leq \tau \leq \bar{\alpha} + \eta$. By setting $P_{\bar{\alpha}}(\tau) = X'(\tau)$ for every such τ, we would get a continuous path that extends $P_{\bar{\alpha}}$ to the interval $[0, \bar{\alpha} + \eta)$. We could work similarly also in case $P_{\bar{\alpha}}(\bar{\alpha}) \in \Delta$. Indeed, our working assumption (2., Sect. 2.3) implies that arbitrarily close to $P_{\bar{\alpha}}(\bar{\alpha})$ we could find at most one proper cornerpoint $X'(\tau)$, for all τ with $\bar{\alpha} \leq \tau \leq \bar{\alpha} + \eta$ and η sufficiently small, to be used to extend $P_{\bar{\alpha}}$. If there is no such a proper cornerpoint, $P_{\bar{\alpha}}$ could be extended by setting $P_{\bar{\alpha}}(\tau) = P_{\bar{\alpha}}(\bar{\alpha})$ for the same values of τ. In any case, we would get a contradiction of our assumption that $\bar{\alpha} = \sup A$.

We now show that there is a unique path $P : [0,1] \to \Delta^+ \cup \Delta$ that is admissible for c and starts at X. Assume that another path $P' : [0,1] \to \Delta^+ \cup \Delta$ admissible for c exists, with $X = P(0) = P'(0)$. Denote by $\bar{\tau}$ the greatest value for which $P(\tau) = P'(\tau)$ for all $\tau \in [0, \bar{\tau}]$. Since P differs from P', $\bar{\tau} < 1$. By the 1-dimensional Stability Theorem, if $P(\bar{\tau}) \notin \Delta$ then $P(\bar{\tau})$ is a proper cornerpoint of $\mathrm{Dgm}(f^*_{c(\bar{\tau})})$ with multiplicity strictly greater than 1, against our assumption that $c(\tau) \in \mathrm{Reg}(f)$ for all $\tau \in [0,1]$. If $P(\bar{\tau}) \in \Delta$ then P and P' contradict the definition of admissible path for c, because of our working assumption (2., Sect. 2.3). Therefore, the path P must be unique.

We say that c *transports* X to $X' = P(1)$ *with respect to* f. Now, we need to define the concept of transporting a matching along a path $c : [0,1] \to \mathrm{Reg}(f) \cap \mathrm{Reg}(g)$ with $c(0) = (a, b)$. Suppose that $\sigma_{(a,b)}(X) = Y$. Let $\sigma_{a,b}$ be a matching between $\mathrm{Dgm}(f^*_{a,b})$ and $\mathrm{Dgm}(g^*_{a,b})$, with (a, b) an element of $\mathrm{Reg}(f) \cap \mathrm{Reg}(g)$. We can naturally associate to $\sigma_{a,b}$ a matching $\sigma_{c(1)} : \mathrm{Dgm}(f^*_{c(1)}) \to \mathrm{Dgm}(g^*_{c(1)})$. We set $\sigma_{c(1)}(X') = Y'$ if and only if c transports X to X' with respect to f and Y to Y' with respect to g. We also say that c *transports* $\sigma_{a,b}$ *to* $\sigma_{c(1)}$ *along* c *with respect to the pair (f,g)*.

Following the same line of proof of Proposition 1, we can also prove the following result.

Proposition 2. *Let* $G(s, x) := (1 - s) \cdot \varphi(x) + s \cdot \psi(x)$ *be a homotopy between* $\varphi, \psi : M \to \mathbb{R}$. *Then, for every* $X \in Dgm(\varphi)$ *of multiplicity 1 and every sufficiently small* $\varepsilon > 0$, *a unique path* $P : [0, \varepsilon] \to \Delta^+ \cup \Delta$ *exists, such that* $P(0) = X$ *and* P *is admissible for* G *at any* $\tau \in [0, \varepsilon]$.

We say that G *transports* X *to* X'.

We are now ready to introduce the coherent 2-dimensional matching distance.

Definition 6. *Choose a point* $(a, b) \in Reg(f) \cap Reg(g)$. *Let* Γ *be the set of all continuous paths* $c : [0,1] \to Reg(f) \cap Reg(g)$ *with* $c(0) = (a, b)$. *Let* S *be the set of all matchings* $\sigma : Dgm(f^*_{c(0)}) \to Dgm(g^*_{c(0)})$. *For every* $\sigma \in S$ *and every* $c \in \Gamma$, *the symbol* $T_c^{(f,g)}(\sigma)$ *will denote the matching obtained by transporting* σ *along* c *with respect to the pair* (f, g). *We define the* coherent 2-dimensional matching distance $CD_{match}(\beta_f, \beta_g)$ *as*

$$CD_{match}(\beta_f, \beta_g) = \max \left\{ \min_{\sigma \in S} \sup_{c \in \Gamma} cost \left(T_c^{(f,g)}(\sigma) \right), \gamma_\infty \right\}, \qquad (2)$$

with

- $cost \left(T_c^{(f,g)}(\sigma) \right)$ *the cost of the matching* $T_c^{(f,g)}(\sigma)$ *with respect to the max-norm;*

- γ_∞ the maximum distance between the cornerpoint at infinity of $f^*_{a,b}$ and the cornerpoint at infinity of $g^*_{a,b}$ varying (a,b) in $P(\Lambda)$ for degrees 0 and m, and 0 for the other degrees.

The following statements hold, proving that CD_{match} is well-defined and satisfies the properties of a pseudo-distance.

Proposition 3. *The definition of $CD_{match}(\beta_f, \beta_g)$ does not depend on the choice of the point $(a,b) \in Reg(f) \cap Reg(g)$.*

Proof. Let us choose another basepoint $(a',b') \in Reg(f) \cap Reg(g)$. We can take a path $c' \in \Gamma$ with $c'(1) = (a',b')$. It is sufficient to observe that $T_c^{(f,g)}(\sigma) = T_{c*c'^{-1}}^{(f,g)}\left(T_{c'}^{(f,g)}(\sigma)\right)$ for every $\sigma \in S$ and every $c \in \Gamma$, where $*$ denotes the concatenation of paths and c'^{-1} is the inverse path of c', i.e., $c'^{-1}(t) := c'(1-t)$.

Proposition 4. *$CD_{match}(\beta_f, \beta_g)$ is a pseudo-distance.*

Proof. It is sufficient to observe that if two matchings $\sigma : \mathrm{Dgm}(f^*_{a,b}) \to \mathrm{Dgm}(g^*_{a,b})$, $\tau : \mathrm{Dgm}(g^*_{a,b}) \to \mathrm{Dgm}(h^*_{a,b})$ are given, then $T_c^{(f,h)}(\tau \circ \sigma) = T_c^{(g,h)}(\tau) \circ T_c^{(f,g)}(\sigma)$ for every $c \in \Gamma$ taking values in $Reg(f) \cap Reg(g) \cap Reg(h)$. This implies that $\mathrm{cost}\left(T_c^{(f,h)}(\tau \circ \sigma)\right) \leq \mathrm{cost}\left(T_c^{(g,h)}(\tau)\right) + \mathrm{cost}\left(T_c^{(f,g)}(\sigma)\right)$. Hence the triangle inequality follows.

The next result shows the stability of the coherent 2-dimensional matching distance.

Theorem 2. *It holds that $CD_{match}(\beta_f, \beta_g) \leq \|f - g\|_\infty$.*

Proof. First of all we recall that for every $(a,b) \in P(\Lambda)$ and every $x \in M$, we have

$$|f^*_{a,b}(x) - g^*_{a,b}(x)| = \min\{a, 1-a\} \cdot |f_{a,b}(x) - g_{a,b}(x)| \leq$$
$$\min\{a, 1-a\} \cdot \max\left\{\left|\frac{f_1(x) - g_1(x)}{a}\right|, \left|\frac{f_2(x) - g_2(x)}{1-a}\right|\right\} \leq$$
$$\max\{|f_1(x) - g_1(x)|, |f_2(x) - g_2(x)|\},$$

in other words, $\|f^*_{a,b} - g^*_{a,b}\|_\infty \leq \|f - g\|_\infty$. Let us consider the closed set C obtained from $P(\Lambda)$ by taking away a union of small balls of radius δ around the singular pairs for f. Let ε be the infimum for $(a,b) \in C$ of the minimum distance between any two proper cornerpoints of $\mathrm{Dgm}(f^*_{a,b})$. Note that, setting $K = \max_{x \in M}\{\|f(p)\|_\infty, \|g(p)\|_\infty\}$, the computation of ε can be accomplished on the compact set $C' = \{(a,b) \in C : |b| \leq K\}$. Indeed, for all $(a,b) \in C$ with $|b| > K$, we have that either $\mathrm{Dgm}(f^*_{a,b}) = \mathrm{Dgm}(f^*_1)$ or $\mathrm{Dgm}(f^*_{a,b}) = \mathrm{Dgm}(f^*_2)$ (analogously for g). Hence, by recalling our working assumption (2., Sect. 2.3) and by construction of C', we have that $\varepsilon > 0$.

Let ε' be the infimum for $(a,b) \in C$ of the minimum distance between any two points of $\mathrm{Dgm}(g^*_{a,b})$. From the fact that $\|f^*_{a,b} - g^*_{a,b}\|_\infty \leq \|f - g\|_\infty$, we know that when we change f according to the homotopy $G_s := (1 - s) \cdot f + s \cdot g$, each point in the normalized persistence diagram moves by a distance not greater than $\|f - G_s\|_\infty \leq \|f - g\|_\infty$, because of the stability of 2-dimensional persistent Betti numbers under the multidimensional matching distance [6, Thm. 4.2]. Hence we have that $\varepsilon' \geq \varepsilon - 2\|f - G_s\|_\infty \geq \varepsilon - 2\|f - g\|_\infty$. Therefore, if $\|f - g\|_\infty$ is small enough, we have that $\varepsilon > 0$ implies $\varepsilon' > 0$.

As a consequence, if $\|f - g\|_\infty$ is small enough, for each $(a, b) \in C$ we can consider the unique matching $\sigma_{a,b} : \mathrm{Dgm}(f^*_{a,b}) \to \mathrm{Dgm}(g^*_{a,b})$ obtained by changing the identity $id_{a,b} : \mathrm{Dgm}(f^*_{a,b}) \to \mathrm{Dgm}(f^*_{a,b})$ according to the change of persistence diagrams induced by the homotopy $G^*_s := (1 - s) \cdot f^*_{a,b} + s \cdot g^*_{a,b}$ (see Proposition 2). Formally, for each proper cornerpoint X of $\mathrm{Dgm}(f^*_{a,b})$, we set $\sigma_{a,b}(X) = X'$ if and only if the homotopy G^*_s transports X to the cornerpoint X' of $\mathrm{Dgm}(g^*_{a,b})$.

We have that $\mathrm{cost}(\sigma_{a,b}) \leq \|f^*_{a,b} - g^*_{a,b}\|_\infty \leq \|f - g\|_\infty$. If $\hat{c}$ is a continuous path from $(\bar{a}, \bar{b})$ to (a, b) with $\hat{c}([0, 1]) \subset C$, it is easy to see that the function $\sigma_{\hat{c}(t)}$ in the variable t describes the transport of $\sigma_{\bar{a}, \bar{b}}$ to $\sigma_{a,b}$ made by $\hat{c}$ with respect to the pair (f, g), because $\sigma_{\hat{c}(t)}$ depends continuously on t.

Now, let c be a continuous path from $(\bar{a}, \bar{b})$ to (a, b) with $c([0, 1]) \subset \mathrm{Reg}(f) \cap \mathrm{Reg}(g)$. The image of c may be not contained in C, but if $(\bar{a}, \bar{b}), (a, b) \in C$, c is relative homotopic to a path $\hat{c} : [0, 1] \to C$ from $(\bar{a}, \bar{b})$ to (a, b). This path $\hat{c}$ transports $\sigma_{\bar{a}, \bar{b}}$ to $\sigma_{a,b}$, too.

Hence we have that $\mathrm{cost}\left(T^{(f,g)}_c(\sigma_{\bar{a}, \bar{b}})\right) = \mathrm{cost}\left(T^{(f,g)}_{\hat{c}}(\sigma_{\bar{a}, \bar{b}})\right) = \mathrm{cost}(\sigma_{a,b}) \leq \|f - g\|_\infty$. From this and from the fact that we can choose an arbitrarily small δ, it follows that $\min_{\sigma \in S} \sup_{c \in \Gamma} \mathrm{cost}\left(T^{(f,g)}_c(\sigma_{\bar{a}, \bar{b}})\right) \leq \|f - g\|_\infty$ for every $c \in \Gamma$.

We conclude by observing that the definition of the coherent 2-dimensional matching distance immediately implies that it is not less informative than the usual 2-dimensional matching distance. Formally,

Proposition 5. $D_{match}(\beta_f, \beta_g) \leq CD_{match}(\beta_f, \beta_g)$.

4   Conclusions

In this contribution we have introduced the notion of coherent 2-dimensional matching distance. Similarly to the classical 2-dimensional matching distance, it is based on defining a suitable family of filtrations associated with lines having a positive slope; however, this new distance only considers matchings that change coherently with respect to the filtrations which are taken into account.

Through the paper we formally proved that the coherent matching distance between 2-dimensional Persistent Betti numbers is well-defined, stable and does not loose discriminative information with respect to the usual 2-dimensional matching distance. We believe that these first results make the coherent matching distance deserving further investigation, such as extending the study of its theoretical properties as well as exploring how to develop computational techniques for its practical evaluation.

References

1. Biasotti, S., Cerri, A., Frosini, P., Giorgi, D.: A new algorithm for computing the 2-dimensional matching distance between size functions. Pattern Recogn. Lett. **32**(14), 1735–1746 (2011)
2. Biasotti, S., Cerri, A., Frosini, P., Giorgi, D., Landi, C.: Multidimensional size functions for shape comparison. J. Math. Imaging Vis. **32**(2), 161–179 (2008)
3. Cagliari, F., Di Fabio, B., Ferri, M.: One-dimensional reduction of multidimensional persistent homology. Proc. Am. Math. Soc. **138**(8), 3003–3017 (2010)
4. Carlsson, G., Singh, G., Zomorodian, A.J.: Computing multidimensional persistence. J. Comput. Geom. **1**(1), 72–100 (2010)
5. Carlsson, G., Zomorodian, A.: The theory of multidimensional persistence. Discr. Comput. Geom. **42**(1), 71–93 (2009)
6. Cerri, A., Di Fabio, B., Ferri, M., Frosini, P., Landi, C.: Betti numbers in multidimensional persistent homology are stable functions. Math. Methods Appl. Sci. **36**(12), 1543–1557 (2013)
7. Cerri, A., Ethier, M., Frosini, P.: A study of monodromy in the computation of multidimensional persistence. In: Gonzalez-Diaz, R., Jimenez, M.-J., Medrano, B. (eds.) DGCI 2013. LNCS, vol. 7749, pp. 192–202. Springer, Heidelberg (2013)
8. Cohen-Steiner, D., Edelsbrunner, H., Harer, J.: Stability of persistence diagrams. Discrete Comput. Geom. **37**(1), 103–120 (2007)
9. d'Amico, M., Frosini, P., Landi, C.: Natural pseudo-distance and optimal matching between reduced size functions. Acta Applicandae Math. **109**(2), 527–554 (2010)
10. Edelsbrunner, H., Letscher, D., Zomorodian, A.J.: Topological persistence and simplification. Discrete Comput. Geom. **28**(4), 511–533 (2002)
11. Frosini, P., Landi, C.: Size functions and formal series. Appl. Algebra Eng. Commun. Comput. **12**(4), 327–349 (2001)
12. Lesnick, M.: The theory of the interleaving distance on multidimensional persistence modules. Found. Comput. Math. **15**(3), 613–650 (2015)

Persistent Homology on Grassmann Manifolds for Analysis of Hyperspectral Movies

Sofya Chepushtanova[1]([⊠]), Michael Kirby[2], Chris Peterson[2], and Lori Ziegelmeier[3]

[1] Wilkes University, Wilkes-Barre, PA, USA
`sofya.chepushtanova@wilkes.edu`
[2] Colorado State University, Fort Collins, CO, USA
`{kirby,peterson}@math.colostate.edu`
[3] Macalester College, Saint Paul, MN, USA
`lziegel1@macalester.edu`

Abstract. The existence of characteristic structure, or shape, in complex data sets has been recognized as increasingly important for mathematical data analysis. This realization has motivated the development of new tools such as persistent homology for exploring topological invariants, or features, in large data sets. In this paper, we apply persistent homology to the characterization of gas plumes in time dependent sequences of hyperspectral cubes, *i.e.* the analysis of 4-way arrays. We investigate hyperspectral movies of Long-Wavelength Infrared data monitoring an experimental release of chemical simulant into the air. Our approach models regions of interest within the hyperspectral data cubes as points on the real Grassmann manifold $G(k, n)$ (whose points parameterize the k-dimensional subspaces of $\mathbb{R}^n$), contrasting our approach with the more standard framework in Euclidean space. An advantage of this approach is that it allows a sequence of time slices in a hyperspectral movie to be collapsed to a sequence of points in such a way that some of the key structure within and between the slices is encoded by the points on the Grassmann manifold. This motivates the search for topological features, associated with the evolution of the frames of a hyperspectral movie, within the corresponding points on the Grassmann manifold. The proposed mathematical model affords the processing of large data sets while retaining valuable discriminatory information. In this paper, we discuss how embedding our data in the Grassmann manifold, together with topological data analysis, captures dynamical events that occur as the chemical plume is released and evolves.

Keywords: Grassmann manifold · Persistent homology · Hyperspectral imagery · Signal detection · Topological data analysis

1 Introduction

Hyperspectral imaging (HSI) technology allows the acquisition of information across the electromagnetic spectrum that is invisible to humans. In a very real

© Springer International Publishing Switzerland 2016
A. Bac and J.-L. Mari (Eds.): CTIC 2016, LNCS 9667, pp. 228–239, 2016.
DOI: 10.1007/978-3-319-39441-1_21

sense, these cameras allow us to "see the unseen" by including wavelengths spanning ultraviolet and far infrared. In contrast, humans can observe a very limited range of the electromagnetic spectrum, *i.e.* wavelengths of approximately 400–700 nm are visible to the human eye.

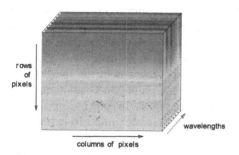

Fig. 1. Illustration of one frame, or data cube, of a hyperspectral movie collected with the Fabry-Pérot Interferometer.

Multi- and hyper-spectral imaging technology has become widely available, and there is an increasing number of canonical data sets available for scientific analysis including, *e.g.* the AVIRIS Indian Pines[1] and the ROSIS University of Pavia[2] data sets. In addition, moving objects may be detected with devices such as the Fabry-Pérot Interferometer [10] which can capture *hyperspectral movies* at frame rates of approximately 5 Hz. See Fig. 1 for an illustration. The resulting 4-way arrays of spatial-spectral-temporal data provide a high fidelity view of our environment and may help in the monitoring of pollution in the air and water. An application that concerns us in this paper is the characterization of gaseous plumes as they are released into the environment.

Traditionally, one of the primary applications of hyperspectral image analysis consists of object detection and classification. The focus is generally on the identification of anomalous pixels in the image and the determination of the composition of the materials in the pixel. A range of mathematical tools have been developed for the analysis of hyperspectral images including, *e.g.* matched subspaces, the RX algorithm, and the adaptive cosine estimator [19]. More recently, manifold learning algorithms have been applied to hyperspectral images to exploit topology and geometry, *i.e.* mathematical shape, or signatures, in data at the pixel level [1,18].

The subspace perspective is also taken in this paper, but in the direction of understanding the topology and geometry of the Grassmann manifold (Grassmannian) associated with hyperspectral images, *i.e.* the manifold parameterizing the k-dimensional subspaces of n-dimensional space. While we are motivated by ideas similar to those found in prior applications of manifold learning algorithms, *e.g.* [1,18], our application data is not at the pixel level. By constructing

---

[1] Available from https://engineering.purdue.edu/~biehl/MultiSpec.

[2] Available from http://www.ehu.es/ccwintco/index.php/.

subspaces of pixels we are able to exploit the rich metric structure of the Grass-mannian based on measuring angles between subspaces. The advantage of this approach is that a set of pixels used to form a subspace is seen to capture the variability in the data missing in a single pixel observation.

An example that illustrates the power of this framework is the application to illumination spaces in the face recognition problem. The variation in illumination on an object may be approximated by a cone captured in a low-dimensional subspace. Subspace angles can be used to compute similarity of illumination spaces and the effect on classification accuracy was striking when applied to the CMU-PIE data set, even on ultra-low resolution images [4]. More recently, tools have been developed to represent points on Grassmannians via subspace means [20], or nested flags of subspaces [12]. In another application to video sequence data, we used the setting of the Grassmannian to extend an algorithm on vector spaces for detection of anomalous activities [25].

In this paper, we address the question of the existence of topological sig-natures in the setting of hyperspectral movies mapped to the Grassmannian. Our approach builds on applying the Grassmannian architecture to hyperspec-tral movies that has shown promise in preliminary work [6,7]. Here, our focus is on application of persistent homology (PH) to the characterization of the evo-lution of chemical plumes as acquired by hyperspectral movie data sets. As in the application to face recognition, we encode a single frame of a hyperspectral movie (or a collection of pixels of a single frame in the movie) as a point on the Grassmann manifold. We speculate that this manifold representation affords a form of compression of information while capturing essential topological struc-ture. We consider the application of this approach to the characterization of chemical signals as measured by the Long-Wavelength Infrared (LWIR) data set [10]. Our goal is to establish the existence of topological signatures that can provide insight into the evolution of complex 4-way data arrays.

The paper outline includes an overview of PH in Sect. 2 and the geometry of the Grassmannian in Sect. 3. Computational experiments are discussed in Sect. 4 and conclusions are given in Sect. 5.

2   Persistent Homology

Homology is an invariant that measures features of a topological space and can be used to distinguish distinct spaces from one another [16]. Persistent homology encodes a parameterized family of these homological features. It is a computa-tional approach to topology that allows one to answer basic questions about the structure of point clouds in data sets at multiple scales [3]. This proce-dure involves (1) interpreting a point cloud as a noisy sampling of a topological space, (2) creating a global object by forming connections between proximate points based on a scale parameter, (3) determining the topological structure made by these connections, and (4) looking for structures that persist across different scales. PH has been used to understand the topological structure of data arising from applications including [8,11,17,21,22,24]. For a detailed discussion of homology, see [16], and for further discussions of persistent homology, see [3,14,15].

One way to associate a family of topological objects with a point cloud is to use the points to construct a family of nested simplicial complexes. The *Vietoris-Rips* complex builds a simplicial complex S_ϵ with vertices as the data points and higher dimensional k-simplices formed whenever $k+1$ points have pairwise distances less than ϵ. The k-dimensional holes of this simplicial complex generate a homology group $H_k(S_\epsilon)$ whose rank, known as the k-th Betti number, counts the number of k-dimensional holes. For instance, $Betti_0$ measures the number of connected components (clusters) of the point cloud, while $Betti_1$ indicates the existence of topological circles (loops), or periodic phenomenon. To avoid picking a specific scale ϵ, persistent homology seeks structures that persist over a range of scales, exploiting the fact that as ϵ grows, so do the simplicial complexes $S_{\epsilon_1} \subseteq S_{\epsilon_2} \subseteq S_{\epsilon_3} \subseteq \dots$ indexed by the parameters $\epsilon_1 \leq \epsilon_2 \leq \epsilon_3 \leq \dots$. Thus, PH tracks homology classes of the point cloud along the scale parameter, indicating at which ϵ a kth order hole appears and for which range of ϵ values it persists. The Betti numbers, as functions of the scale ϵ, can be visualized in a distinct *barcode* for each dimension k [15].

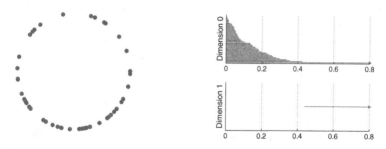

Fig. 2. $Betti_0$ (top right) and $Betti_1$ (bottom right) barcodes corresponding to point cloud data sampled from the unit circle (left).

Figure 2 is an example of the $k = 0$ and $k = 1$ barcodes generated for a point cloud sampled from a circle. Each horizontal bar begins at the scale where a topological feature first appears and ends at the scale where the feature no longer remains. The kth Betti number at any given parameter value ϵ is the number of bars that intersect the vertical line through ϵ. Short-lived features are often considered as noise while those features persisting over a large range of scale represent true topological characteristics. In the case of $Betti_0$, at small values of ϵ there will be a distinct bar for each point, as the simplicial complex S_ϵ consists of isolated vertices. At large values of ϵ, only one bar remains, as all data will eventually connect into a single component. For the circle, $Betti_0 = Betti_1 = 1$ which correspond to the number of connected components and number of loops, respectively, shown by the longest (persistent) horizontal bars in each plot. We use JavaPlex, a library for computing PH and TDA in this paper [23].

3   The Geometry of the Grassmann Manifold

The (real) Grassmann manifold $G(k, n)$ is a parameterization of all k-dimensional subspaces of n-dimensional space [13]. A point on $G(k, n)$ can be represented by a tall $n \times k$ matrix Y with the property that $Y^T Y = I_k$ where Y is an element of the equivalence class $\lfloor Y \rfloor$ consisting of all matrices of the form YQ with $Q \in O(k)$, the orthogonal group that consists of $k \times k$ orthogonal matrices [13].

Hyperspectral data is a 3-way cube $x \times y \times \lambda$ that can be mapped to points in a Grassmannian in a variety of ways. Here, we select a subset of k frequencies λ_i. For each of the k frequencies we propose to "vectorize" the $xy = n$ spatial components to form an $n \times k$ matrix X. It is assumed that the construction is such that $k < n/2 - 1$ so subspaces don't overlap trivially. To map X to a matrix Y representing a point on the Grassmann we compute any orthogonal basis for the column space of X. For instance, the $n \times k$ matrix U in the *thin* singular value decomposition $X = U \Sigma V^T$ provides one option as a representation of a point on the Grassmanian $G(k, n)$.

The mapping described above allows us to construct a sequence of points on $G(k, n)$, each one taken from the same spatial location in the 3-way array of hyperspectral pixels or from the same frame of a hyperspectral movie. The pairwise distances between the points in this sequence are computed in terms of the principal angles between the subspaces. The implementation of the Grassmannian framework is, in part, motivated by the rich metric structure of a variety of distance measures including the chordal, geodesic, and Fubini-Study distances, which are all functions of the k principal angles between the subspaces [2,9].

The experiments in this paper use the (pseudo)distance between two subspaces measured by the smallest principal angle. This (pseudo)distance has been effective in other numerical experiments [4,6], and in fact, we observed, in the experiments in this paper, that using it resulted in stronger topological signals than other distance measures. Once a distance matrix for the points on the Grassmannian is computed, we apply PH to determine topological structure. In particular, we explore $Betti_0$ barcodes to estimate the number of connected components and $Betti_1$ barcodes to detect topological circles. The goal is to associate physical properties in the HSI image that relate to these structures.

4   Experimental Results

In this section, we apply PH to Long-Wavelength Infrared (LWIR) multispectral movies, each of an explosive release of a chemical and resulting toxic plume which travels across the horizon of the scene [10]. The simulants released included Triethyl Phosphate (TEP) and Methyl Salicylate (MeS) in quantities of 75 kg and 150 kg, respectively. The LWIR data sets are captured using an interferometer in the 8–11 μm range of the electromagnetic spectrum. A single *frame*, or data cube, of this movie consists of 256×256 pixels collected at 20 IR bands. A given movie is a sequence of data cubes consisting of *pre-burst* and *post-burst* frames.

The purpose of this paper is not to propose a new algorithm for detecting chemical plumes but rather to investigate the topological features associated

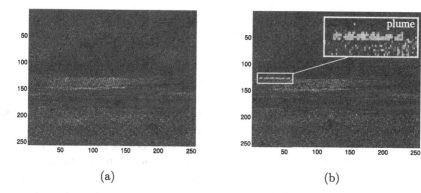

Fig. 3. The ACE detector on LWIR data cubes: (a) ACE values of a pre-burst cube indicating that no plume is detected; (b) ACE values of a post-burst cube with a plume detected. We have magnified the plume region to illustrate the performance of the ACE detector.

with a known plume. The data processing workflow consists of the following steps: (1) band selection, (2) identification of the region containing the chemical plume, (3) mapping data to the Grassmannian, (4) computing (pseudo)distances on the Grassmannian using the smallest principal angle, (5) determination of PH $Betti_0$ and $Betti_1$ barcodes, and, finally, (6) interpretation of the structure in the data as encoded by the topological invariants. We describe more detail of steps (1) and (2) below.

Band Selection. We applied the sparse support vector machine (SSVM) algorithm for optimal *in situ* band selection, *i.e.* the SSVM identifies wavelengths that best discriminate the plume from the natural background [5]. In another approach, we visually choose bands which have the strongest plume signal in data cubes which have had the background removed and thus, have visible plume.

Plume detection. The location of the chemical plume in the post-burst cubes is determined using the well-known adaptive-cosine-estimator (ACE) [19]. The ACE detector is one of the benchmark hyperspectral detection algorithms which computes the squared cosine of the angle between the whitened test pixel and the whitened target's spectral signature. Based on a chosen threshold, an ACE score indicates if the chemical is present in the test pixel. Figure 3a depicts an image corresponding to a cube without a plume, and Fig. 3b depicts a cube with a chemical plume detected by the ACE.

4.1 Experiment on Triethyl Phosphate Movie

We first consider the 561 frame multispectral movie of the data collection event of chemical Triethyl Phosphate (TEP) being released into the air. The data consists of the raw, unpreprocessed data including background clutter. It was determined that the wavelengths $\{9.53, 8.30, 10.68\}$ (nm) were optimal for discriminating TEP from background using the SSVM band selection algorithm.

In this experiment, we determine $Betti_0$ barcodes using all 561 TEP cubes, where $4 \times 8 \times 3$ subcubes have been extracted from regions of each data cube along the plume location region.

The $Betti_0$ barcode in Fig. 4a arises from the 561 Grassmanian points corresponding to the left horizon $4 \times 8 \times 3$ region in each data cube, limited by pixel rows 124 to 127 and pixel columns 34 to 41. This region belongs to the area when a plume forms and first becomes visible at frame 112 as detected by the ACE. At scale $\epsilon = 1.5 \times 10^{-3}$, there are 31 bars corresponding to 31 connected components on $G(3, 32)$, with 28 isolated points from frames 111 to 142, one cluster containing frames $\{134, 135, 137\}$, one cluster containing frame 519, and another containing all other frames. At scale $\epsilon = 2 \times 10^{-3}$, we have 19 bars corresponding to 19 connected components on $G(3, 32)$, with 18 isolated frames from 112 to 129, and one cluster containing all the rest. These bars persist for a large range of parameter value (to just beyond 3×10^{-3}), indicating a large degree of separation. At $\epsilon = 4 \times 10^{-3}$, we have 13 clusters with 11 isolated frames 112, 114 to 118, 120 to 123 and 125, one cluster of frames $\{119, 124\}$, and another containing everything else; see also [7]. Cubes following frame 111 are where the plume first occurs with the highest concentration of chemical, and the composition of the plume changes quickly as time progresses. PH detects separation of these cubes from pre-plume cubes and those cubes where plume no longer remains at multiple scales.

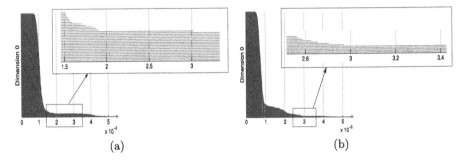

Fig. 4. (a) $Betti_0$ barcode generated on $4 \times 8 \times 3$ left horizon (plume formation) region limited by pixel rows 124–127 and columns 34–41, through all 561 TEP cubes, mapped to $G(3, 32)$. (b) $Betti_0$ barcode generated on $4 \times 8 \times 3$ horizon region limited by pixel rows 124–127 and columns 75–82, through all 561 TEP cubes, mapped to $G(3, 32)$.

After the plume is released, the plume drifts to the right in the multispectral movie as time progresses. We now consider a plume patch corresponding to a horizon region located to the right of the original plume location discussed above. That is, a $4 \times 8 \times 3$ patch is drawn from pixel rows 124 to 127 and pixel columns 75 to 82 for each of the 561 data cubes in the TEP movie. This data is embedded in $G(3, 32)$, and PH is implemented to uncover the structure of the data. Figure 4b contains the 0-dimensional barcode. Analyzing connected components as ϵ varies, we observe that they differ from those found in the previous experiment, see Fig. 4a. At scale $\epsilon = 1.5 \times 10^{-3}$, we have 52 connected components on $G(3, 32)$ corresponding to 47 isolated points from 119 to 141, 145 to 165, and 170 to 172. The other points are connected into four smaller clusters {142,143,144}, {166,167}, {168,169}, and {173,174}, and one cluster containing all the other points. At scale $\epsilon = 2 \times 10^{-3}$, there are 30 connected components on the Grassmannian, including 25 isolated points from 119 to 127, 129 to 140, 149, and 151 to 156; four clusters each containing {128,136–138}, {141–150}, {157,158}, {162–164}; and one cluster containing all the rest. Further, at scale $\epsilon = 3 \times 10^{-3}$, the barcode plot has 5 bars that persist over a large range of values, namely, up to a little beyond 4×10^{-3}: 4 isolated points from frame 121 to 124 and one cluster containing all the rest.

We observe that for this region, PH separates points from frame 119 and later, in contrast to the frames separated from frame 112 in the previous experiment (Fig. 4a). Note that points corresponding to frames 112 to 118 are "plume-free" as the plume does not reach this region until frame 119. It is also interesting to note that points corresponding to frames 121 to 124 are kept isolated for a large range of scales, *i.e.* they are far away from each other and the rest of the points. PH, under the Grassmannian framework, treats these frames as being the most distinct in this region.

4.2 Experiments Detecting a Loop in Methyl Salicylate Movie

The next two experiments consider the multispectral movie of the data collection event of chemical Methyl Salicylate (MeS) being released into the air, consisting of 829 frames. Here we use 3 out of 20 wavelength bands {10.57,10.68,10.94} (nm) that were determined by visual inspection of a background-removed data cube where plume was present. These bands, in particular, were selected as strong plume signal was visible at these corresponding wavelengths. In this movie, the plume first becomes visible at frame 32.

In the first experiment, we construct a sliding window along the horizon, where the plume is released, in both a frame with and without a plume present (frames 32 and 1, respectively) to compare the topological structure of each. This sliding window is constructed by selecting $4 \times 8 \times 3$ patches of each frame limited by rows 125–128 and columns 190–245 where each new point samples 8 columns, incrementing by one. Each patch is then embedded into $G(3, 32)$ and the topological structure is analyzed with PH. In this experiment and the next, our focus is on the $Betti_1$ information which measures the number of loops present in the data.

Observe in Fig. 5 that no persistent topological circle is present in the $Betti_1$ barcode of frame 1, while a persistent loop is present in the $Betti_1$ barcode of frame 32. This is interpreted as follows. In frame 32, where a plume is present, the sliding window first constructs points in $G(3, 32)$ of the natural background, then traverses through points that contain plume, finally returning to points of the natural background. This creates a closed loop in $G(3, 32)$. This behavior is captured in the topological structure of the plume cube. On the other hand, the sliding window in frame 1 only has points in $G(3, 32)$ of the natural background, and thus, no persistent loop is formed in this space.

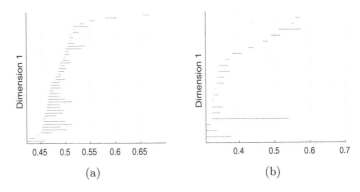

Fig. 5. Data constructed by sliding a window along the horizon region of a single frame of the MeS movie, embedded into $G(3, 32)$ and analyzed with PH. (a) $Betti_1$ barcode of frame 1. (b) $Betti_1$ barcode of frame 32. Observe that a persistent loop is present.

We mention that this experiment was done on background removed frames. In analysis with raw data, loops were not as prevalent with this framework. However, the next experiment does in fact use raw data in our analysis.

In the second experiment, we consider the first one hundred frames of the MeS cubes and focus on a "plume location" patch of size $4 \times 8 \times 3$, limited by pixel rows 125 to 128 and pixel columns 217 to 224, embedded into $G(3, 32)$ for each cube.

Figure 6a displays the $Betti_0$ and $Betti_1$ barcodes from applying PH to this Grassmannian data. A fairly persistent bar appears in the $Betti_1$ barcode that begins at $\epsilon = 0.00979$ and ends at $\epsilon = 0.0141$. This represents a loop through the data in $G(3, 32)$. All other bars are considered as noise. Let us inspect this loop further. It begins once all of the data has been connected into a single component (refer to $\epsilon = 0.00979$ in the $Betti_0$ barcode). The maximum pairwise distance–measured by the smallest angle between subspaces–for this data is 0.0308. This loop persists until just under half this distance.

We conclude the following from this experiment. The first few frames start with a fixed background. Then, the plume begins to form, spreading through the plume patch until the plume no longer remains in the $4\times8\times3$ sampled region. The remaining cubes then return to a fixed background, reflecting periodic behavior in the data. This collection of cubes traces out a closed loop, encoded in the Grassmann manifold $G(3, 32)$. PH captures this loop in the persistent $Betti_1$ bar. Figure 6b displays a schematic of one possibility in the equivalence class of the edges that form this loop. While not all data cubes are present, we notice that those cubes immediately following 31 connect to one another sequentially. This is when the chemical is first released and begins to evolve. Cubes before this frame (where no plume is present) do not follow sequentially and connect with later cubes which no longer contain plume in the sampled plume patch. That is, the time dependent information of 'pre-plume' and 'post-plume' cubes–which simply contain information about the natural background and not the evolving plume–is not as important as 'plume' cubes.

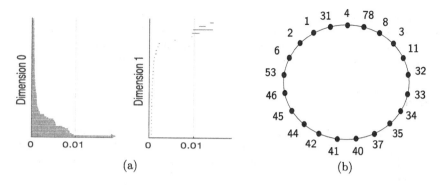

Fig. 6. (a) $Betti_0$ and $Betti_1$ barcodes generated on $4 \times 8 \times 3$ plume locations limited by pixel rows 125–128 and columns 217–224, of the first one hundred frames of the MeS movie, mapped to $G(3, 32)$. (b) Schematic of the edges forming the persistent $Betti_1$ feature.

5   Conclusion

We propose a geometric and topological model for capturing dynamical changes in hyperspectral movies. The HSI data cubes (or a sequence of pixel patches) are viewed as a sequence of points on the Grassmann manifold. The tools of persistent homology are then applied to capture topological novelty in the setting of the Grassmann manifold. This approach models cubes as points, a technique that permits the processing of potentially large amounts of data while retaining basic dynamical structure.

The dynamic structure recorded by the multispectral movie of the gas plume consisting of the simulant Triethyl Phosphate was illuminated in the $Betti_0$ barcodes. Frames containing the plume were identified as topological singletons, *i.e.* isolated points on the manifold for large ranges of scale. Grassmann points before the release, as well as long after the release, appeared as clusters of points. At a location to the right of this region, we see that later frames had a similar behavior, indicating that the geometric model of the Grassmannian allows the dynamics of the scene to be effectively characterized in a topological sense.

In the next two experiments, we use the $Betti_1$ barcode on the movie of the release of Methyl Salicylate mapped to the Grassmannian to reveal that a closed loop is present on the manifold, again reflecting the evolution of the plume. First, we consider a sliding window of pixels along the plume location region and observe that a loop is present in a frame with a plume unlike a frame without a plume. Second, we consider a patch of pixels in each of the first one hundred frames and observe a closed loop that encompasses frames immediately following the release of the chemical in a sequential manner. We mention that in other HSI movies in the LWIR data set, when the amount of chemical released was not as much as in the MeS cubes, the signal of this loop was not as strong. These experiments illustrate that the use of the Grassmann manifold together with PH provide insight into the presence and concentration of chemical contamination in a hyperspectral movie.

Acknowledgments. This paper is based on research partially supported by the National Science Foundation grants DMS-1228308, DMS-1322508. Any opinions, findings, and conclusions or recommendations expressed in this material are those of the authors and do not necessarily reflect the views of the National Science Foundation.

References

1. Bachmann, C.M., Ainsworth, T.L., Fusina, R.A.: Exploiting manifold geometry in hyperspectral imagery. IEEE Trans. Geosci. Remote Sens. **43**(3), 441–454 (2005)
2. Björck, Å., Golub, G.H.: Numerical methods for computing angles between linear subspaces. Math. Comput. **27**(123), 579–594 (1973)
3. Carlsson, G.: Topology and data. Bull. Am. Math. Soc. **46**(2), 255–308 (2009)
4. Chang, J.-M., Kirby, M., Kley, H., Peterson, C., Draper, B.A., Beveridge, J.R.: Recognition of digital images of the human face at ultra low resolution via illumination spaces. In: Yagi, Y., Kang, S.B., Kweon, I.S., Zha, H. (eds.) ACCV 2007, Part II. LNCS, vol. 4844, pp. 733–743. Springer, Heidelberg (2007)
5. Chepushtanova, S., Gittins, C., Kirby, M.: Band selection in hyperspectral imagery using sparse support vector machines. In: Proceedings of the SPIE, vol. 9088, pp. 90881F–90881F-15 (2014)
6. Chepushtanova, S., Kirby, M.: Classification of hyperspectral imagery on embedded Grassmannians. In: Proceedings of the 2014 IEEE WHISPERS Workshop, Lausanne, Switzerland, June 2014
7. Chepushtanova, S., Kirby, M., Peterson, C., Ziegelmeier, L.: An application of persistent homology on Grassmannians manifolds for the detection of signals in hyperspectral imagery. In: Proceedings of the IEEE International Geoscience and Remote Sensing Symposium (IGARSS), Milan, Italy, July 2015
8. Chung, M.K., Bubenik, P., Kim, P.T.: Persistence diagrams of cortical surface data. In: Prince, J.L., Pham, D.L., Myers, K.J. (eds.) IPMI 2009. LNCS, vol. 5636, pp. 386–397. Springer, Heidelberg (2009)
9. Conway, J.H., Hardin, R.H., Sloane, N.J.A.: Packing lines, planes, etc.: packings in grassmannian spaces. Exp. Math. **5**, 139–159 (1996)
10. Cosofret, B.R., Konno, D., Faghfouri, A., Kindle, H.S., Gittins, C.M., Finson, M.L., Janov, T.E., Levreault, M.J., Miyashiro, R.K., Marinelli, W.J.: Imaging sensor constellation for tomographic chemical cloud mapping. Appl. Opt. **48**, 1837–1852 (2009)
11. Dabaghian, Y., Memoli, F., Frank, L., Carlsson, G.: A topological paradigm for hippocampal spatial map formation using persistent homology. PLoS Comput. Biol. **8**(8), e1002581 (2012)
12. Draper, B., Kirby, M., Marks, J., Marrinan, T., Peterson, C.: A flag representation for finite collections of subspaces of mixed dimensions. Linear Algebra Appl. **451**, 15–32 (2014)
13. Edelman, A., Arias, T.A., Smith, S.T.: The geometry of algorithms with orthogonality constraints. SIAM J. Matrix Anal. Appl. **20**(2), 303–353 (1998)
14. Edelsbrunner, H., Harer, J.: Persistent homology - a survey. Contemp. Math. **453**, 257–282 (2008)
15. Ghrist, R.: Barcodes: The persistent topology of data. Bull. Am. Math. Soc. **45**(1), 61–75 (2008)
16. Hatcher, A.: AlgebraiC Topology. Cambridge University Press, Cambridge (2002)

17. Heath, K., Gelfand, N., Ovsjanikov, M., Aanjaneya, M., Guibas, L.J.: Image webs: computing and exploiting connectivity in image collections. In: 2010 IEEE Conference on Computer Vision and Pattern Recognition (CVPR), pp. 3432–3439. IEEE (2010)
18. Ma, L., Crawford, M.M., Tian, J.: Local manifold learning-based-nearest-neighbor for hyperspectral image classification. IEEE Trans. Geosci. Remote Sens. **48**(11), 4099–4109 (2010)
19. Manolakis, D.: Signal processing algorithms for hyperspectral remote sensing of chemical plumes. In: IEEE International Conference on Acoustics, Speech and Signal Processing, ICASSP 2008, pp. 1857–1860, March 2008
20. Marrinan, T., Draper, B., Beveridge, J.R., Kirby, M., Peterson, C.: Finding the subspace mean or median to fit your need. In: 2014 IEEE Conference on Computer Vision and Pattern Recognition (CVPR), pp. 1082–1089. IEEE (2014)
21. Perea, J.A., Harer, J.: Sliding windows and persistence: An application of topological methods to signal analysis. Foundations of Comput. Math., pp. 1–40 (2013)
22. Singh, G., Memoli, F., Ishkhanov, T., Sapiro, G., Carlsson, G., Ringach, D.L.: Topological analysis of population activity in visual cortex. J. Vis. **8**(8), 11 (2008)
23. Adams, H., Tausz, A., Vejdemo-Johansson, M.: JavaPlex: a research software package for persistent (co)homology. In: Hong, H., Yap, C. (eds.) ICMS 2014. LNCS, vol. 8592, pp. 129–136. Springer, Heidelberg (2014)
24. Topaz, C.M., Ziegelmeier, L., Halverson, T.: Topological data analysis of biological aggregation models. PLoS ONE **10**(5), e0126383 (2015). http://dx.doi.org/10.1371
25. Wang, K., Thompson, J., Peterson, C., Kirby, M.: Identity maps and their extensions on parameter spaces: applications to anomaly detection in video. In: Proceedings Science and Information Conference, pp. 345–351 (2015)

Persistence Based on LBP Scale Space

Ines Janusch[(✉)] and Walter G. Kropatsch

Pattern Recognition and Image Processing Group,
Institute of Computer Graphics and Algorithms, TU Wien, Vienna, Austria
{ines,krw}@prip.tuwien.ac.at

Abstract. This paper discusses the connection between the texture operator LBP (local binary pattern) and an application of LBPs to persistent homology. A shape representation - the LBP scale space - is defined as a filtration based on the variation of an LBP parameter. A relation between the LBP scale space and a variation of thresholds used in the segmentation of a graylevel image is discussed. Using the LBP scale space a characterization of (parts of) shapes is demonstrated based on simple shape primitives, the observations may also be generalized for smooth curves. The LBP scale space is augmented by associating it with polar coordinates (with the origin located at the LBP center). In this way a procedure of shape reconstruction based on the LBP scale space is defined and its reconstruction accuracy is demonstrated in an experiment. Furthermore, this augmented LBP scale space representation is invariant to translation and rotation of the shape.

Keywords: LBP · Persistence · Scale space · Filtration · Shape analysis · Shape reconstruction · Segmentation

1 Introduction

Biasotti et al. note in [1] that a digital model of an object is quantitatively similar to the object, while a description is only qualitatively similar. The authors further quote a clear distinction between *representation* and *description*:

> "*An object representation contains enough information to reconstruct (an approximation to) the object, while a description only contains enough information to identify an object as a member of some class.*" [1, p. 5]

Following this definition we study in this paper a novel shape representation based on topological persistence and on local binary patterns (LBPs): the LBP scale space. We show that an approximate reconstruction of the shape is possible using the LBP scale space. The shape representation is given as a vector of the persistence of LBP types around a center pixel. It may as such also be used to classify and compare shapes and can thus also be seen as a topological shape descriptor.

Shape description based on topological persistence is for example defined by the size functions described by Verri et al. [2] which represent the persistent Betti number β_0. Carlsson et al. presented persistence barcodes for shape

© Springer International Publishing Switzerland 2016
A. Bac and J.-L. Mari (Eds.): CTIC 2016, LNCS 9667, pp. 240–252, 2016.
DOI: 10.1007/978-3-319-39441-1_22

description and classification [3]. For shape retrieval the shapes may for example be compared based on their persistence diagram using the matching distance as presented by Cerri et al. [4]. Barcodes encode the persistent homology of a data set in the form of a parametrized version of a Betti number [5]. Barcodes visualize the lifetime for which a features persist and therefore encode multisets of intervals in $\mathbb{R}$. While a persistence diagram is "a multiset of points (u, v) whose abscissa and ordinate are, respectively, the level at which a new k-homology class is created and the level at which it is annihilated through the filtration" [4, p. 2]. This filtration produces a sequence of nested spaces. While the filtration grows, topological features appear (birth) or disappear again (death). The interval between the birth and the death of such a feature, its lifetime, is its persistence. A filtration is for example given by the level cuts of a Morse function on a manifold, as it is used to derive a Reeb graph (another topological shape representation) [6]. The simplest way to obtain a Reeb graph of an image or a 3D shape is to use a height function as a Morse function. In the same way persistence diagrams or barcodes can be determined using a filtration based on a height function. Such topological shape representations based on a filtration are in general dependent on the filtration. Especially height functions are not invariant to rotations of the shape and therefore lack in representational power.

The proposed LBP scale space represents a shape based on changes of the local topology captured by LBPs of increasing radii[1]. Since the LBP takes a circle around a center pixel into consideration, this representation is invariant to translation and rotation of the shape. The LBP scale space can be further extended by associating polar coordinates with the observed local topology. In this way the LBP scale space provides the possibility of reconstruction of the shape (for a discrete LBP scale space up to the sampling of the scale space) based solely on this shape descriptor.

The rest of the paper is structured as follows: Sect. 2 describes the way in which the LBP texture operator captures local topology. In Sect. 3 filtrations using LBPs are presented. The LBP operator is used to derive a shape descriptor based on its persistence - the LBP scale space. Two experiments were conducted: a study of the behaviour of the LBP scale space for special shapes (primitives) and an approach to shape reconstruction based on the LBP scale space. These experiments and the results are discussed in Sect. 4. Section 5 concludes the paper and gives an outlook to future work.

2   Capturing Topology Using LBPs

Although originally proposed as a tool of texture classification, LBPs have in the past also been studied as tool of topological shape and image analysis. For a given grayscale digital image I, the local-binary-pattern codification of I: $LBP(I)$ again yields a grayscale digital image. The grayvalues of $LBP(I)$ are now LBP codes that are used to represent the texture element at each pixel in I. However,

---

[1] Note that our scale space therefore differs from a scale space obtained through Gaussian smoothing with diverse variances.

the LBP codes not only capture local texture information, they also describe the local topology observed.

2.1   Introduction to LBPs

LBPs were first introduced for texture classification [7] and since then became popular texture operators. An LBP computation around a pixel $p = (x, y)$ studies the grayvalues along a subsampled circle of a specific radius r around p. Each position in a bit pattern (corresponding to the sampling points along the circle) is set to 1 if the grayvalue at the sampling point $(g(x_i))$ is larger than or equal to the grayvalue of the center pixel $(g(p))$ and to 0 otherwise (Fig. 1a and b):

$$s(x_i) = \begin{cases} 1 & \text{if } |g(p) - g(x_i)| \geq 0 \\ 0, & \text{otherwise} \end{cases} \tag{1}$$

The two parameters P and r determine the LBP computation: P fixes the number of sampling points along a circle of radius r around the center pixel, for which the LBP operator is computed [8]. Figure 1c shows different parameter configurations. By varying r we may examine concentric circles around p, the LBP is given by $c(x, y, r)$.

2.2   Connected Components of a Graylevel Image

A binary segmentation of a graylevel image is easily obtained as level sets using a threshold t or an interval around the threshold t:

$$|g(x, y) - t| \leq \epsilon \tag{2}$$

Such a threshold interval is also used in the LBP computation of robust local binary patterns (RLBPs) [9]:

$$s(x_i) = \begin{cases} 1 & \text{if } |g(p) - g(x_i) + t| \geq 0 \\ 0, & \text{otherwise} \end{cases} \tag{3}$$

According to Eq. (2) the image is then segmented into several connected components $C_i, i = 1, ..., n$. One such connected component C_a consists of either:

$x_1 \geq c$	$x_2 < c$	$x_3 < c$
$x_8 \geq c$	p	$x_4 < c$
$x_7 \geq c$	$x_6 \geq c$	$x_5 \geq c$

1	0	0
1	p	0
1	1	1

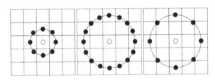

(a)   comparison with neighbours

(b) bit pattern

(c) $(P,r) = (8,1)$; $(P,r) = (16,2)$; $(P,r) = (8,2)$ - according to [8].

Fig. 1. (a) and (b) LBP computation for center pixel p and (c) variations of the parameters P (sampling points) and r (radius).

1. grayvalues that are inside the defined interval around $t \pm \epsilon$ so called *plateaus*: $g(x, y) \in C_a : t - \epsilon \leq g(x, y) \leq t + \epsilon$,
2. grayvalues that are larger than the threshold interval $t + \epsilon$ so called *maxima*: $g(x, y) \in C_a : g(x, y) > t + \epsilon$,
3. or grayvalues that are smaller than the threshold interval $t - \epsilon$ so called *minima*: $g(x, y) \in C_a : g(x, y) < t - \epsilon$.

The input image is thus segmented into connected components, each of them belonging to one of the mentioned categories: *plateau*, *minimum* and *maximum*. The connected components (which may have holes) are surrounded by closed boundaries $b(x, y)$.

2.3   Local Topology Based on LBP Types

In previous work we defined a shape descriptor based on persistence of LBP types around critical points of a shape [10]. For this purpose we analysed a binary shape based on LBPs. We computed LBP types which describe the local topology of the foreground region around the center pixel p. The LBP types are defined by the number of transitions from 0 to 1, respectively vice versa (*bit-switches*) in the LBP code.

- (local) maximum (no bit-switches: the bit pattern contains only 0s),
- (local) minimum (no bit-switches: the bit pattern contains only 1s),
- plateau (no bit-switches: the bit pattern contains only 1s, but all pixels of the region have the same gray value),
- slope (two bit-switches - compare uniform patterns [8]),
- saddle point (four or more bit-switches) [11].

In a segmented image these bit-switches BS correspond to the intersections of the boundary of the connected component which holds p and the LBP circle of radius r:

$$BS = b(x, y) \cap c(x, y, r) \qquad (4)$$

Note that the LBP types not necessarily correspond to the types of connected components as defined in the previous section.

3   LBP Based Persistence

In persistence, those features which persist for a parametrised family of spaces over a range of parameters are considered signals of interest. Short-lived features are treated as noise [5]. The persistence of a feature is given as its lifetime, the span in between the birth and the death of a feature according to a filtration of a space which is "a filtered space, a nested sequence of subspaces that begins with the empty and ends with the complete space" [12, p. 5].

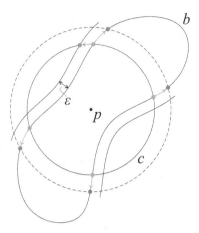

Fig. 2. Increasing the radius r for a fixed center p shifts the intersections with b along the boundary b (blue arrows). Varying ϵ moves the intersections along the LBP circle (orange arrows). (Color figure online)

3.1   Filtration Based on LBPs

Using LBPs we may perform a filtration either:

1. by varying the radius r of the LBP computation c for a fixed boundary b,
2. or by varying the parameter ϵ of the segmentation thus varying b, for a fixed r.

Varying the radius r of the LBP computation c corresponds to a movement along the boundary b (blue arrows in Fig. 2), while varying ϵ (varying the boundary b) leads to a movement along the circle defined by r and p (orange arrows in Fig. 2). By continuously increasing the radius r the intersection points are moving towards or away from each other along the boundary b. For a certain radius the circle will not intersect the boundary anymore but touch it in one point. By further increasing the radius a pair of intersections will disappear or,

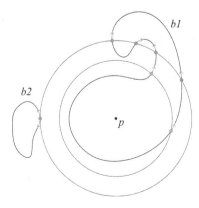

Fig. 3. Configurations of intersections observed when varying the radius allow assumptions about a shape's connected components.

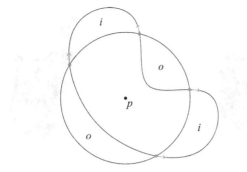

Fig. 4. Characterisation of shape parts based on the LBP scale space.

ultimately the whole shape will be inside the circle of radius r, this corresponds to an LBP of type maximum.

The persistence of an LBP type can be measured by the movement of the intersection points along the boundary b: For a certain radius r the circle of the LBP computation c intersects the closed boundary b $2n$ times (apart from osculation points). These intersections divide the shape in $n + 1$ regions and $2n$ boundary segments (in a persistence diagram: a "birth" for each region / segment). By increasing r the intersection points are moving along b. Once two intersection points coincide the LBP type changes (in the persistence diagram the "death" of the respective segment / region).

Assumptions about a connected component's structure can be made by varying the radius r of the LBP computation c with fixed center p. By analysing the configurations of intersections (bit-switches BS) observed when increasing r a conclusion about characteristics of the shape's boundary can be drawn (see Fig. 3). Moreover, further connected components in the vicinity can be detected. A connected component is divided into several parts that are either covered by the LBP circle of a certain radius or not. These parts can be characterized as follows (see also Fig. 4):

1. *inside a shape (interior)*: The shape is simply connected and corresponds to the bounded connected component of the Jordan curve theorem. Parts inside a shape are identified by intersection points that are moving closer to each other along the boundary for increasing LBP radii until they converge in osculation points.
2. *outside a shape (exterior)*: This connected component is unbounded, indicated by intersection points that diverge along the boundary for increasing LBP radii.
3. *holes*: A hole in the foreground connected component can show no intersections with the LBP circle because it is fully contained inside the LBP circle or not at all covered by the LBP circle. A hole that is intersected by the LBP circle needs special consideration: the shape is divided in parts inside and outside the LBP circle. Since the hole is intersected by the LBP circle, the hole is reduced to concavities along the boundaries of these individual shape parts - the topology changes.

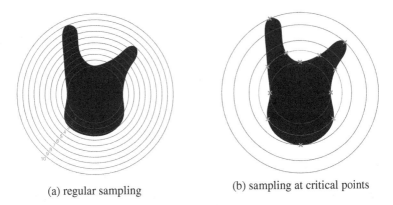

(a) regular sampling   (b) sampling at critical points

Fig. 5. LBP scale space in the (a) a discrete case showing an example sampling scheme and (b) in the continuous case - the osculation points marked along the red LBP circles are critical points at which the topology changes. (Color figure online)

3.2   LBP Scale Space

The LBP scale space is a novel shape representation proposed in this paper. For a chosen pixel or point of a shape or connected component (as LBP center) we compute the LBP over a range of scales (range of LBP radii). We may start with a radius of 0 and increase it either continuously or in the discrete case according to a predefined sampling scheme (for example increasing always by 1 to cover all integer radii, which corresponds to the full pixel resolution). Figure 5 illustrates the LBP scale space for regular sampling and sampling at critical points.

We are interested in the local topology captured by the LBPs for varying radii. We therefore consider the number of bit-switches observed for each of the LBP radii analysed. This shape representation can be stored as a vector or matrix (depending on the sampling and whether the radii need to be explicitly stored as well). The LBP scale space may be used as a shape descriptor for classification or recognition purposes, as we showed for a similar shape descriptor in [10]. Moreover, the LBP scale space enables the reconstruction of a shape based on this representation. For this purpose, we extend the LBP scale space by polar coordinates. This procedure is discussed in more detail in the experiments Sect. 4.2.

4   Experiments

The conducted experiments include a thorough study of characteristics of the proposed LBP scale space on simple shapes (primitives) and an application to shape representation and reconstruction based on the LBP scale space. As test dataset for the shape reconstruction we use a dataset of binary shapes.

4.1   Shape Analysis - Special Cases

We study two simple shapes (primitives): we start with the simplest shape - a circle, before moving on to an already more generalised shape - an ellipse. For this experiment we study the shape primitives using our LBP scale space with the center pixel of the LBP computation located inside the shape. Based on the LBP scale space derived for varying LBP center locations inside the shape, we categorize parts of the shape according to the LBP scale space. This experiment is related to a shape descriptor based on the LBP scale space that we presented in previous work [10], since it may identify locations within the shape which are well suited as well as parts that may not be suitable at all as centers for this shape description.

This study although conducted on circles and ellipses is not limited to these shapes: By computing the medial axis of a shape and splitting the shape into parts at skeleton branching points we can derive a normalised shape representation by straightening the skeleton segments. Any shape may therefore be represented by circles (according to the medial axis) or more general by ellipses. In addition, we show that some observations made for the shape primitives can be generalised to smooth curves.

Circle: For a circle with radius r_c we can distinguish the following two cases:

1. LBP center coincides with center of the shape (circle):
   In this case there are either no intersection points between the LBP circle and the shape (the LBP radius is smaller or larger than r_c) or the whole boundary of the shape intersects with the LBP circle (the LBP radius equals r_c).
2. LBP center at any location within the shape, other than the circle's center:
   We observe one intersection point t_1 for the LBP circle with radius equal to the smallest distance to the boundary and this boundary itself. When increasing the LBP radius we observe two intersection points which move away from t_1 until they converge at the second osculation point of the boundary and the LBP circle t_2. We thus observe the following sequence of number of intersections: *0-1-2-1-0*.

Ellipse: For an ellipse there are four cases to distinguish. We start with the most general case (1) since all others (2–4) are special cases of it:

1. The LBP center is located inside the ellipse but not on the ellipse's axes:
   We observe one intersection point between the LBP circle and the shape's boundary (LBP radius equal to the smallest distance of the LBP center to the shape boundary). For increasing radii we observe two intersection points which move away from each other and for further increasing radii three intersections (one of them a degenerate intersection). Increasing the radius even further yields four intersections. The intersection points in each case in pairs move away from each other and towards each other along the shape's boundary, until two of them converge. Here we observe three intersections,

for increasing radii again two intersections until these two as well converge in one osculation point for the maximum radius. The sequence of number of intersections therefore is: *0-1-2-3-4-3-2-1-0*.

2. The LBP center is located at the major axis of the ellipse:
   The smallest distance to the shape's boundary is *smaller than or equal* to the radius of the circle of curvature for a major apex: In this case we observe one intersection point between LBP circle and shape's boundary (at one major apex). For increasing radii we observe two intersection points which move away from one major apex towards the other. The intersection points converge again in the second major apex. Thus, we observe the following sequence of number of intersections: *0-1-2-1-0*.

   The smallest distance to the shape's boundary is *larger than* the radius of the circle of curvature for a major apex: We observe two osculation points of the shape's boundary with the LBP circle. For increasing radii we observe four intersection points, which converge pairwisely towards the major apexes. Once two of them converge in an apex, we observe three intersections. For further increasing radii we observe two intersections until these two converge in the second major apex. The sequence of number of intersections is as follows: *0-2-4-3-2-1-0*.

3. the LBP center is located at the minor axis of the ellipse:
   We observe one osculation point at a minor apex in which the LBP circle intersects the shape's boundary. For increasing radii we observe two intersections. When increasing the radius further a third intersection at the second minor apex can be observed, followed by four intersections. The intersection points pairwisely move away from the minor apexes until they converge pairwisely in the major apexes and only two intersections remain - the maximum LBP radius is reached. The sequence of number of intersections is: *0-1-2-3-4-2-0*.

4. the LBP center coincides with the center of the ellipse:
   We observe two osculation points of the shape's boundary with the LBP circle located at the minor apexes of the ellipse. For increasing radius four intersection points are observed which move away from the minor axis towards the major axis along the shape's boundary. The intersection points converge in the major apexes. This yields the following sequence of number of intersections: *0-2-4-2-0*.

Smooth Curves: Some observations made for circles and ellipses in this experiment apply also for smooth curves in general. For a continuously increasing radius we observe intersection points at the shape's boundary that move along the boundary and cover the boundary completely when considering the complete LBP scale space. Osculation points are degenerate intersection points since two intersection points coincide at such an osculation point. We can determine the osculation points as the points along the shape's boundary for which the LBP circle's and the boundary's tangent in that point coincide. These osculation points are critical points and describe a birth or death of a component. LBP circles of radii in the interval in between those radii associated with osculation points yield true (non-degenerate) intersections. The interval of radii spanned by such critical points thus describes the lifetime (persistence) of a component.

4.2 Shape Reconstruction

We compute an LBP scale space for a fixed center point inside a shape. The LBP scale space consists of the intersections (bit-switches see Eq. 4) of the shape boundary with the LBP circle for radii ranging from 0 to the radius for which no intersection appears anymore (largest Euclidean distance from the LBP center to the shape boundary). We further associate angles measured around the LBP center with the intersections. In this way we obtain polar coordinates of all intersections and can restore the boundary and thus the shape based on this LBP scale space. For a discrete LBP scale space the quality of the reconstruction is of course dependent on the sampling of the LBP radii and the angle measurement as well as on the chosen center point.

This LBP scale space representation is invariant to translation and rotation of the shape. The rotation leads only to a change in the angles associated with the intersection points. All angles measured for a shape are altered by a constant α which complies with the angle of the rotation applied to the whole shape (for the discrete case deviations from α may be observed because of the sampling).

We tested the accuracy of the LBP scale space reconstructions on the Kimia99 dataset [13]. The reconstruction obtained using a discrete LBP scale space is highly dependent on the chosen LBP center. This is well visible in the comparison of reconstructions based on different locations of center points (see Fig. 6).

For the experiments on the whole dataset, we used two different LBP centers. We defined the LBP center for each shape as (1) the location of minimum eccentricity [14] and (2) the location of maximum distance transform. We used a regular sampling scheme of 1 pixel radius increases, thus covering all integer radii in the range of the LBP scale space of a shape. Based on the LBP scale space obtained for a shape, we reconstructed the shape again. An input shape together with the reconstruction based on the LBP scale space is shown in Fig. 6. For all shapes in the dataset we evaluated the reconstruction quality for the boundary only and for the whole shape. For this purpose we computed the

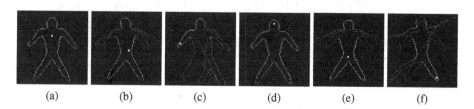

(a)	(b)	(c)	(d)	(e)	(f)

Fig. 6. Reconstructions - green, input - red, matching boundaries - yellow. LBP center - white: (a) max. distance transform, (b) min. eccentricity, (c)–(f) chosen manually. (Color figure online)

reconstruction error considering the set of pixels: M_b which are the boundary pixels of the reconstruction matching the boundary pixels of the input shape respectively M_{shape} all shape pixels of the reconstruction matching the input shape - the true positives. We computed the precision using the set of boundary pixels of the input shape b_{in} and as well as the set of pixels of the whole input shape $shape_{in}$:

1. precision boundary: M_b/b_{in}
2. precision shape: $M_{shape}/shape_{in}$

Figure 7 shows the precision of the reconstruction of the boundary and the whole shape for all 99 shapes in the dataset. The precision for the boundary is low, due to the fact that deviations from the exact position of boundary pixels even by only 1 pixel in the reconstruction are considered an inaccurate reconstruction. The boundary precision is at maximum 0.67 for the center at the location of maximum distance transform (Fig. 7: boundary1) and maximum 0.72 for the center at the location of minimum eccentricity (Fig. 7: boundary2). Since the goal of this approach is to reconstruct the whole shape not only its boundary, we used morphological operations, to close gaps in the boundary and to fill the shape region. The precision for the whole shape is considerably higher: up to 0.99 for the LBP center at the location of maximum distance transform (Fig. 7: shape1) and 1.00 for the LBP center at the location of minimum eccentricity (Fig. 7: shape2). As visible in Fig. 7 the precision of the reconstructed shape is low for some shapes of the dataset - in these cases the boundary reconstruction showed larger gaps, which did not allow to reconstruct a connected boundary

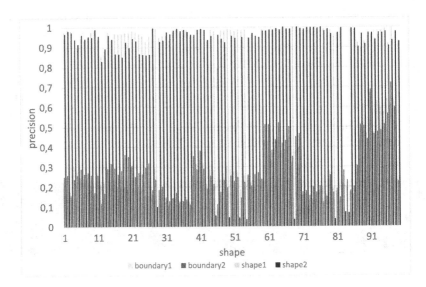

Fig. 7. Precision of the reconstructed boundary (boundary 1 and 2) and the reconstructed shape (shape 1 and 2) for the 99 shapes in the dataset and two different LBP centers.

that could be filled. In general the reconstruction of the whole shape works well: a precision of 0.9 is reached for 92 % of the shapes for the LBP center at the maximum distance transform and for 81 % of the shapes for the LBP center at the location of the minimum eccentricity.

5   Conclusion and Future Work

The novel shape representation - the LBP scale space - is based on persistence, studying a shape by creating a filtration using the LBP operator over a range of scales (radii). The experiments presented in this paper show that the LBP scale space may be easily augmented using polar coordinates around the LBP center. Based on this extended shape representation not only classification of a shape but also a reconstruction is possible.

In future work we would like to perform experiments including noise flawed input shapes, since we can employ the persistence information to exclude noise from the reconstruction. The LBP scale space may also be extended to the 3D space using spheres instead of circles. Of course the intersection points used for the LBP scale space in 2D are also extended to intersection curves between the LBP sphere and the 3D shape then. Moreover the presented approach may be used in future work as a tool of quality control in obtaining binary image segmentations, similar to the MSER (maximally stable extremal regions) approach [15]. The segmentation is defined by the threshold t as well as the interval defined by ϵ. By fixing the LBP parameter r and varying the boundary b we may for example determine parameters t and ϵ that yield a segmentation of a given number of connected components with stable boundaries regarding the persistence of LBP types when varying the radius r of the LBP computation c slightly.

Acknowledgments. We thank the anonymous reviewers for their constructive comments.

References

1. Biasotti, S., De Floriani, L., Falcidieno, B., Frosini, P., Giorgi, D., Landi, C., Papaleo, L., Spagnuolo, M.: Describing shapes by geometrical-topological properties of real functions. ACM Comput. Surv. (CSUR) **40**(4), 12 (2008)
2. Verri, A., Uras, C., Frosini, P., Ferri, M.: On the use of size functions for shape analysis. Biol. Cybern. **70**(2), 99–107 (1993)
3. Carlsson, G., Zomorodian, A., Collins, A., Guibas, L.J.: Persistence barcodes for shapes. Int. J. Shape Model. **11**(02), 149–187 (2005)
4. Cerri, A., Di Fabio, B., Medri, F.: Multi-scale approximation of the matching distance for shape retrieval. In: Ferri, M., Frosini, P., Landi, C., Cerri, A., Di Fabio, B. (eds.) CTIC 2012. LNCS, vol. 7309, pp. 128–138. Springer, Heidelberg (2012)
5. Ghrist, R.: Barcodes: the persistent topology of data. Bull. Am. Math. Soc. **45**(1), 61–75 (2008)

6. Biasotti, S., Giorgi, D., Spagnuolo, M., Falcidieno, B.: Reeb graphs for shape analysis and applications. Theoret. Comput. Sci. **392**(13), 5–22 (2008)

7. Ojala, T., Pietikäinen, M., Harwood, D.: A comparative study of texture measures with classification based on featured distributions. Pattern Recogn. **29**(1), 51–59 (1996)

8. Pietikäinen, M., Hadid, A., Zhao, G., Ahonen, T.: Computer vision using local binary patterns. Computational Imaging and Vision. Springer, London (2011)

9. Chen, J., Kellokumpu, V., Zhao, G., Pietikäinen, M.: RLBP: robust local binary pattern. In: Proceedings of the British Machine Vision Conference (2013)

10. Janusch, I., Kropatsch, W.G.: Shape classification according to LBP persistence of critical points. In: Normand, N., Guédon, J., Autrusseau, F. (eds.) DGCI 2016. LNCS, vol. 9647, pp. 166–177. Springer, Heidelberg (2016). doi:10.1007/978-3-319-32360-2_13

11. Gonzalez-Diaz, R., Kropatsch, W.G., Cerman, M., Lamar, J.: Characterizing configurations of critical points through LBP. In: Computational Topology in Image Context (2014)

12. Edelsbrunner, H.: Persistent homology: theory and practice (2014)

13. Sebastian, T., Klein, P., Kimia, B.: Recognition of shapes by editing shock graphs. In: International Conference on Computer Vision, vol. 1, pp. 755–762. IEEE Computer Society (2001)

14. Kropatsch, W.G., Ion, A., Haxhimusa, Y., Flanitzer, T.: The eccentricity transform (of a digital shape). In: Kuba, A., Nyúl, L.G., Palágyi, K. (eds.) DGCI 2006. LNCS, vol. 4245, pp. 437–448. Springer, Heidelberg (2006)

15. Matas, J., Chum, O., Urban, M., Pajdla, T.: Robust wide-baseline stereo from maximally stable extremal regions. Image Vis. Comput. **22**(10), 761–767 (2004)

On Some Local Topological Properties of Naive Discrete Sphere

Nabhasmita Sen, Ranita Biswas[(✉)], and Partha Bhowmick

Department of Computer Science and Engineering,
Indian Institute of Technology, Kharagpur, India
nabhasmita.sgsits@gmail.com, biswas.ranita@gmail.com,
bhowmick@gmail.com

Abstract. Discretization of sphere in the integer space follows a particular discretization scheme, which, in principle, conforms to some topological model. This eventually gives rise to interesting topological properties of a discrete spherical surface, which need to be investigated for its analytical characterization. This paper presents some novel results on the local topological properties of the *naive model* of discrete sphere. They follow from the bijection of each *quadraginta octant* of naive sphere with its projection map called *f-map* on the corresponding functional plane and from the characterization of certain *jumps* in the *f*-map. As an application, we have shown how these properties can be used in designing an efficient reconstruction algorithm for a naive spherical surface from an input voxel set when it is sparse or noisy.

Keywords: Discrete sphere · Functional plane · 3D imaging · Digital topology · Digital geometry

1 Introduction

Sphere is one of the important geometric primitives. It finds numerous applications in different areas of science and engineering, starting from the age-old manufacturing industry and ending at today's rapid prototyping and 3D imaging. Being non-linear, its discretization in the integer space remains a well-studied research problem, since the topological properties like gap-freeness, tunnel-freeness, tiling, surface connectivity, and minimality are associated with it [1,2,5,9,10,13,21,23,26].

For reconstruction of 2D/3D discrete objects, several techniques can be found in the literature [16,20]. Since formulation of an appropriate topology ensures well-composed discrete sets [19], topological properties are used in many techniques to repair/reconstruct discrete surfaces [11,15,18,22,24,25]. However, all the existing works are related either with discrete triangulated objects in the Euclidean space or with general voxelized surfaces. To the best of our knowledge, there is no existing work related with topological properties of a specific voxelized surface such as sphere. In this paper, we present the first study on some local topological properties of (naive) discrete sphere and show an interesting application of these properties for reconstruction of a spherical surface.

© Springer International Publishing Switzerland 2016
A. Bac and J.-L. Mari (Eds.): CTIC 2016, LNCS 9667, pp. 253–264, 2016.
DOI: 10.1007/978-3-319-39441-1_23

1.1 Motivation and Main Results

Reconstruction of a discrete object, such as sphere, is a pertinent problem in 3D imaging and allied areas. In this paper, we derive and analyze some of the local topological properties of discrete sphere that provide some insights about its composition and can be used for related applications such as spherical surface reconstruction.

At this point, it is worth mentioning that among the different models of discrete sphere, the naive sphere contains the minimum number of voxels ensuring the separability of its interior and exterior. Hence, it is a subset of any other valid model of discrete sphere [5]. Further, in topological terms, it is 2-minimal and marked by its readiness to decomposition in separate parts based on functional plane [6]. These unique points of naive sphere make us choose it as our model of study.

We first analyze the basic properties of a naive spherical surface. From these basic properties, we derive its other local topological properties. Since a naive sphere is 48-symmetric and each of its quadraginta octants (*q-octants*, in short) has a unique functional plane [5,6], we use, w.l.o.g., the projection of its 1st q-octant on its functional plane (i.e., *xy*-plane). We call this projection as *f-map* and analyze it for studying the topology in 3×3 neighborhood.

Figure 1 shows an input *f*-map and the reconstructed *f*-map by our algorithm along with the corresponding naive spherical surface in 3D. In the result, each yellow voxel definitely belongs to the naive sphere and hence called a *definite voxel*. Each blue voxel, on the other hand, is a *semi-definite voxel*, as itself or its 2-adjacent voxel with a higher *z*-value belongs to the naive sphere. In this example, there are 41 voxels in the initial set (taken randomly from a naive sphere of radius 30). Our algorithm is able to reconstruct 57 definite and 54 semi-definite voxels for this surface. In practice, especially when the input voxel set is not sparse, our algorithm runs in a time linear in the area of the reconstructed surface. This is one of its strong points in comparison with other reconstruction algorithms that are mostly based on Hough transform.

2 Preliminaries

In this section, we explain some fundamental concepts, and fix some basic notions, notations, and definitions to be used in the sequel. For more details, we refer to [17].

For two (real or integer) points $p(i, j, k)$ and $p'(i', j', k')$, we define the distance between them along each coordinate axis. For the coordinate $w \in \{'x', 'y', 'z'\}$, it is given by

$$d_w(p, p') = \begin{cases} |i - i'| & \text{if } w = 'x' \\ |j - j'| & \text{if } w = 'y' \\ |k - k'| & \text{if } w = 'z'. \end{cases}$$

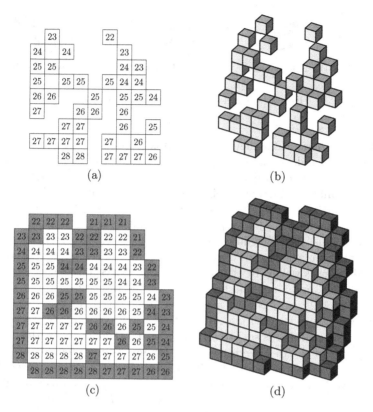

Fig. 1. A snapshot of our work. (a) A small input instance of an incomplete f-map, M. (b) Representation of M in $\mathbb{Z}^3$. (c) Reconstructed f-map based on topological properties. (d) Representation of the reconstructed f-map in $\mathbb{Z}^3$.

These inter-point distances, in turn, define the respective x-, y-, and z-distances between a point $p(i, j, k)$ and a (real) surface Γ as follows.

$$d_w(p, \Gamma) = \begin{cases} \min\{d_w(p, p') : p' \in \Gamma_w(p)\} & \text{if } \Gamma_w(p) \neq \emptyset \\ \infty & \text{otherwise} \end{cases}$$

where, $\Gamma_w(p) = \{p' \in \Gamma : d_v(p, p') = 0 \; \forall v \in \{`x`, `y`, `z`\} \smallsetminus \{w\}\}$.

The above definitions are used to define the isothetic distance between two points, or between a point and a surface. Between two points $p(i, j, k)$ and $p'(i', j', k')$, isothetic distance is taken as the Minkowski norm [17], given by

$$d_\infty(p, p') = \max\{d_x(p, p'), d_y(p, p'), d_z(p, p')\}.$$

Between a point $p(i, j, k)$ and a surface Γ, it is defined as

$$d_\perp(p, \Gamma) = \min\{d_x(p, \Gamma), d_y(p, \Gamma), d_z(p, \Gamma)\}.$$

Next, we define some terms related to voxels and their adjacency. A *voxel* is an integer point in 3D space, and equivalently, a 3-cell [17]. Two distinct voxels

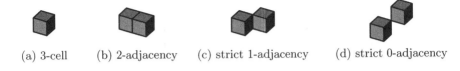

(a) 3-cell   (b) 2-adjacency   (c) strict 1-adjacency   (d) strict 0-adjacency

Fig. 2. A voxel as 3-cell, and its adjacency.

are said to be *0-adjacent* if they share a vertex (0-cell), *1-adjacent* if they share an edge (1-cell), and *2-adjacent* if they share a face (2-cell). Figure 2 shows an illustration. Note that 0-adjacent (resp., 1-adjacent) voxels are not considered as adjacent while considering 1-neighborhood (resp., 2-neighborhood) connectivity. Thus, for $l \in \{0, 1, 2\}$, two voxels $p(i, j, k)$ and $p'(i', j', k')$ are l-adjacent if $d_\infty(p, p') = 1$ and $d_x(p, p') + d_y(p, p') + d_z(p, p') \leqslant 3 - l$. In Fig. 3(a), a single voxel is marked in blue, its 2-neighbors are marked in green, 1-neighbors in yellow, and 0-neighbors in saffron. Note that the 0-, 1-, and 2-neighborhood notations, as adopted by us as well as by the authors in [26], correspond respectively to the classical 26-, 18-, and 6-neighborhood notations used in [12].

For $l \in \{0, 1, 2\}$, an *l-path* in a 3D discrete object A (or the discrete space $\mathbb{Z}^3$) is a sequence of voxels from A such that every two consecutive voxels are l-adjacent. The object A is said to be *l-connected* if there is an l-path connecting any two points of A. An *l-component* is a maximal l-connected subset of A.

Let B be a subset of a discrete object A. If $A \smallsetminus B$ is not l-connected, then B is *l-separating* in A [12]. If such an l-separating set B contains a 3-cell c such that both $A \smallsetminus B$ and $A \smallsetminus (B \smallsetminus \{c\})$ are individually l-separating in A, then c is said to be *l-simple* in B w.r.t. A; otherwise, B is *l-minimal* in A. If a subset B of A is not l-separating in A, then B has *l-gaps*. 2-gaps, in particular, are called *tunnels* [7]. Figure 3(b) shows how a tunnel is formed in a 2-minimal surface when any of its voxels is removed.

3   Naive Sphere and Its Topological Properties

There are several models of discrete sphere and hypersphere, which are built with different topological constraints; see, for example, [14,26], and the references therein. Our work is focused on the naive model of discrete sphere [2]. A *naive sphere* $S(r)$ is a 2-minimal set of 3D integer points (equivalently, voxels) such that $\max\{d_\perp(p, \mathsf{S}(r)) : p \in S(r)\}$ is minimized. Here $\mathsf{S}(r)$ denotes a real sphere of radius r, considered as a positive integer in our work. Also, its center is an integer point, which is taken as $(0, 0, 0)$ for simplicity and without loss of generality.

A detailed number-theoretic analysis of naive sphere with integer specification and an integer algorithm for its generation can be found in [4]. As shown in [3,4],

$$S(r) = \left\{ \begin{array}{l} p \in \mathbb{Z}^3 : r^2 - \max(X) \leqslant s < r^2 + \max(X) \\ \land \left((s \neq r^2 + \max(X) - 1) \lor (\mathrm{mid}(X) \neq \max(X)) \right) \end{array} \right\}$$

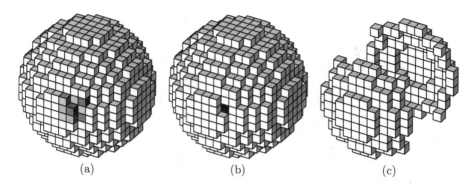

Fig. 3. (a) A naive sphere of radius 15. All voxels are shown in white except p (blue) and its adjacent voxels (green = 2-adjacent, yellow = 1-adjacent, saffron = 0-adjacent). (b) A tunnel formed when any voxel is withdrawn from the naive sphere. (c) 16 q-octants (8 in front, 8 in back) whose functional plane is the xy-plane. (Color figure online)

where $p = (i, j, k)$, $s = i^2 + j^2 + k^2$, and $X = \{|i|, |j|, |k|\}$.

In this section, we put forward some interesting properties characterizing the local neighborhood of a voxel in a naive sphere. First, we explain some basic properties and then we derive some additional properties from the basic ones. These properties are later used for efficient reconstruction of a sparse or noisy voxel set generated from a naive spherical surface.

3.1   Basic Properties

As already mentioned, $S(r)$ is a 2-minimal surface that best-approximates $\mathsf{S}(r)$. It does not contain any tunnel, but contains 0- and 1-gaps. A naive sphere is made up of 48 basic symmetric parts called *quadraginta octants*, or *q-octants* in short [4]. Figure 4(a) shows the first q-octant of $S(r)$ with radius 30.

Given a discrete object $A \subseteq \mathbb{Z}^3$, we say that a coordinate plane, say, xy, is functional for A, if for every voxel $v = (i, j, k) \in A$ there is no other voxel in A with the same first two coordinates. For a plane in general orientation, the functional plane (FP) is unique, and it is one of the coordinate planes. For a sphere, on the contrary, it is not so; rather, for each q-octant, the concept is analogous with plane. To explain this, we denote by $S^t(r)$ the tth q-octant of $S(r)$, where $t = 1, 2, \ldots, 48$. A characterization of the q-octants of $S(r)$ can be found in [4]. The functional plane of $S^t(r)$ is the coordinate plane on which its projection has a bijection with $S^t(r)$.

Each coordinate plane serves as the FP of 16 specific q-octants. Figure 3(c) shows the voxels of q-octants whose FP is the xy-plane. For a detailed analysis on FPs of discrete sphere, we refer to [6].

We define *f-map* as the projection of a q-octant on its FP. Since the projection is bijective for naive sphere, the *f-map*, at each pixel position, holds the maximum coordinate value of the corresponding voxel. From now on, we refer

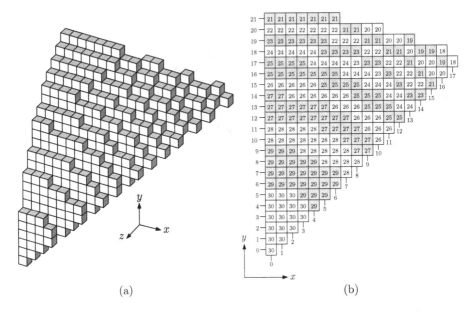

(a)   (b)

Fig. 4. (a) $S^1(r)$ and (b) $F^1(r)$ for $r = 30$. Note that each segment in $F^1(r)$ with constant f-value is a discrete annulus (shown in white or yellow). (Color figure online)

to this maximum coordinate as f-coordinate. Figure 4(b) shows an example of the f-map of $S^1(r)$. As xy-plane is the FP of $S^1(r)$, each position of its f-map contains the z-value of the corresponding voxel. The voxels with same z-value form discrete annuli and are shown in alternate colors. We denote the f-map of $S^1(r)$ by $F^1(r)$.

We start with the following theorems on two basic properties of the voxels in the 1st q-octant, which can be generalized for all other octants as well.

Theorem 1. *Unit increment (decrement) in x- or y-value in $F^1(r)$ results in at most unit decrement (increment) in z-value.*

Proof. By [4, Theorem 4], the ith circular arc (voxels with $x = i$) in $S^1(r)$ forms a 1-connected monotone path with increasing y-value and non-increasing z-value. This ensures the connectivity of each arc corresponding to a column in $F^1(r)$. Similarly, as the naive surface is 2-separating, for a fixed y-value, unit increment in x-value results in at most unit decrement in z-value.   □

Theorem 2. *Unit increment (decrement) in x-value and unit decrement (increment) in y-value in $F^1(r)$ results in at most unit increment (decrement) in z-value.*

Proof. The z-coordinate induces discrete annuli in $S^1(r)$ (see Fig. 4). The upper and the lower boundaries of the projection of each such annulus in $F^1(r)$ are monotone paths, with increasing x-value and non-increasing y-value from left to right. This owes to the fact that in $S^1(r)$, the x-value at a point does not exceed the y-value. So, a unit increment (decrement) in x-value and a unit decrement (increment) in y-value in $F^1(r)$ points to either the same or the next annulus, whence the result.   □

<table>
<tr><td>3</td><td>2</td><td>1</td></tr>
<tr><td>4</td><td>p</td><td>0</td></tr>
<tr><td>5</td><td>6</td><td>7</td></tr>
</table>
(a)

<table>
<tr><td>29</td><td>29</td><td>29</td></tr>
<tr><td>30</td><td>30</td><td>29</td></tr>
<tr><td>30</td><td>30</td><td>30</td></tr>
</table>
(b)

<table>
<tr><td>27</td><td>27</td><td>26</td></tr>
<tr><td>27</td><td>27</td><td>27</td></tr>
<tr><td>28</td><td>27</td><td>27</td></tr>
</table>
(c)

<table>
<tr><td>22</td><td>21</td><td>20</td></tr>
<tr><td>22</td><td>22</td><td>21</td></tr>
<tr><td>23</td><td>22</td><td>22</td></tr>
</table>
(d)

Fig. 5. 3×3 neighborhood of a pixel p in $F^1(r)$. (a) Direction codes w.r.t. p. (b–d) Examples from $F^1(r)$ of Fig. 4(b).

3.2   Derived Properties

We explain here some local properties in 3×3 neighborhood of $F^1(r)$, which are proved using the basic properties stated in Theorems 1 and 2. For ease of understanding, for a pixel p in $F^1(r)$, we refer to its 3×3 neighborhood using direction codes, as shown in Fig. 5(a). We refer to the z-value of pixel p in $F^1(r)$ as $f(p)$, and that of some in the neighborhood of p by $f(p + d_\alpha)$, where α is the direction. For example, in Fig. 5(b), $f(p) = 30$, $f(p + d_0) = 29$, $f(p + d_5) = 30$. The set of points in 3×3 neighborhood of p and contained in $F^1(r)$ is given by $N(p) = \{p\} \cup \{p + d_\alpha : (p + d_\alpha \in F^1(r)) \wedge (\alpha = 0, 1, \ldots, 7)\}$. We denote by $f(N(p))$ the set of f-values of $N(p)$. With these basic notations, we now introduce the properties that are true in the 3×3 neighborhood of every point p in the f-map of $S^1(r)$ and can be generalized for any other q-octant of $S(r)$.

Theorem 3. If $(p + d_5) \in N(p)$, then $f(p + d_5) = \max\{f(N(p))\}$; and if $(p + d_1) \in N(p)$, then $f(p + d_1) = \min\{f(N(p))\}$.

Proof. Observe that $p + d_5$ has the lowest x- and y-values, and $p + d_1$ has the highest x- and y-values. By Theorem 1, increase in x- or y-value decreases the z-value in $S^1(r)$. Therefore, z-value is maximum for $p + d_5$, and minimum for $p + d_1$, whence the proof.   □

Notice that all three examples in Fig. 5 conform to the above theorem.

Theorem 4. $\min\{f(N(p))\} \geqslant f(p) - 2$ and $\max\{f(N(p))\} \leqslant f(p) + 2$.

Proof. By Theorem 1, the maximum change of z-value occurs in a diagonal direction and it can be 2. So, if we move diagonally from p, then $f(p + d_1)$ can be as low as $f(p) - 2$, and $f(p + d_5)$ can be as high as $f(p) + 2$.   □

For example, in Fig. 5(d), $\min\{f(N(p))\} = f(p) - 2$ at $p + d_1$. Note that the above theorem indicates the occurrence of *jump* that arises in naive planes [8] as well as in naive spheres, as shown recently in [6]. A jump is formed when two voxels are not adjacent on the naive surface but their projections are two 0-adjacent pixels on the functional plane.

Theorem 5. *If* $\max\{f(N(p))\} - \min\{f(N(p))\} = 3$, *then* $f(p) \in$ $\{\min\{f(N(p))\} + 1, \max\{f(N(p))\} - 1\}$.

Proof. By Theorem 3, $f(p+d_5) = \max\{f(N(p))\}$ and $f(p+d_1) = \min\{f(N(p))\}$. By Theorem 4, diagonal z-value change is no more than 2. So, with $\max\{f(N(p))\} - \min\{f(N(p))\} = 3$, we get $f(p) = \min\{f(N(p))\} + 1$ or $\max\{f(N(p))\} - 1$. □

The following theorems can be used to get $f(p)$ when a partial information is available in $N(p)$.

Theorem 6. *For* $0 \leqslant \alpha \leqslant 3$ *and* $4 \leqslant \beta \leqslant 7$, *if* $f(p+d_\alpha) = f(p+d_\beta)$, *then each of these two is equal to* $f(p)$.

Proof. From Theorems 1, 2, and 4, it can be inferred that the f-value never increases from p at directions 0, 1, 2, and 3, and never decreases at directions 4, 5, 6, and 7. Therefore, if $f(p+d_\alpha) = f(p+d_\beta)$, where $0 \leqslant \alpha \leqslant 3$ and $4 \leqslant \beta \leqslant 7$, then p also holds the same f-value. □

Theorem 7. *If* $f(p + d_\alpha) = c_1$, $f(p + d_\beta) = c_2$, *and* $|c_1 - c_2| = 2$, *where* $\alpha \in \{0, 2, 3\}$ *and* $\beta \in \{4, 6, 7\}$, *then* $f(p) = \frac{1}{2}(c_1 + c_2)$.

Proof. From Theorems 1 and 2, we can infer that from p at directions 0, 2, and 3, the f-value remains same or decrements by one, and at directions 4, 6, and 7, the f-value remains same or increments by one. Therefore, if the difference between $f(p + d_\alpha)$ and $f(p + d_\beta)$ is two, then $f(p)$ has to lie exactly between these two values. □

Theorem 8. *If* $f(p + d_\alpha) = c_1$, $f(p + d_\beta) = c_2$, *and* $|c_1 - c_2| = 1$, *where* $\alpha, \beta \in \{0, 2, 3\}$, *then* $f(p) = \max(c_1, c_2)$; *or where* $\alpha, \beta \in \{4, 6, 7\}$, *then* $f(p) = \min(c_1, c_2)$.

Proof. From Theorems 1 and 2, we can infer that from p at directions 0, 2, and 3, the f-value remains same or decrements by one, and at directions 4, 6, and 7, the f-value remains same or increments by one. Therefore, if $\alpha, \beta \in \{0, 2, 3\}$ or $\alpha, \beta \in \{4, 6, 7\}$, i.e., both are either towards non-incrementing or towards non-decrementing directions, and if the difference between $f(p + d_\alpha)$ and $f(p + d_\beta)$ is unity, then $f(p)$ has to be the maximum of these two values for the non-incrementing direction $(\alpha, \beta \in \{0, 2, 3\})$ and the minimum of these two values for the non-decrementing direction $(\alpha, \beta \in \{4, 6, 7\})$. □

4   Reconstruction

In this section, we show how the local topological properties of a naive sphere can be used to reconstruct its surface when holes or noise are present in input data. The input to the algorithm is a set of integer points/voxels belonging to the first q-octant of a naive spherical surface. We have also assumed that $N(p)$

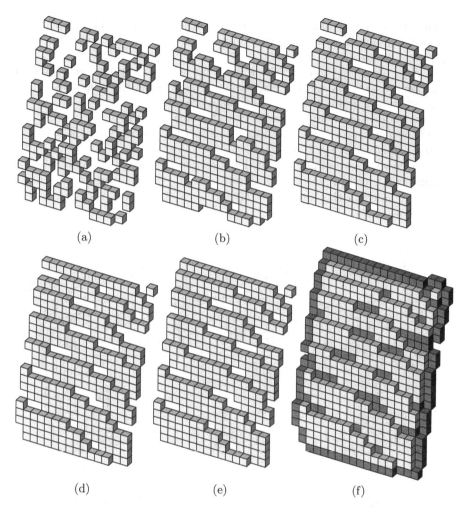

Fig. 6. Demonstration of our algorithm. (a) An input instance containing 145 voxels taken from a naive spherical surface. (b) Reconstructed surface after 1st iteration. (c) Reconstructed surface after 2nd iteration. (d) Reconstructed surface after 3rd iteration. (e) Reconstructed surface after last, i.e., 6th iteration. (f) Reconstructed surface after inclusion of semi-definite voxels.

always consists of 9 points, i.e., the points in the input set are all taken from the interior of the q-octant and not from its border. This assumption helps us to also generate semi-definite voxels around the final definite point set of the reconstructed naive set.

Note that, while generating the definite voxels to fill the gaps appearing in the input surface, only Theorems 6 to 8 are used, as these are the only ones which can definitely predict $f(p)$. Few of the other properties are used to predict the semi-definite voxels after generation of the final definite result set, and can also be used to detect noisy voxels in the input set. The steps of the reconstruction algorithm are as follows.

1. Generate the f-map of the input by taking projection on the functional plane.
2. Scan the f-map row-wise (or column-wise) to detect pixels with undefined $f(p)$, i.e., holes.
3. For each detected hole p, check whether any appropriate pair fits to Theorems 6 to 8, and get the exact value of $f(p)$ as per that theorem.
4. Repeat steps 2 and 3 as long as new holes are filled.
5. Produce a one-pixel empty boundary to the final f-map.
6. Scan the f-map; for any empty location, select the lowest z-value if there are two possible values for $f(p)$.

   Note that existence of any one of the voxels at directions 0, 2, 3, 4, 6, and 7 ensures that we can get two possible values for $f(p)$. So, every semi-definite voxel included in this step has a 50 % chance of actually belonging to the naive spherical surface.

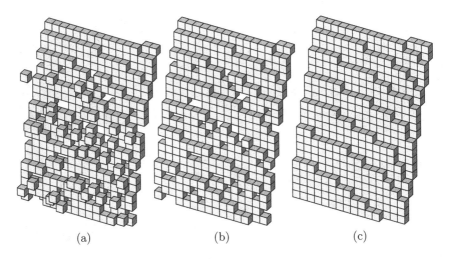

(a)   (b)   (c)

Fig. 7. Applying the local topological properties to detect and restore noisy voxels. (a) An input instance containing noisy voxels in a naive spherical surface. (b) After removal of noise. (c) Reconstruction with defined and semi-definite voxels.

Each iteration of this algorithm scans through the entire f-map to detect empty locations, i.e., it scans through total $(n + m)$ number of points, where n is the number of points in the input set, and m is the number of empty locations in the f-map. Inclusion of definite voxels at each stage helps to predict the values for undefined voxels in the next stage; and in the worst case, the holes may be interconnected in such a way that each iteration is able to produce only one definite voxel. Therefore, the number of iterations depends on the configuration, size, and shape of the hole regions. The time complexity of the algorithm therefore varies from $O(n + m)$ in the best case to $O((n + m)m)$ in the worst case. However, as we have experimented, the number of iterations in practice is low and the algorithm stops after a small number of iterations, thus practically giving us a linear time complexity when m is in the order of $n/2$ or less.

In the above-stated experimentation, we have assumed that the input set of points belongs to a particular q-octant and all definitely belong to a naive sphere whose radius and center are not known. As the properties stated in the previous section can easily be generalized to work for other q-octants as well, a set of points from different q-octants of a naive sphere can be partitioned by their corresponding functional planes. This can be done by analyzing the projections on the three coordinate planes. After partitioning by functional planes, each part can be used for reconstruction based on the local topological properties and their corresponding f-maps.

We have run the algorithm for multiple input instances and found encouraging results. Figure 6 shows us how the algorithm reconstructs the input surface step by step. In this example, the input surface contains 145 voxels, all of which belongs to a particular naive sphere. After six iterations, our algorithm is able to reconstruct the surface as in Fig. 6(f), which contains 269 definite voxels and 115 semi-definite voxels. Another result shown in Fig. 7 presents how the local topological properties can effectively be used to discard noisy voxels and restore the correct ones to reconstruct a naive spherical surface.

5   Concluding Remarks

We have studied various local properties of naive sphere and have demonstrated a method to utilize these properties for reconstruction of a naive spherical surface containing missing voxels. One point to note is that these properties are not supposed to work while the input data is very sparse and we are not able to find any local neighbors of the holes. In that case, some different methodologies like Hough transform can be employed. Nevertheless, local topological properties in larger neighborhoods may yield better scope for reconstruction, which we foresee interesting and challenging in continuation of the proposed work. Further, these local properties of naive sphere, as stated in this paper, may also be compared with local properties of other naive discrete surfaces, e.g., naive plane, for analyzing how they differ across different parameters.

References

1. Andres, E.: Discrete circles, rings and spheres. Comput. Graph. **18**(5), 695–706 (1994)
2. Andres, E., Jacob, M.-A.: The discrete analytical hyperspheres. IEEE Trans. Vis. Comput. Graph. **3**(1), 75–86 (1997)
3. Biswas, R., Bhowmick, P.: On finding spherical geodesic paths and circles in $\mathbb{Z}^3$. In: Barcucci, E., Frosini, A., Rinaldi, S. (eds.) DGCI 2014. LNCS, vol. 8668, pp. 396–409. Springer, Heidelberg (2014)
4. Biswas, R., Bhowmick, P.: From prima quadraginta octant to lattice sphere through primitive integer operations. Theor. Comput. Sci. **624**, 56–72 (2016)
5. Biswas, R., Bhowmick, P., Brimkov, V.E.: On the connectivity and smoothness of discrete spherical circles. In: Barneva, R.P., et al. (eds.) IWCIA 2015. LNCS, vol. 9448, pp. 86–100. Springer, Heidelberg (2015)

6. Biswas, R., Bhowmick, P., Brimkov, V.E.: On the polyhedra of graceful spheres and circular geodesics. Discrete Appl. Math. (in press). doi:10.1016/j.dam.2015.11.017

7. Brimkov, V.E.: Formulas for the number of $(n-2)$-gaps of binary objects in arbitrary dimension. Discrete Appl. Math. **157**(3), 452–463 (2009)

8. Brimkov, V.E., Barneva, R.P.: Graceful planes and lines. Theoret. Comput. Sci. **283**(1), 151–170 (2002)

9. Brimkov, V.E., Barneva, R.P.: On the polyhedral complexity of the integer points in a hyperball. Theoret. Comput. Sci. **406**(1–2), 24–30 (2008)

10. Chamizo, F., Cristobal, E.: The sphere problem and the L-functions. Acta Math. Hung. **135**(1–2), 97–115 (2012)

11. Chen, L., Rong, Y.: Linear time recognition algorithms for topological invariants in 3D. In: ICPR 2008, pp. 1–4 (2008)

12. Cohen-Or, D., Kaufman, A.: Fundamentals of surface voxelization. Graph. Models Image Process. **57**(6), 453–461 (1995)

13. Fiorio, C., Jamet, D., Toutant, J.-L.: Discrete circles: an arithmetical approach with non-constant thickness. In: Vision Geometry XIV, SPIE, vol. 6066, p. 60660C (2006)

14. Fiorio, C., Toutant, J.-L.: Arithmetic discrete hyperspheres and separatingness. In: Kuba, A., Nyúl, L.G., Palágyi, K. (eds.) DGCI 2006. LNCS, vol. 4245, pp. 425–436. Springer, Heidelberg (2006)

15. Fourey, S., Malgouyres, R.: Intersection number and topology preservation within digital surfaces. Theoret. Comput. Sci. **283**(1), 109–150 (2002)

16. Kazhdan, M.: Reconstruction of solid models from oriented point sets. In: SGP 2005, Article 73 (2005)

17. Klette, R., Rosenfeld, A.: Digital Geometry: Geometric Methods for Digital Picture Analysis. Morgan Kaufmann, San Francisco (2004)

18. Latecki, L., Conrad, C., Gross, A.: Preserving topology by a digitization process. J. Math. Imaging Vis. **8**(2), 131–159 (1998)

19. Latecki, L., Eckhardt, U., Rosenfeld, A.: Well-composed sets. Comput. Vis. Image Underst. **61**(1), 70–83 (1995)

20. Latecki, L.J., Rosenfeld, A.: Recovering a polygon from noisy data. Comput. Vis. Image Underst. **86**(1), 32–51 (2002)

21. Maehara, H.: On a sphere that passes through n lattice points. Eur. J. Comb. **31**(2), 617–621 (2010)

22. Malgouyres, R., Lenoir, A.: Topology preservation within digital surfaces. Graph. Models **62**(2), 71–84 (2000)

23. Montani, C., Scopigno, R.: Graphics Gems, pp. 327–334 (1990)

24. Siqueira, M., Latecki, L.J., Tustison, N.J., Gallier, J.H., Gee, J.C.: Topological repairing of 3D digital images. J. Math. Imaging Vis. **30**(3), 249–274 (2008)

25. Stelldinger, P., Latecki, L.J., Siqueira, M.: Topological equivalence between a 3D object and the reconstruction of its digital image. IEEE TPAMI **29**(1), 126–140 (2007)

26. Toutant, J.-L., Andres, E., Roussillon, T.: Digital circles, spheres and hyperspheres: from morphological models to analytical characterizations and topological properties. Discrete Appl. Math. **161**(16–17), 2662–2677 (2013)

DIG: Discrete Iso-contour Geodesics for Topological Analysis of Voxelized Objects

Gurman Bhalla and Partha Bhowmick[(⊠)]

Department of Computer Science and Engineering,
Indian Institute of Technology, Kharagpur, India
gurman.bhalla@gmail.com, bhowmick@gmail.com

Abstract. Discretized volumes and surfaces—used today in many areas of science and engineering—are approximated from the real objects in a particular theoretical framework. After a discretization produces a triangle mesh (2-manifold surface), a well-formed voxel set can be prepared from the mesh by voxelization of its constituent triangles based on some digitization principle. Since there exist different topological models of digital plane, choosing the appropriate model to meet the desired requirement appears to be of paramount importance. We introduce here the concept of discrete iso-contour geodesics (DIG) and show how they can be constructed on a voxelized surface with the assurance of certain topological requirements, when the voxelization conforms to the naive model with judicious inclusion of Steiner voxels from the graceful model, as and when needed. We also show some preliminary results on its practical application towards extraction of high-level topological features of 3D objects, which can subsequently be used for various shape-analytic applications.

Keywords: Digital geometry · Discrete topology · Iso-contour geodesics · Shape analysis · Voxelization

1 Introduction

Voxelization today is not only important in the field of object discretization and representation but also gaining remarkable progress in additive manufacturing through rapid prototyping (RP) techniques like stereo-lithography, 3D printing, and fused deposition modeling [11,19–21,30]. Hence, the collection of work related to voxelization, as seen in today's literature, can be divided into two categories—one covering the theories and algorithmic solutions for object discretization and another dealing with different RP techniques using digital technology. The latter category mostly relies on a *digital building matter* in the sense that the building block is a digital unit or *voxel*, as opposed to the analog (continuous) material used in conventional RP [6,17,18,28,32].

Whether the subject relates to analytical discretization or relates to physical manufacturing, the underlying theory or methodology of voxelization has a strong impact on the consistency or on the solidity of the resultant product.

A. Bac and J.-L. Mari (Eds.): CTIC 2016, LNCS 9667, pp. 265–276, 2016.
DOI: 10.1007/978-3-319-39441-1_24

In either case, these characteristics can be analyzed well in the purview of discrete geometry and topology, as a collection of voxels is usually obtained by a particular process in a certain theoretical framework [23,27]. In our work, we focus on this with a two-fold objective—first to show how a surface should be voxelized for its readiness to discrete iso-contour geodesic (DIG) construction and then to demonstrate the usefulness of DIG in extraction of high-level topological information from a voxelized object.

1.1 Existing Work

We give here a brief review of the development and the state-of-the art practices related to voxelization and also to computation of discrete geodesics and iso-contours.

Voxelization. The early work on physical modeling of a surface or volume element can be seen in [19,30,31] and in the articles referred to therein. Those work, however, did not address the topological issues related to voxelization. The theoretical frameworks along with the topological issues came up gradually in a later stage. For example, in [9], some of the topological properties were discussed, which included holes, cavities, simple points, separability, and penetration.

With the growing need for digitization and cutting-edge technology, different techniques for voxelization have been proposed off and on, taking into account different apparatus, computational models, cost factors, and product requirement. A low-cost methodology based on z-buffer and multi-view depth information is developed in [22]. To incorporate an anti-aliasing effect during voxel rendition, a multi-resolution technique is proposed in [10]. For adding more features available in graphics workstations, such as texture mapping and frame-buffer blending functions, a hardware-accelerated approach is shown to be effective in [16]. The idea of exploiting programmable graphics hardware is also used in [13] for voxelization of a polygonal model after mapping it into three sheet buffers and then synthesizing into a single worksheet recording the volumetric representation of the target.

Voxelization is also useful for simplification and repair of a polygonal model, as shown in [29], with 3D morphological operations on the scan-converted voxel set. Further, with the emergence of GPU functionalities, a variety of applications with voxelized objects have come up in recent time. For example, in [24], a GPU-accelerated approach is proposed for creation of multi-valued solid volumetric models with different solid slice functions and material description in order to make it useful for different applications like collision detection, medical simulation, volume deformation, 3D printing, and computer art. In [15], a filtering algorithm is designed to build a density estimate for deduction of normals from the voxelized model, which is shown to be useful in simulation of translucency effects and particle interactions. In fact, very recently, many such real-time simulations and applications are shown to be efficiently realizable when a voxelized dataset is used; these include urban modeling [34], octree-based sparse voxelization for 3D animation [12], fluid simulation with dynamic obstacles [38], discrete radiosity [25], light refraction and transmittance in complex scenes [7], etc.

Geodesics. The literature on geodesics and iso-contours, as on today, is predominantly focused on closed orientable 2-manifold surfaces, i.e., objects with triangulated-mesh representation in the Euclidean space. Hence, the techniques are mostly from differential and computational geometry; see, for example, [1, 8, 26, 33, 35–37]. As geodesics find various applications in remeshing, non-rigid registration, surface parametrization, shape editing and segmentation, the notion of approximate geodesic distance is also proposed recently in [36] as a practical alternative for the exact solution [35].

In the domain of voxel complexes, however, no significant work can be found on discrete geodesics, barring a few [8, 12]. In [8], the concept of visibility—a well-known concept in computational geometry—is defined in the discrete space based on digital straightness. In [12], as sparse (i.e., highly disconnected) voxel set is used, the geodesic metric is based on Euclidean norm.

1.2   Our Contribution

As briefed in Sect. 1.1, a multitude of work have been carried out on voxelization of 2-manifolds and on geodesics in the Euclidean space. However, geodesics on voxelized (i.e., 3-manifold) surfaces and their topological properties have not been studied so far. This motivates us to look into this interesting problem. We introduce here the concept of *discrete iso-contour geodesics* (DIG) that can be constructed on a well-formed voxelized surface and then demonstrate their usefulness in shape-analytic applications similar to those in the Euclidean domain.

Henceforth in this paper, a voxelized curve (DIG in our case) or a voxelized surface means discrete approximation of its real counterpart by a set of voxels in a certain topological model. In order to ensure that the concerned object is well-defined in the voxelized space, the connectivity and related topological issues come up alongside, which are addressed and fixed in this paper.

2   Voxelization of 2-Manifolds

We discuss here some definitions and concepts related to topology of voxel complexes and the underlying metric space, which are relevant to our work, following the convention as in [23]. For easy understanding and for easily relating the theoretical results with the experimental results on voxelized objects, we discuss the topological concepts in terms of voxels and relations among them, which can be equivalently represented and explained in graph-theoretic terms as well [23, 27].

2.1   Voxel Topology and Metric Space

A *voxel* is a 3-cell, i.e., an axis-parallel cube-shaped 3-manifold of unit length. Two voxels are said to be 0-, 1-, or 2-*adjacent* if they share a vertex (0-cell), an edge (1-cell), or a face (2-cell), respectively (also called 26-, 18-, 6-neighborhoods [9]). Note that 0-adjacent (1-adjacent) voxels are not treated as adjacent while considering 1-adjacency (2-adjacency).

A voxelized object A means a set of voxels. A k-*path* $(k = 0, 1, 2)$ in A is a sequence of voxels from A such that every two consecutive voxels are k-adjacent. If a k-path exists between every two voxels of A, then A is said to be k-*connected*. A k-*component* is a maximal k-connected subset of A.

A subset A' of A is k-*separating* (w.r.t. A) if $A \setminus A'$ is not k-connected. In addition, if $A \setminus A'$ has exactly two k-components and A' has a voxel v such that $A' \setminus \{v\}$ is still k-separating, then v is called a *simple voxel* in A'. If now A' contains no simple voxel, then A' is k-*minimal* and so has no *tunnel*; and if A' has any tunnel, then it is not 2-separating (see [3] for further details).

The *supercover* $K(X)$ of a set $X \subseteq \mathbb{R}^3$ is the set of all voxels intersected by X. The *standard* and the *naive* voxelizations of X are 0- and 1-minimal subsets of $K(X)$. If X is a real plane or its part (such as a triangle in our work), then its naive set $N(X)$ is *functional* in at least one coordinate plane and hence has one-to-one correspondence with its projection on that coordinate plane. For example, if xy-plane is functional for X, then $N(X)$ has one-to-one correspondence with its projection on the xy-plane. Some examples are given in Fig. 1.

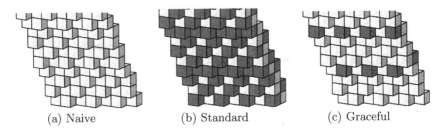

(a) Naive   (b) Standard   (c) Graceful

Fig. 1. Instances of three models of digital plane for $(a, b, c) = (4, -5, 8)$, in the domain $x \in [-3, 3], y \in [-7, 7]$. White voxels belong to the naive plane and the blue belong to (b) standard or (c) graceful. Notice that the xy-plane is functional here only for the naive plane. (Color figure online)

We define x-, y-, and z-*distance* between two (real or integer) points, p and p', as $d_x(p, p') = |i - i'|$, $d_y(p, p') = |j - j'|$, and $d_z(p, p') = |k - k'|$, respectively, where d_z is not applicable in 2D, $p = (i, j)$ and $p' = (i', j')$ in 2D, and $p = (i, j, k)$ and $p' = (i', j', k')$ in 3D. The x-distance between $p(i, j, k)$ and a curve/surface X is $d_x(p, X) = d_x(p, p')$ if $\exists\, p'(x', y', z') \in X$ such that $(y', z') = (j, k)$; otherwise, $d_x(p, X) = \infty$. The distances $d_y(p, X)$ and $d_z(p, X)$ are defined in a similar way. Let D denote the set $\{d_x(\cdot), d_y(\cdot)\}$ in 2D and $\{d_x(\cdot), d_y(\cdot), d_z(\cdot)\}$ in 3D. Then the *isothetic distance* between two points p and p' is the *Minkowski norm*, $d_\infty(p, p') = \max\{\delta : \delta \in D\}$, and that between p and a curve/surface X is $d_\perp(p, X) = \min\{\delta : \delta \in D\}$.

As shown in [2], each voxel of a naive plane (triangle in our case) has an isothetic distance of at most $\frac{1}{2}$ from the corresponding real plane. Hence, the naive voxelization results in the best possible approximation of a manifold with the guarantee of one-to-correspondence with the projection (pixel set) on its functional plane.

2.2 Homeomorphism of Voxels and 2-Manifolds

Let S be a closed and orientable (2-manifold) surface in the 3D Euclidean space, such that exactly two 2-manifolds (triangles) are incident on each of its 1-manifolds (edges) and at least three 1-manifolds are incident on each of its 0-manifolds (vertices). To derive a 3-manifold representation (naive voxelization) of S in $\mathbb{Z}^3$, we choose a *scale factor* $\xi > 0$ and apply an isotropic scaling on the 0-manifolds of S. Next, for every 2-manifold $t \in S$, we make its naive voxelization to obtain $N(t, \xi)$, and hence obtain the naive voxelization of S (scaled by the factor ξ) as $N(S, \xi) = \bigcup_{t \in S} N(t, \xi)$. Being closed and orientable, S is a compact surface without any boundary and the outward normal to each 2-manifold $t \in S$ is uniquely determinable, whence the functional plane(s) of each $N(t, \xi)$ is also fixed.

To define the topological space for S, let v be a voxel in $N(S, \xi)$. Then v is obtained in the naive voxelization of one or more 2-manifolds in S. So, we define $T(v) = \{t : t \in S \wedge v \in N(t, \xi)\}$ and $S = \{T(v) : v \in N(S, \xi)\}$. If Γ_S denotes the topology defined on S, then the corresponding topological space becomes (S, Γ_S). Now, to obtain the topological space $(\mathcal{V}, \Gamma_{\mathcal{V}})$ for $N(S, \xi)$, we define $\mathcal{V} = \{V(v) : v \in N(S, \xi)\}$, where $V(v) = N(T(v), \xi) := \bigcup_{t \in T(v)} N(t, \xi)$. Henceforth, for brevity, we denote (S, Γ_S) and $(\mathcal{V}, \Gamma_{\mathcal{V}})$ simply by S and $\mathcal{V}$, respectively [14].

We show the homeomorphism of S with $\mathcal{V}$ shortly in Theorem 1. For this, we define a basis B_S for S and another $B_{\mathcal{V}}$ for $\mathcal{V}$, as follows.

(i) Each $\beta_S^{(i)} \in B_S$ contains (as its element) every set $\{T(v) : (v \in N(T(v), \xi)) \wedge (x(v) = i)\}$, where $x(v)$ denotes the x-coordinate of the center of the voxel v.

(ii) If $(\beta_S^{(i)}, \beta_S^{(j)}) \in B_S^2$ and $\beta_S^{(i)} \cap \beta_S^{(j)} \neq \emptyset$, then $\beta_S^{(i)} \cap \beta_S^{(j)} \in B_S$.

(iii) Each $\beta_{\mathcal{V}}^{(i)} \in B_{\mathcal{V}}$ contains every set $\{V(v) : x(v) = i\}$.

(iv) If $(\beta_{\mathcal{V}}^{(i)}, \beta_{\mathcal{V}}^{(j)}) \in B_{\mathcal{V}}^2$ and $\beta_{\mathcal{V}}^{(i)} \cap \beta_{\mathcal{V}}^{(j)} \neq \emptyset$, then $\beta_{\mathcal{V}}^{(i)} \cap \beta_{\mathcal{V}}^{(j)} \in B_{\mathcal{V}}$.

Theorem 1. *The topological spaces S and $\mathcal{V}$ are homeomorphic for a sufficiently large value of ξ.*

Proof. Let $f : S \to \mathcal{V}$. So, if $T(v), T(v') \in S$, then $f(T(v)) = V(v) \in \mathcal{V}$ and $f(T(v')) = V(v') \in \mathcal{V}$. As ξ is sufficiently large, $T(v) \neq T(v')$ and $V(v) \neq V(v')$ for any (v, v') with $v \neq v'$. So, $V(v) = V(v') \iff T(v) = T(v') \iff v = v'$, wherefore f is bijective.

Now, to show that f is continuous, let v be a voxel with $x(v) = i$. Then $T(v)$ belongs to an element of $\beta_S^{(i)}$, and $f(T(v)) := V(v)$ belongs to an element of $\beta_{\mathcal{V}}^{(i)}$, which imply $f(\beta_S^{(i)}) \subset \beta_{\mathcal{V}}^{(i)}$, or, $\beta_S^{(i)} \subset f^{-1}(\beta_{\mathcal{V}}^{(i)})$, whence $f^{-1}(\beta_{\mathcal{V}}^{(i)})$ is an open set, thus showing f continuous. Similarly, $g := f^{-1} : \mathcal{V} \to S$ is also continuous, since $V(v)$ belongs to an element of $\beta_{\mathcal{V}}^{(i)}$ and $g(V(v))$ to an element of $\beta_S^{(i)}$, or, $g(\beta_{\mathcal{V}}^{(i)}) \subset \beta_S^{(i)}$, or, $\beta_{\mathcal{V}}^{(i)} \subset g^{-1}(\beta_S^{(i)})$, or, $g^{-1}(\beta_S^{(i)})$ is an open set. As a result, there exists a bijective continuous open map from S to $\mathcal{V}$, and hence the homeomorphism. $\qquad\square$

3   DIG: Topology and Construction

Given a seed voxel $s \in N(S,\xi)$ and a positive integer τ, we define a *discrete iso-contour geodesic* (DIG) as the 0-minimal path whose each voxel v has an intersection with S and a geodesic distance τ from s. We denote this DIG by $\Pi(S,\xi,s,\tau)$. The geodesic distance $d_g(s,v)$ from s to v is given by the length n of the shortest 0-path $\langle v_i : (0 \leqslant i \leqslant n) \wedge v_i \in K(S,\xi) \rangle$ from $s := v_0$ to $v := v_n$, $K(S,\xi)$ being the supercover of S.

If a DIG is made of voxels only from the naive voxelization, then it may not be 0-minimal. However, on replacing some of its voxel pairs by some special voxels from the *graceful triangles* corresponding to S, it becomes 0-minimal. In analogy with other geometric problems, we term these special voxels as *Steiner voxels*, since they are added to the naive set to make it graceful. Detailed study and analysis related to graceful planes may be seen in [4,5]. As shown in [4], a graceful plane is the thinnest possible voxelized plane on which primitives like lines, triangles, and arbitrary polygons are always connected sets of voxels. Here we show its usefulness for construction of DIG as well.

Let t be a 2-manifold in S with its functional plane $F(t)$, and let $N(t,\xi)$ and $G(t,\xi)$ be the respective naive and graceful planes. Let p and q be two distinct voxels in $N(t,\xi)$, and p' and q' be their respective projections on $F(t)$. Then $p' \neq q'$, due to the one-to-one correspondence between $N(t,\xi)$ and its projection on $F(t)$. Further, if p and q are 0-adjacent (resp., 1- or 2-adjacent) to each other, then p' and q' are also 0-adjacent (resp., 1-adjacent). However, 0-adjacency of p' and q' does not ascertain the connectedness of p and q in $N(t,\xi)$—a typical topological characteristic of naive plane that arises due to *jump* [4]. Figure 2 shows two jump configurations for each functional plane (FP). This is resolved in $G(t,\xi)$ by inserting a Steiner voxel in between the two voxels forming a jump in $N(t,\xi)$ so that the two voxels corresponding to two 0-adjacent pixels on the FP are 0-adjacent in $G(t,\xi)$ as well. The Steiner voxel is chosen from the supercover $K(t,\xi)$ and hence has an intersection with t. Observe that the *tandem* formed by the Steiner voxel and its 2-adjacent jump voxel maps to a single pixel on the FP.

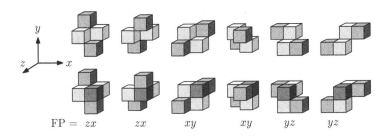

Fig. 2. Examples of adding a *Steiner voxel* to a *naive triangle*. **Top:** A *jump* (green voxel pair) on a naive triangle along with its two common 1-adjacent voxels (white). **Bottom:** A *tandem* made by a jump voxel and a *Steiner voxel* (blue) added from the graceful triangle. (Color figure online)

We have the following lemma.

Lemma 1 (Path Projection). *Let t be a 2-manifold in S, and P be a 0-path in $N(t, \xi)$. Then the projection of P on $F(t)$ is 0-minimal if and only if no voxel of P forms a tandem with some voxel in $G(t, \xi)$.*

Proof. Let p, q, r be three consecutive voxels in P, and p', q', r' be the respective pixels in the projection P' of P on $F(t)$. Clearly, p', q', r' are three distinct pixels on $F(t)$, since $P \subset N(t, \xi)$ and $N(t, \xi)$ has one-to-one correspondence with its projection on $F(t)$. So, q does not make any tandem with p or r in $G(t, \xi)$. Hence, p' and r' are adjacent if and only if q' is simple in P', or equivalently, q forms a tandem with some Steiner voxel in $G(t, \xi)$.   □

An example of obtaining a 0-minimal path based on Lemma 1 is shown in Fig. 3. Notice that here the path is basically a DIG. By Lemma 1, if the projection P' of P is not 0-minimal in $F(t)$, then there are one or more tandems. Each such tandem is formed by pairing a voxel $q \in N(S, \xi)$ with a voxel $u \in G(S, \xi) \setminus N(S, \xi)$. These local repairs in the constitution of P result in the desired 0-minimality of P' without breaking the connectedness of P in the voxel topology. In particular, we have the following lemma.

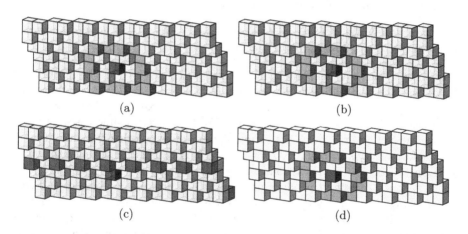

(a)   (b)

(c)   (d)

Fig. 3. An example showing DIG construction with the seed voxel s shown in red. (a) Green voxels are at geodesic distance 2 from s. (b) Back projection (green voxels and one yellow voxel) to the naive plane from the 0-minimal path on the functional plane; the two green voxels adjacent to the yellow voxel form the jump. (c) Graceful plane with the Steiner voxels shown in blue. (d) DIG (green voxels) after replacing one of the jump voxels by a Steiner voxel. (Color figure online)

Lemma 2 (Steiner Repair). *Let P be a 0-path in $N(t, \xi)$, P' be its projection on $F(t)$, and q' be the projection of $q \in P$. If q' is a simple pixel, then q and one of its adjacent voxels can be replaced by a single Steiner voxel to ensure the 0-minimality in both P and P'.*

Proof. As q' is a simple pixel, its preceding pixel p' and succeeding pixel r' in P' are 0-adjacent. Hence, by Lemma 1, one of their pre-images (p or r) forms a tandem with some Steiner voxel $u \in G(t, \xi)$. Let, w.l.o.g., that tandem be (p, u). Then replacing (p, q) by u ensures the local minimality in both P and P'.   □

We now introduce the following lemma for the theorem that explains the construction of DIG using $N(S, \xi)$ and the Steiner voxels as needed.

Lemma 3 (Geodesic Distance). *For any 2-manifold t in S, the geodesic distance between two voxels in $N(t, \xi)$ is given by the isothetic distance between their projections on $F(t)$.*

Proof. $N(t, \xi)$ has one-to-one correspondence with its projection on $F(t)$. Hence, by definitions of isothetic distance and geodesic distance, the proof follows.   □

Let $B(S, \xi, s, \tau)$ denote the set of voxels from $N(S, \xi)$ having geodesic distance τ from s. This is obtained by breadth-first-search in $N(S, \xi)$ with s as the start vertex in the underlying graph. Let $B(S, \xi, s, \tau)'$ denote the collection of its piecewise projections on the respective functional planes of the participating 2-manifolds of S. Let $\Pi(S, \xi, s, \tau)'$ denote the piecewise projections of $\Pi(S, \xi, s, \tau)$ in a similar manner. We have now the following theorem.

Theorem 2 (DIG). *$\Pi(S, \xi, s, \tau)'$ is contained in $B(S, \xi, s, \tau)'$ and is 0-minimal on the respective functional planes.*

Proof. Let t be any 2-manifold in S. Let $\Pi_t(S, \xi, s, \tau)$ be the portion of $\Pi(S, \xi, s, \tau)$ corresponding to t. Also, let $\Pi_t(S, \xi, s, \tau)'(\subseteq \Pi(S, \xi, s, \tau)')$ be the piecewise projection of $\Pi(S, \xi, s, \tau)$ on $F(t)$. Each voxel $q \in \Pi_t(S, \xi, s, \tau)$ has the geodesic distance $d_g(s, q) = \tau$ from s. We have two possible cases: either s belongs to $N(t, \xi)$ or it belongs to the naive set of some other 2-manifold in S. For the former, $d_g(s, q) = d_\infty(p, q)$ by Lemma 3. For the latter, let p be the voxel lying on a/the geodesic path from s to q and common to $N(t, \xi)$ and $N(t_1, \xi)$, where t_1 is a 2-manifold incident on one of the three 1-manifolds of t. Then, $d_g(s, q) = d_g(s, p) + d_\infty(p, q)$ by Lemma 3. Hence, in either case, if q (along with one of its adjacent voxels) is replaced by a Steiner voxel u (Lemma 2), then for u, we have $d_u(s, u) = d_g(s, q)$. This ensures the containment of $\Pi_t(S, \xi, s, \tau)'$ in $B(S, \xi, s, \tau)'$, and hence the result follows.   □

4   Concluding Remarks

We have tested the algorithm for DIG construction on the naive voxel sets of different objects at different scales. On examining and analyzing these test results, it becomes evident that a collection of DIG, constructed with regular geodesic distances of $\tau, 2\tau, 3\tau, \ldots$, from a seed point randomly chosen in $N(S, \xi)$, can aid in inferring on interesting geometric and topological features and the complexity of the object. A couple of such empirical observations are presented here.

See Fig. 4, which shows a collection of DIG, with uniformly changing values of τ, constructed on a regular icosahedron made of 20 equilateral triangles. Notice that the DIG for $\tau = 5$ is almost squarish, since it lies in some triangles whose functional planes are same. As the value of τ increases to $10, 15, \ldots$, the functional planes of the concerned triangles gradually vary, thereby changing the shape of DIG more and more. If the object is more roundish, such as a regular polyhedron with a larger number of faces, a DIG also becomes more regular and symmetric. The position of the seed point does not have any significant role, and neither the orientation of the object S, as far as the scale factor ξ is not uncompromisingly small.

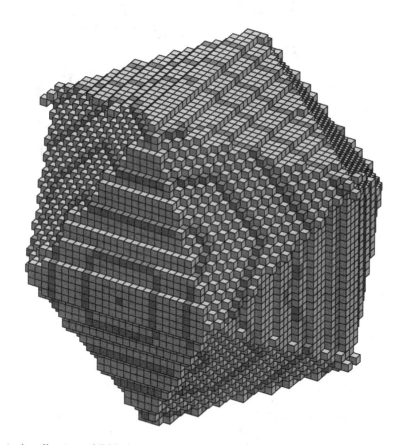

Fig. 4. A collection of DIG for $\tau = 5, 10, 15, \ldots$, constructed by our algorithm on the naive set of a regular icosahedron (the seed point s is shown in red). (Color figure online)

The collection of DIG can also be used to detect tunnels in an object, and hence to compute its genus, which is a strong topological feature commonly used for shape analysis. Figure 5 shows a typical set of results on the naive

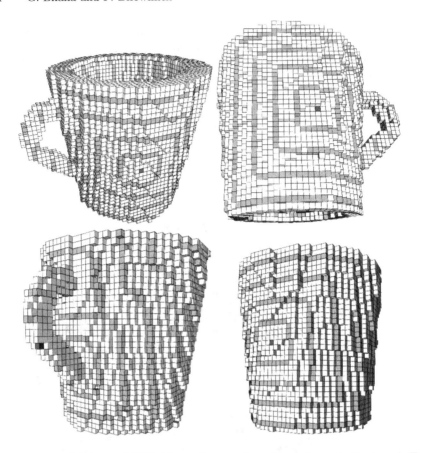

Fig. 5. Four collections of DIG (shown in green) on the naive set of a 'mug'. Each collection is generated with a seed point shown in red. (Color figure online)

voxel set of an object with genus one. With four different seed points widely varying in position, the final results in all cases lead to the occurrence of a 'handle' in the object. This is inferred from the fact that in the handle, the DIGs can be paired based on their geodesic distances from s. The two DIGs in every pair are geodesically equidistant from s, and hence implies two articulation points from the main 'body'. If the two DIGS in the farthest pair among these pairs are connected with each other within a geodesic distance of 2τ, then the connecting part belongs to the handle and hence indicates the occurrence of a tunnel; otherwise, it signifies two articulated projections connected through the main body.

The notion of DIG introduced in this paper clearly shows its theoretical merit as well as practical uses in different tasks and applications related to voxelized objects. Setting the value of the scale factor ξ for voxelization of a 2-manifold S in order to ensure its topological equivalence with S through homeomorphism remains an open problem. We foresee this problem important in the theoretical context of DIG construction and also for shape analysis. Apart from genus and articulation points that are briefly discussed in this paper, many other shape features like regularity, concavity, convexity, and symmetry can also possibly be analyzed through DIG constitution in the voxel space, which if done, would further establish its potential in digital geometry and topology.

References

1. Balasubramanian, M., Polimeni, J.R., Schwartz, E.L.: Exact geodesics and shortest paths on polyhedral surfaces. IEEE TPAMI **31**, 1006–1016 (2009)
2. Biswas, R., Bhowmick, P.: On different topological classes of spherical geodesic paths and circles in $\mathbb{Z}^3$. Theoret. Comput. Sci. **605**, 146–163 (2015)
3. Brimkov, V.E.: Formulas for the number of $(n - 2)$-gaps of binary objects in arbitrary dimension. Discrete Appl. Math. **157**, 452–463 (2009)
4. Brimkov, V.E., Barneva, R.P.: Graceful planes and lines. Theoret. Comput. Sci. **283**, 151–170 (2002)
5. Brimkov, V.E., Coeurjolly, D., Klette, R.: Digital planarity–a review. Discrete Appl. Math. **155**, 468–495 (2007)
6. Chandru, V., Manohar, S., Prakash, C.E.: Voxel-based modeling for layered manufacturing. IEEE Comput. Graph. App. **15**, 42–47 (1995)
7. Chang, H.H., Lai, Y.C., Yao, C.Y., Hua, K.L., Niu, Y., Liu, F.: Geometry-shader-based real-time voxelization and applications. Vis. Comput. **30**, 327–340 (2014)
8. Coeurjolly, D., Miguet, S., Tougne, L.: 2D and 3D visibility in discrete geometry: an application to discrete geodesic paths. PRL **25**, 561–570 (2004)
9. Cohen-Or, D., Kaufman, A.: Fundamentals of surface voxelization. Graph. Models Image Process. **57**, 453–461 (1995)
10. Dachille, F., Kaufman, A.E.: Incremental triangle voxelization. In: Graphics Interface Conference, pp. 205–212 (2000)
11. Desimone, J.M., Ermoshkin, A., Samulski, E.T.: Method and apparatus for three-dimensional fabrication. US Patent 20140361463 (2014)
12. Dionne, O., de Lasa, M.: Geodesic binding for degenerate character geometry using sparse voxelization. IEEE TVCG **20**, 1367–1378 (2014)
13. Dong, Z., Chen, W., Bao, H., Zhang, H., Peng, Q.: Real-time voxelization for complex polygonal models. In: PG 2004, pp. 43–50 (2004)
14. Edelsbrunner, H., Harer, J.L.: Computational Topology. American Mathematical Society, Providence (2009)
15. Eisemann, E., Décoret, X.: Single-pass GPU solid voxelization for real-time applications. In: GI 2008, pp. 73–80 (2008)
16. Fang, S., Fang, S., Chen, H., Chen, H.: Hardware accelerated voxelization. Comput. Graph. **24**, 433–442 (2000)
17. Hiller, J., Lipson, H.: Design and analysis of digital materials for physical 3D voxel printing. Rapid Prototyping J. **15**, 137–149 (2009)
18. Hiller, J., Lipson, H.: Tunable digital material properties for 3D voxel printers. Rapid Prototyping J. **16**, 241–247 (2010)
19. Hull, C.: Apparatus for production of three-dimensional objects by stereolithography. US Patent 4575330 (1986)
20. Jee, H.J., Sachs, E.: A visual simulation technique for 3D printing. Adv. Eng. Softw. **31**, 97–106 (2000)

21. Kamrani, A.K., Nasr, E.A.: Engineering Design and Rapid Prototyping. Springer, Boston (2009)
22. Karabassi, E.A., Papaioannou, G., Theoharis, T.: A fast depth-buffer-based voxelization algorithm. J. Graph. Tools 4(4), 5–10 (1999)
23. Klette, R., Rosenfeld, A.: Digital Geometry: Geometric Methods for Digital Picture Analysis. Morgan Kaufmann, San Francisco (2004)
24. Liao, D.: GPU-accelerated multi-valued solid voxelization by slice functions in real time. In: VRCAI 2008, pp. 18:1–18:6 (2008)
25. Malgouyres, R.: A discrete radiosity method. In: Braquelaire, A., Lachaud, J.-O., Vialard, A. (eds.) DGCI 2002. LNCS, vol. 2301, p. 428. Springer, Heidelberg (2002)
26. Mitchell, J.S.B., Mount, D.M., Papadimitriou, C.H.: The discrete geodesic problem. SIAM J. Comput. 16, 647–668 (1987)
27. Mukhopadhyay, J., Das, P.P., Chattopadhyay, S., Bhowmick, P., Chatterji, B.N.: Digital Geometry in Image Processing. CRC, Boca Ration (2013)
28. Nanya, T., Yoshihara, H., Maekawa, T.: Reconstruction of complete 3D models by voxel integration. J. Adv. Mech. Des. Syst. Manuf. 7, 362–376 (2013)
29. Nooruddin, F.S., Turk, G.: Simplification and repair of polygonal models using volumetric techniques. IEEE TVCG 9, 191–205 (2003)
30. Pomerantz, I., Cohen-Sabban, J., Bieber, A., Kamir, J., Katz, M., Nagler, M.: Three dimensional modelling apparatus. US Patent 4961154 (1990)
31. Prakash, C., Manohar, S.: Volume rendering of unstructured grids–a voxelization approach. Comput. Graph. 19, 711–726 (1995)
32. Steingart, Robert, C., Tzu-Wei, D.: Fabrication of non-homogeneous articles via additive manufacturing using three-dimensional voxel-based models. US Patent 8509933 (2013)
33. Surazhsky, V., Surazhsky, T., Kirsanov, D., Gortler, S.J., Hoppe, H.: Fast exact and approximate geodesics on meshes. ACM TOG 24, 553–560 (2005)
34. Truong-Hong, L., Laefer, D.F., Hinks, T., Carr, H.: Combining an angle criterion with voxelization and the flying voxel method in reconstructing building models from lidar data. Comput. Aided Civ. Infrastruct. Eng. 28, 112–129 (2013)
35. Xin, S.Q., Wang, G.J.: Improving Chen and Han's algorithm on the discrete geodesic problem. ACM TOG 28, 104:1–104:8 (2009)
36. Xin, S.Q., Ying, X., He, Y.: Constant-time all-pairs geodesic distance query on triangle meshes. In: I3D 2012, pp. 31–38 (2012)
37. Ying, X., Wang, X., He, Y.: Saddle vertex graph (SVG): a novel solution to the discrete geodesic problem. ACM TOG 32, 170:1–170:12 (2013)
38. Zhang, Z., Morishima, S.: Application friendly voxelization on GPU by geometry splitting. In: Christie, M., Li, T.-Y. (eds.) SG 2014. LNCS, vol. 8698, pp. 112–120. Springer, Heidelberg (2014)

Solving Distance Geometry Problem with Inexact Distances in Integer Plane

Piyush K. Bhunre[(⊠)], Partha Bhowmick, and Jayanta Mukhopadhyay

Department of Computer Science and Engineering, Indian Institute of Technology,
Kharagpur, India
kbpiyush@gmail.com, {pb,jay}@cse.iitkgp.ernet.in

Abstract. Given the pairwise distances for a set of unknown points in a known metric space, the distance geometry problem (DGP) is to compute the point coordinates in conformation with the distance constraints. It is a well-known problem in the Euclidean space, has several variations, finds many applications, and so has been attempted by different researchers from time to time. However, to the best of our knowledge, it is not yet fully addressed to its merit, especially in the discrete space. Hence, in this paper we introduce a novel variant of DGP where the pairwise distance between every two unknown points is given a tolerance zone with the objective of finding the solution as a collection of integer points. The solution is based on characterization of different types of annulus intersection, their equivalence, and cardinality bounds of integer points. Necessary implementation details and useful heuristics make it attractive for practical applications in the discrete space.

1 Introduction

Given the inter-point distances for a finite set of points with unknown coordinates in a particular metric space, the *distance geometry problem* (DGP) is to embed the points in that space so as to satisfy the distance constraints. The basic problem in the Euclidean space was first introduced by Menger in 1920s, and later formally established by Blumenthal in 1950s [2,14,17]. The general version of the problem is as follows. Given a positive integer k and a simple undirected weighted graph $G = (V, E)$ with weight function $w : E \mapsto \mathbb{R}^+$, determine if there is a function $f : V \mapsto \mathbb{R}^k$ such that for all $(u, v) \in E$, $\|f(u) - f(v)\| = w(u, v)$, where $\| \cdot \|$ denotes the Euclidean norm in k-dimensional space. The problem is NP-complete for $k = 1$ (i.e., embedding in real line) and NP-hard for $k > 1$ [19]. Following are the variations of DGP as per the input distance set.

1. *Exact Distances:* The point-set embedding should be such that the corresponding distance set exactly matches the input distance set. Here all the distances are given, and so the distance set is a complete set.
2. *Inexact or Bounded Distances:* Each pairwise distance in the solution need not be exact, but lies in an interval.
3. *Sparse Distances:* The distance set is incomplete; in addition, the distances could be of exact or of bounded type.

© Springer International Publishing Switzerland 2016
A. Bac and J.-L. Mari (Eds.): CTIC 2016, LNCS 9667, pp. 277–289, 2016.
DOI: 10.1007/978-3-319-39441-1_25

The second and the third variations of the problem are more realistic than the first one, since the exact distances are often not available due to limitations of the measurement device. The DGP is related to many important research problems, such as Molecular Distance Geometry Problem and Molecular Conformation [8,11–13,15,16,20], Sensor Network Localization [4,5,18], and Graph Drawing [1,3,6,9]. In all these problems, the DGP is dealt in the real space. When the pairwise distances are exact, it can be solved in polynomial time [7,20]. For bounded and sparse distances, the scenario becomes complex and difficult to solve [14,20].

In our work, we have addressed the DGP for inexact or bounded distances in the integer plane and have devised an efficient technique for solving it whenever a given input instance admits a feasible solution. Although all the solution points have integer coordinates, the proposed technique can be extended for finding solutions in finer grids. To start with, in Sect. 2, we formulate the problem in the integer plane. In Sect. 3, we present a characterization of digital annulus intersection for efficient computation of the solution in the integer plane. The proposed heuristics and algorithm are discussed in Sect. 4. Further research directions and concluding notes are put in Sect. 5.

2   Preliminaries and Problem Definition in $\mathbb{Z}^2$

We fix here few definitions and notations used in the sequel. A real circle with center at $c \in \mathbb{R}^2$ and radius r is denoted by $\mathcal{C}^{\mathbb{R}}(c, r)$. A real annulus is defined as $\mathcal{A}^{\mathbb{R}}(c, a, b) = \{p \in \mathbb{R}^2 : a \leqslant \|c - p\| \leqslant b\}$, where $c \in \mathbb{R}^2$ is the center of the annulus, and a and b are the respective inner and outer radii. A digital annulus centered at $c \in \mathbb{R}^2$, and with inner and outer radii a and b respectively, is defined as the set of all integer points in $\mathcal{A}^{\mathbb{R}}(c, a, b)$, and so given by $\mathcal{A}^{\mathbb{Z}}(c, a, b) = \{p \in \mathbb{Z}^2 : a \leqslant \|c - p\| \leqslant b\}$.

Let $P = \{p_1, p_2, \ldots, p_n\}$ be a set of points with unknown coordinates in $\mathbb{Z}^2$ such that the distance between every two points p_i and p_j is known to lie in a given interval $[a_{ij}, b_{ij}]$. The objective is to determine the integer coordinates of all points in P. We consider an alternative form of the distance bounds where the distance between p_i and p_j lies in a known interval $[d_{ij} - \epsilon, d_{ij} + \epsilon]$ for some $\epsilon > 0$. We also assume that the given intervals are such that there is at least one solution to the problem.

To simplify our strategy, we first fix the reference frame with p_1 as the origin, p_2 as an integer point on the $+x$-axis, and p_3 lying left of $\overrightarrow{p_1 p_2}$. With $a_{12} \leqslant \|p_1 - p_2\| \leqslant b_{12}$, we get $p_2 \in \mathcal{A}^{\mathbb{Z}}(p_1, a_{12}, b_{12})$, and $p_3 \in \mathcal{A}^{\mathbb{Z}}(p_1, a_{13}, b_{13}) \cap \mathcal{A}^{\mathbb{Z}}(p_2, a_{23}, b_{23})$ (Fig. 1a). Now, for $i \geqslant 4$, if we try to embed p_i based on its distance intervals from the previous $i - 1$ points, then we need to compute $\bigcap_{j=1}^{i-1} \mathcal{A}^{\mathbb{Z}}(p_j, a_{ji}, b_{ji})$, which is expensive. Hence, as a faster solution, we use $\bigcap_{j=1}^{3} \mathcal{A}^{\mathbb{Z}}(p_j, a_{ji}, b_{ji})$ for each $p_i (i \geqslant 4)$ (Fig. 1b). For this, we have the following lemma.

Lemma 1. *All the solution points corresponding to p_i for $i = 4, 5, \ldots, n$ belong to $\bigcap_{j=1}^{3} \mathcal{A}^{\mathbb{Z}}(p_j, a_{ji}, b_{ji})$, which, if empty, does not yield any solution for P.*

Proof. Follows from the problem statement and our consideration.   □

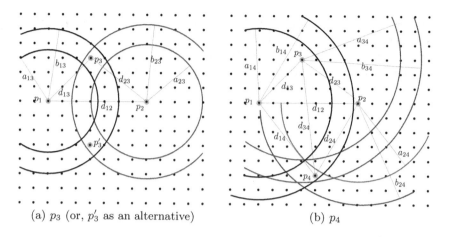

(a) p_3 (or, p'_3 as an alternative)   (b) p_4

Fig. 1. Embedding of p_3 followed by p_4 in $\mathbb{Z}^2$.

3 Characteristics of Annulus Intersection

We first provide some basic concepts of discrete geometry in the integer plane. For more details we refer to [10]. Let $p = (i, j) \in \mathbb{Z}^2$. The 4-*neighbors* of p in $\mathbb{Z}^2$ is defined as $N_4(p) = \{(i', j') \in \mathbb{Z}^2 : |i - i'| + |j - j'| = 1\}$, and the 8-*neighbors* as $N_8(p) = \{(i', j') \in \mathbb{Z}^2 : \max(|i - i'|, |j - j'|) = 1\}$. A subset A of $\mathbb{Z}^2$ is said to be 4-*connected* (resp., 8-*connected*) if either A is singleton or there exists a sequence of points in A between every two points of A such that every two consecutive points in that sequence are 4-neighbors (resp., 8-neighbors) of each other. When A is not connected, its maximally connected subsets are called *connected components*.

Let, without loss of generality, p_1 and p_2 be the farthest pair, and the distance d_{12} between them be denoted by d for simplicity. We examine the possible locations of an unknown point p, which is at a distance d_1 and d_2 from p_1 and p_2 respectively, where $a_1 \leqslant d_1 \leqslant b_1$ and $a_2 \leqslant d_2 \leqslant b_2$. So, by our supposition, $\max(b_1, b_2) \leqslant d$. Also, $b_1 + b_2 \geqslant d$, failing which there will be no such point p. Then p can be chosen as some point common to the digital annuli $\mathcal{A}^{\mathbb{Z}}(p_1, a_1, b_1)$ and $\mathcal{A}^{\mathbb{Z}}(p_2, a_2, b_2)$ if their intersection is non-empty. When a solution exists, i.e., the width of the annulus is sufficiently large, the intersection will be either a single or a pair of 8-connected components (Fig. 2). We define $\mathcal{I}^{\mathbb{Z}}$ as the set of integer points belonging to both $\mathcal{A}^{\mathbb{Z}}(p_1, a_1, b_1)$ and $\mathcal{A}^{\mathbb{Z}}(p_2, a_2, b_2)$. The non-empty intersection $\mathcal{I}^{\mathbb{Z}}$ is classified as follows.

- *Type 1* ($a_1 + a_2 > d$): $\mathcal{I}^{\mathbb{Z}}$ comprises two connected components lying in two different sides of the line $p_1 p_2$. (As a degeneracy, it may have a single component when $a_1 + a_2$ tends to d.)
- *Type 2* (($a_1 + a_2 \leqslant d) \wedge (a_1 + b_2 \geqslant d) \wedge (b_1 + a_2 \geqslant d)$): $\mathcal{I}^{\mathbb{Z}}$ is a single connected component. (As a degeneracy, it may have two connected components as in Type 1.)

- *Type 3* $(((a_1 + b_2 < d) \wedge (b_1 + a_2 > d)) \vee ((b_1 + a_2 < d) \wedge (a_1 + b_2 > d)))$: $\mathcal{I}^{\mathbb{Z}}$ is a single connected component.
- *Type 4* $((a_1 + a_2 < d) \wedge (a_1 + b_2 < d) \wedge (b_1 + a_2 < d))$: $\mathcal{I}^{\mathbb{Z}}$ is a single connected component bounded by the outer circles of the real annuli.

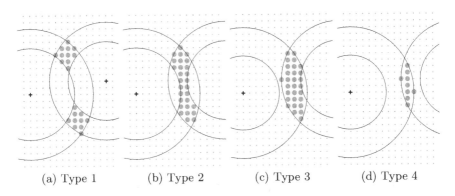

(a) Type 1   (b) Type 2   (c) Type 3   (d) Type 4

Fig. 2. Possible types of intersection between two annuli such that no radius exceeds the distance between the annulus centers.

Note that there can be other types of intersection if $\max(b_1, b_2) > d$. Hence, to avoid this, we consider p_1 and p_2 as the farthest pair. we have the following lemma related to intersection types.

Lemma 2. *Type 3 and Type 4 intersections are computationally equivalent to Type 2.*

Proof. For Type 3 (Fig. 3(a-b)), we can replace $\mathcal{A}^{\mathbb{Z}}(p_1, a_1, b_1)$ by $\mathcal{A}^{\mathbb{Z}}(p_1, a_1', b_1)$ with $a_1' = d - b_2$, whence it gets converted to Type 2, thereby equalizing $\mathcal{A}^{\mathbb{Z}}(p_1, a_1, b_1) \cap \mathcal{A}^{\mathbb{Z}}(p_2, a_2, b_2)$ to $\mathcal{A}^{\mathbb{Z}}(p_1, a_1', b_1) \cap \mathcal{A}^{\mathbb{Z}}(p_2, a_2, b_2)$; similarly, we can replace $\mathcal{A}^{\mathbb{Z}}(p_2, a_2, b_2)$ by $\mathcal{A}^{\mathbb{Z}}(p_2, a_2', b_2)$ with $a_2' = d - b_1$, which also results in Type 2. In case of Type 4 (Fig. 3c), $\mathcal{A}^{\mathbb{Z}}(p_1, a_1, b_1)$ and $\mathcal{A}^{\mathbb{Z}}(p_2, a_2, b_2)$ can be replaced by $\mathcal{A}^{\mathbb{Z}}(p_1, a_1', b_1)$ and $\mathcal{A}^{\mathbb{Z}}(p_2, a_2', b_2)$ respectively, with $a_1' = d - b_2$ and $a_2' = d - b_1$, thereby again resulting in Type 2. □

Thus, Type 3 and Type 4 intersections can be converted to their equivalent Type 2 intersection, which simplifies the associated computation.

3.1   Computing Annulus Intersection in $\mathbb{Z}^2$

We first ensure that an annulus intersection is of Type 1 or of Type 2 or its equivalent. For computing the integer points in $\mathcal{I}^{\mathbb{Z}} := \mathcal{A}^{\mathbb{Z}}(p_1, a_1, b_1) \cap \mathcal{A}^{\mathbb{Z}}(p_2, a_2, b_2)$, we first find a seed point $s \in \mathcal{I}^{\mathbb{R}} := \mathcal{A}^{\mathbb{R}}(p_1, a_1, b_1) \cap \mathcal{A}^{\mathbb{R}}(p_2, a_2, b_2)$. The seed point s is taken as an/the intersection point of the circles $\mathcal{C}^{\mathbb{R}}(p_1, \frac{a_1 + b_1}{2})$

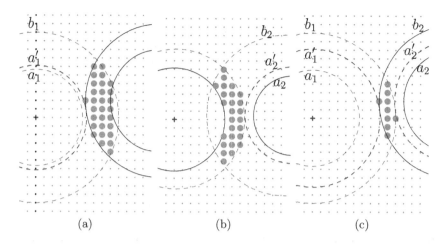

(a)   (b)   (c)

Fig. 3. Converting Type 3 and Type 4 intersections into equivalent Type 2 intersection by removing the empty sub-annulus incident on the inner circle (shown dashed and blue). (Color figure online)

and $\mathcal{C}^{\mathbb{R}}(p_2, \frac{a_2+b_2}{2})$. Note that $\mathcal{C}^{\mathbb{R}}(p_1, \frac{a_1+b_1}{2})$ and $\mathcal{C}^{\mathbb{R}}(p_2, \frac{a_2+b_2}{2})$ intersect at two points if $d' = \frac{a_1+b_1}{2} + \frac{a_2+b_2}{2} > d$, where d is the distance between p_1 and p_2. If $d' = d$, then they touch at a single point, and they do not intersect or touch at all when $d' < d$. As the intersection is of Type 1 or Type 2, there always exists a seed point with minimum distance $\frac{b_1-a_1}{2}$ and $\frac{b_2-a_2}{2}$ from the boundaries of the annuli (Fig. 4). This gives us the following lemma.

Lemma 3. *If* $\min(b_1 - a_1, b_2 - a_2) \geqslant \sqrt{2}$ *and the intersection of* $\mathcal{A}^{\mathbb{Z}}(p_1, a_1, b_1)$ *and* $\mathcal{A}^{\mathbb{Z}}(p_2, a_2, b_2)$ *is of Type 1 or of Type 2, then there exists an integer point* $q_0 \in \mathcal{A}^{\mathbb{Z}}(p_2, a_1, b_1) \cap \mathcal{A}^{\mathbb{Z}}(p_2, a_2, b_2)$ *and it can be determined in constant time.*

Proof. The seed point s belongs to a unit square U whose vertices are integer points. If the width of each annulus is at least $\sqrt{2}$, then the minimum distance of s from each annulus boundary is at least $\frac{\sqrt{2}}{2} = \frac{1}{\sqrt{2}}$. Hence, the real circle $\mathcal{C}^{\mathbb{R}}(s, \frac{1}{\sqrt{2}})$ is completely enclosed by both the annuli. Since s lies on or inside U, at least one vertex of U is within a distance of $\frac{1}{\sqrt{2}}$ from s, and so lies on or inside $\mathcal{C}^{\mathbb{R}}(s, \frac{1}{\sqrt{2}})$, and hence inside the annulus intersection.   □

By Lemma 3, we get an integer point $q_0 \in \mathcal{I}^{\mathbb{Z}}$. Starting from q_0, we get all other points in $\mathcal{I}^{\mathbb{Z}}$ as a connected component, using breadth first search (BFS) in the integer plane.

Henceforth in this paper, for notional and notational simplicity, we consider 2ϵ as the width of each annulus. So, the digital annulus centered at a point p_i is denoted by $\mathcal{A}^{\mathbb{Z}}(p_i, d_i - \epsilon, d_i + \epsilon)$, where d_i is its mean radius.

Theorem 1. *If* $d := \|p_1 - p_2\| \geqslant \max(d_1, d_2) + \epsilon$, *then the cardinality of* $\mathcal{I}^{\mathbb{Z}}$ *is* $O(\epsilon^2)$ *for Type 1 intersection and* $O(\epsilon\sqrt{\epsilon d})$ *for Type 2 intersection.*

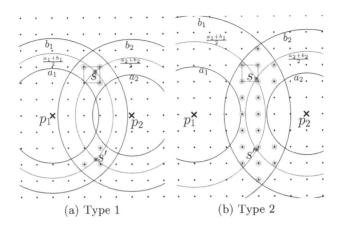

(a) Type 1   (b) Type 2

Fig. 4. Finding a seed points s (red) for computing the annulus intersection. (Color figure online)

Proof. We have $d > d_1 + \epsilon$ and $d > d_2 + \epsilon$. If $d > d_1 + d_2 + 2\epsilon$, then $\mathcal{I}^{\mathbb{Z}}$ is empty. So, $d \leqslant d_1 + d_2 + 2\epsilon$. Without loss of generality, we take $p_1 = (0, 0)$ and $p_2 = (d, 0)$, as illustrated in Fig. 5.

Let $\mathcal{I}^+ = \{(x, y) \in \mathcal{I}^{\mathbb{R}} : y \geqslant 0\}$, $x_{\min} = \min\{x : (x, y) \in \mathcal{I}^+\}$, $x_{\max} = \max\{x : (x, y) \in \mathcal{I}^+\}$, $y_{\min} = \min\{y : (x, y) \in \mathcal{I}^+\}$, and $y_{\max} = \max\{y : (x, y) \in \mathcal{I}^+\}$. Then there exists a rectangle with edge-lengths $w = (x_{\max} - x_{\min})$ and $h = (y_{\max} - y_{\min})$ such that it encloses $\mathcal{I}^+$. Hence, the number of integer points in $\mathcal{I}^+$ is $O(wh)$, and from the symmetry of the problem, the cardinality of $\mathcal{I}^{\mathbb{Z}}$ is $O(wh)$. We determine w and h by case analysis.

Case I ($\mathcal{I}^{\mathbb{R}}$ is Type 1): Here, $d \leqslant d_1 + d_2 - 2\epsilon$, and $\mathcal{I}^{\mathbb{R}}$ consists of two connected components: one lies above and the other lies below the line $p_1 p_2$ (Fig. 5(a)). Then $x_{\min}$ is the abscissa of the point of intersection of $\mathcal{C}^{\mathbb{R}}(p_1, d_1 - \epsilon)$ and $\mathcal{C}^{\mathbb{R}}(p_2, d_2 + \epsilon)$. By solving the equations $x^2 + y^2 = (d_1 - \epsilon)^2$ and $(x - d)^2 + y^2 = (d_2 + \epsilon)^2$, we get $x_{\min} = \frac{d^2 + (d_1 - \epsilon)^2 - (d_2 + \epsilon)^2}{2d}$. Similarly, the abscissa of the point of intersection of $\mathcal{C}^{\mathbb{R}}(p_1, d_1 + \epsilon)$ and $\mathcal{C}^{\mathbb{R}}(p_2, d_2 - \epsilon)$ gives $x_{\max} = \frac{d^2 + (d_1 + \epsilon)^2 - (d_2 - \epsilon)^2}{2d}$. So by using $d \geqslant \max(d_1, d_2) + \epsilon$, we get

$$w = x_{\max} - x_{\min} = \frac{4\epsilon(d_1 + d_2)}{2d} \leqslant 4\epsilon\left(1 - \frac{\epsilon}{d}\right) \leqslant 4\epsilon. \tag{1}$$

By symmetry of the problem, we can interchange x and y to obtain a similar bound for h as 4ϵ. Hence, for each of the two components of the annulus intersection, there exists an enclosing square of length 4ϵ. Hence, the cardinality of $\mathcal{I}^{\mathbb{Z}}$ becomes $O(\epsilon^2)$.

Case II ($\mathcal{I}^{\mathbb{R}}$ is Type 2): Here, $d_1 + d_2 - 2\epsilon < d$, because the inner circles $\mathcal{C}^{\mathbb{R}}(p_1, d_1 - \epsilon)$ and $\mathcal{C}^{\mathbb{R}}(p_2, d_2 - \epsilon)$ do not intersect or touch each other. By following the procedure used in Case I, it can be shown that $x_{max} - x_{min} \leqslant 4\epsilon$, or, $w = O(\epsilon)$.

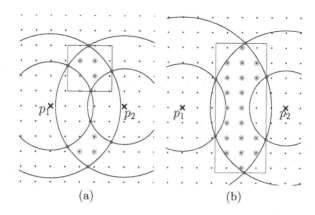

Fig. 5. Cardinality of intersection for annulus width 2ϵ. (a) Type 1 annulus intersection is bounded by a rectangle with (axis-parallel) edges of length $O(\epsilon)$ and cardinality $O(\epsilon^2)$. (b) Type 2 annulus intersection is bounded by a rectangle with edges of length $O(\epsilon)$ and $O(\sqrt{\epsilon d})$, and cardinality $O(\epsilon\sqrt{\epsilon d})$.

Let $c = (c_x, c_y)$ be the topmost point of the annulus intersection, and hence it is the intersection point of the outer real circles of the two annuli. By symmetry, $c' = (c_x, -c_y)$ is the bottommost point, which is the reflection of c with respect to x-axis. Then solving the equations $c_x^2 + c_y^2 = (d_1 + \epsilon)^2$ and $(d - c_x)^2 + c_y^2 = (d_2 + \epsilon)^2$, we get $c_x = \frac{d^2 + (d_1 + \epsilon)^2 - (d_2 + \epsilon)^2}{2d}$ and c_y as follows.

$$
\begin{aligned}
c_y^2 &= (d_1 + \epsilon)^2 - \left(\frac{d^2 + (d_1 + \epsilon)^2 - (d_2 + \epsilon)^2}{2d}\right)^2 = \frac{\left((d_1 + d_2 + 2\epsilon)^2 - d^2\right)\left(d^2 - (d_1 - d_2)^2\right)}{4d^2} \\
&\leqslant \frac{\left((d + 4\epsilon)^2 - d^2\right)\left(d^2 - (d_1 - d_2)^2\right)}{4d^2}, \text{ since } d_1 + d_2 - 2\epsilon < d \\
&= 4\epsilon^2 \left(1 + \frac{d}{2\epsilon}\right)\left(1 - \left(\frac{d_1 - d_2}{d}\right)^2\right) \\
&\leqslant 4\epsilon^2 \left(\frac{d}{2\epsilon} + \frac{d}{2\epsilon}\right), \text{ since } d \geqslant \max(d_1, d_2) + \epsilon, d > d_1 + d_2 - 2\epsilon, \frac{d}{2\epsilon} > 1 \\
&= 4\epsilon d \implies c_y = O(\sqrt{\epsilon d}).
\end{aligned}
\tag{2}
$$

By Eq. 2, $h = O(\sqrt{\epsilon d})$; as $w = O(\epsilon)$, the number of points in $\mathcal{I}^{\mathbb{Z}}$ is $O(\epsilon\sqrt{\epsilon d})$. $\square$

By Lemma 3 and Theorem 1, we have the following corollary.

Corollary 1. $\mathcal{I}^{\mathbb{Z}}$ is computable in $O(\epsilon^2)$ time for Type 1 and in $O(\epsilon\sqrt{\epsilon d})$ time for Type 2 intersections.

4   Proposed Algorithm and Implementation Issues

The first three points p_1, p_2, and p_3 are known, or are fixed with an appropriate coordinate system, as explained in Sect. 2. That is, for $i = 1, 2, 3$, the candidate solutions are $\mathcal{I}_i^{\mathbb{Z}} = \{p_i\}$. For each other point p_i, a set of candidate solutions is obtained as $\mathcal{I}_i^{\mathbb{Z}} := \bigcap_{j=1}^3 \mathcal{A}^{\mathbb{Z}}(p_j, a_{ji}, b_{ji})$. One or more points from $\mathcal{I}_i^{\mathbb{Z}}$ would belong to a/the global solution with all n points. In particular, we have the following theorem on the collection of these sets of candidate solutions.

Theorem 2. *If a given set of distance constraints admits a global solution, then it is always contained in the collection $\mathcal{K} := \{\mathcal{I}_i^{\mathbb{Z}} : i = 1, 2, \ldots, n\}$ whose computational time is $O(\epsilon^2 n)$ in the best case and $O(\epsilon^{\frac{3}{2}} d^{\frac{1}{2}} n)$ in the worst case.*

Proof. The containment of global solution in $\mathcal{K}$ follows from Lemma 1. For proving the computational part, we use Corollary 1. For each p_i with $i = 4, \ldots, n$, computation of $\mathcal{A}^{\mathbb{Z}}(p_1, a_{1i}, b_{1i}) \cap \mathcal{A}^{\mathbb{Z}}(p_2, a_{2i}, b_{2i})$ takes $O(\epsilon^2)$ time for Type 1 intersection and $O(\epsilon \sqrt{\epsilon d})$ time for Type 2 intersection. The additional time needed for validation against $\mathcal{A}^{\mathbb{Z}}(p_3, a_{3i}, b_{3i})$ is subsumed within the aforesaid time complexity. Summing up this over i from 4 to n, we get the result. □

By Lemma 3, we always have an integer point $q_{i,0}$ in $\mathcal{I}_i^{\mathbb{Z}}$ if we set $\epsilon \geqslant \frac{1}{\sqrt{2}}$ and $\mathcal{I}_i^{\mathbb{Z}}$ is of Type 1 or of Type 2, for $i = 1, 2, \ldots, n$. Further, by Theorem 1, if $d \geqslant \max(d_1, d_2) + \epsilon$, then with $\epsilon = \frac{1}{\sqrt{2}}$, we get $|\mathcal{I}_i^{\mathbb{Z}}| = O(1)$ for Type 1 intersection and $O(\sqrt{d})$ for Type 2 intersection. So by Corollary 1, $\mathcal{I}_i^{\mathbb{Z}}$ can be computed in $O(1)$ time in the best case and in $O(\sqrt{d})$ time in the worst case. This gives the following corollary.

Corollary 2. *For an appropriately small value of ϵ $(= \frac{1}{\sqrt{2}}$ for definiteness), the time complexity of the algorithm varies from $O(n)$ to $O(n\sqrt{d})$, where d is the maximum point-pair distance.*

When positions of p_1, p_2, and p_3 are not known, we can choose p_1 as the origin and p_2 lying in $\mathcal{A}^{\mathbb{Z}}(p_1, a_{12}, b_{12})$. There are $O(\epsilon d)$ possible choices for p_2. Next, the third point p_3 can be chosen from the intersection of $\mathcal{A}^{\mathbb{Z}}(p_1, a_{13}, b_{13})$ and $\mathcal{A}^{\mathbb{Z}}(p_2, a_{23}, b_{23})$. There are $O(\epsilon^{\frac{3}{2}} \sqrt{d})$ ways to choose p_3. The rest of the points can be computed using the proposed technique. For certain choices of p_2 and p_3, a global solution (satisfying all the distance bounds) may not exist. So, we may need to consider all possible choices of p_2 and p_3 in order to compute a global solution. There are $O(\epsilon^{\frac{5}{2}} d^{\frac{3}{2}})$ possible pairs of possible solutions for p_2 and p_3 in the worst case. Hence, total time required for computing a global solution is $O(\epsilon^4 d^2 n)$ in the worst case. We also need $O(n^2)$ time to verify the distance bounds for a solution.

4.1   Search for Feasible Solution

By Theorem 2, the collection $\mathcal{K}$ contains all feasible solutions. However, it may contain some infeasible combinations too, alongside. To distinguish the former from the latter, one has to search at least $O(\epsilon^{2n})$ possible combinations. These combinations can be represented in a rooted tree with the root node containing the solution for p_1 and a node at the ith level representing a set of solutions for p_i that satisfies the distance constraints with its predecessor points from p_{i-1} to p_1. The tree traversal is similar to the branch-and-prune technique proposed

---

Algorithm 1. DDGA(n, a, b, P_1, P_2, P_3)

1 Create empty list $\mathcal{I}_i^{\mathbb{Z}}$ of candidate solutions for ith point, for $i = 1, 2, \cdots, n$.
2 $\mathcal{I}_1^{\mathbb{Z}} \leftarrow \{P_1\}$, $\mathcal{I}_2^{\mathbb{Z}} \leftarrow \{P_2\}$, $\mathcal{I}_3^{\mathbb{Z}} \leftarrow \{P_3\}$
3 **for** $i = 4, 2, \cdots, n$ **do** $\triangleright$ set of candidate solutions for ith point
4 $\quad\quad \mathcal{I}_i^{\mathbb{Z}} \leftarrow \mathcal{A}^{\mathbb{Z}}(P_1, a_{1i}, b_{1i}) \cap \mathcal{A}^{\mathbb{Z}}(P_2, a_{2i}, b_{2i}) \cap \mathcal{A}^{\mathbb{Z}}(P_3, a_{3i}, b_{3i})$.
5 $\quad\quad n_i \leftarrow |\mathcal{I}_i^{\mathbb{Z}}|$ $\triangleright$ number of candidate solutions for ith point

6 $N \leftarrow \textbf{ShiftOrigin}(\bigcup_{i=1}^{n} \mathcal{I}_i^{\mathbb{Z}})$ $\triangleright$ make sure all points are in $[1, N] \times [1, N]$
7 Create a matrix $B[1..N; 1..N]$, initialized to 0s $\triangleright$ to accumulate votes
8 **for** $i = 1, 2, \cdots n$ **do** $\triangleright$ casting mutual votes of $\mathcal{I}_i^{\mathbb{Z}}$ and $\mathcal{I}_j^{\mathbb{Z}}$
9 $\quad\quad$ **for** $j = i+1, i+2, \cdots, n$ **do**
10 $\quad\quad\quad\quad$ **for** $k = 1, 2, \cdots, n_i$ **do**
11 $\quad\quad\quad\quad\quad\quad (\alpha_i, \beta_i) \leftarrow \mathcal{I}_i^{\mathbb{Z}}[k]$ $\triangleright$ get the kth point of $\mathcal{I}_i^{\mathbb{Z}}$
12 $\quad\quad\quad\quad\quad\quad$ **for** $l = 1, 2, \cdots, n_j$ **do**
13 $\quad\quad\quad\quad\quad\quad\quad\quad (\alpha_j, \beta_j) \leftarrow \mathcal{I}_j^{\mathbb{Z}}[l]$ $\triangleright$ get the lth point of $\mathcal{I}_j^{\mathbb{Z}}$
14 $\quad\quad\quad\quad\quad\quad\quad\quad d \leftarrow \|(\alpha_i, \beta_i) - (\alpha_j, \beta_j)\|$ $\triangleright$ distance between the points
15 $\quad\quad\quad\quad\quad\quad\quad\quad$ **if** $a_{ij} \leqslant d \leqslant b_{ij}$ **then**
16 $\quad\quad\quad\quad\quad\quad\quad\quad\quad\quad B[\alpha_i, \beta_i] \leftarrow B[\alpha_i, \beta_i] + 1$ $\triangleright$ casting a vote for (α_i, β_i)
17 $\quad\quad\quad\quad\quad\quad\quad\quad\quad\quad B[\alpha_j, \beta_j] \leftarrow B[\alpha_j, \beta_j] + 1$ $\triangleright$ casting a vote for (α_j, β_j)

18 **for** $i = 4, 5, \cdots, n$ **do** $\triangleright$ find a point in $\mathcal{I}_i^{\mathbb{Z}}$ having maximum vote
19 $\quad\quad (\alpha, \beta) \leftarrow \mathcal{I}_i^{\mathbb{Z}}[1]$ $\triangleright$ first point in the list $\mathcal{I}_i^{\mathbb{Z}}$
20 $\quad\quad$ **for** $k = 2, 3, \cdots, n_i$ **do** $\triangleright$ search in $\mathcal{I}_i^{\mathbb{Z}}$
21 $\quad\quad\quad\quad (\alpha_i, \beta_i) \leftarrow \mathcal{I}_i^{\mathbb{Z}}[k]$
22 $\quad\quad\quad\quad$ **if** $B[\alpha, \beta] < B[\alpha_i, \beta_i]$ **then**
23 $\quad\quad\quad\quad\quad\quad (\alpha, \beta) \leftarrow (\alpha_i, \beta_i)$

24 $\quad\quad P_i \leftarrow (\alpha, \beta)$ $\triangleright$ final solution point with maximum vote
25 **return** $\{P_i : i = 1, 2, \cdots, n\}$

---

in [13] and the interval branch-and-prune technique proposed in [12] for finding an approximate DGP solution in $\mathbb{R}^2$. In our approach, the search is performed in $\mathbb{Z}^2$, where we need to verify only a finitely many possible points for a valid solution. A possible solution to the DGP problem corresponds to a path of length $n - 1$ from the root to a leaf node. In order to find a solution, in the worst case, we may need to explore and verify all paths, which would take exponential time when ϵ is not small. In such a case, a voting scheme may be adopted as an efficient heuristic to determine the best solution point in each $\mathcal{I}_i^{\mathbb{Z}}$ in the sense that it satisfies the maximum number of distance constraints. For any $p \in \mathcal{I}_i^{\mathbb{Z}}$ and $q \in \mathcal{I}_j^{\mathbb{Z}}$, if p and q satisfy the distance constraints, then each of them receives a mutual vote. In order to compute the vote for all candidate integer points, we use an accumulator B (implemented as a 2D array), such that for each digital point (α, β), $B(\alpha, \beta)$ stores the number of votes received by (α, β). The main steps are shown in Algorithm 1.

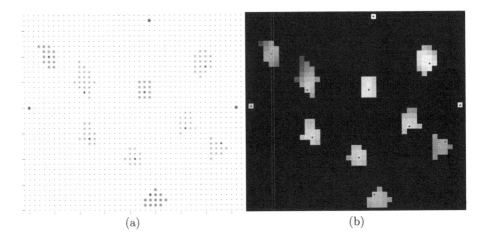

(a)   (b)

Fig. 6. Visualization of a digital solution to DGP with inexact distances. See Appendix for the input. (a) p_1, p_2, and p_3 shown in red; the connected components making the candidate solutions of all other points shown in distinct colors; each point having the highest vote in a component shown as a blue dot. (b) The votes for all candidate solutions are treated as intensity values to visualize the accumulator matrix B as an image. Brighter pixels correspond to higher votes and pixels having highest votes are marked by colored dots. (Color figure online)

4.2   Improvement of Efficiency

If all the annulus intersections computed during the execution of the algorithm are of Type 1, then the algorithm has the speediest execution. So, in order to increase Type 1 intersections, we choose p_1 and p_2 as the farthest pair. Similarly, we choose p_3 such that $\min(b_{13}, b_{23}) \geqslant \min(b_{1i}, b_{2i})$, $\forall i \neq 1, 2$. Further, $\mathcal{I}_i^{\mathbb{Z}}$ can be computed by following an appropriate ordering of computation of the intersection of the three annuli $\mathcal{A}^{\mathbb{Z}}(p_1, a_{1i}, b_{1i})$, $\mathcal{A}^{\mathbb{Z}}(p_2, a_{2i}, b_{2i})$, and $\mathcal{A}^{\mathbb{Z}}(p_3, a_{3i}, b_{3i})$, so that $\mathcal{I}_i^{\mathbb{Z}}$ is Type 1 intersection; otherwise, we defer its computation. In particular, we first run the algorithm for the points that have Type 1 intersection with respect to p_1, p_2 and p_3, and find the solution by voting scheme. We take S_1 as the list of points whose solutions are found by this, and S_2 as the list of the remaining points, i.e., the ones having Type 2 intersection. The solutions for the points in S_2 are determined by using few suitable points from S_1.

4.3   Test Results

The proposed algorithm for solving inexact DGP is implemented and tested on randomly generated problem instances. Experimentation shows that the algorithm is able to solve inexact DGP efficiently. A solution to a small problem instance with $n = 12$ points is visualized in Fig. 6. Notice that only one solution is reported here; there are, in fact, multiple solutions when a connected component contains more than one point having the same maximum vote.

5   Concluding Notes

We have formulated the inexact DG problem in $\mathbb{Z}^2$ and have done a theoretical analysis in order to solve it efficiently. For a large value of ϵ, we have proposed an efficient voting scheme in order to bring down the exponential time complexity to a low-order polynomial. The idea can be extended for solving inexact DGP in 3D integer space also. Besides, many other issues and possibilities have opened up, which requires a deeper analysis of the problem in the discrete space. Some of these, which we foresee as potential research problems, are as follows.

1. The number of worst-case occurrences in the execution of the algorithm can possibly be minimized by changing the sequence in which the annuli are considered while computing their intersections. The rationale is that if the centers of three annuli are well-separated, then their intersection is well-formed, easy to compute, and lead to speedier execution.
2. In many applications, the distances may not be known for some pairs of points. We can then apply the algorithm to compute the candidate solutions for those points whose distances from p_1, p_2, and p_3 are known; then these partial solutions can be used in a suitable order to find the remaining points.
3. For a small ϵ, no integer solution may exist for certain points. To handle this, we can go for a higher resolution by subdividing $\mathbb{Z}^2$ and reporting the solution point coordinates as rational numbers.
4. We have not commented on the behavior of the heuristic used when ϵ is large. As ϵ increases, uncertainty also increases, and the solution space increases exponentially and depends on the distribution of the pairwise distances. We envisage a wide scope of research to analyze this aspect of the problem.

A   Appendix

Top matrix: $(d)_{n\times n}$ contains actual distances among $n = 12$ points generated randomly. An instance of inexact DGP is made with $[a_{ij}, b_{ij}] = [(d_{ij}-\epsilon), (d_{ij}+\epsilon)]$ for $\epsilon = 2$.

Bottom matrix: $(\hat{d})_{n\times n}$ (rounded off to two decimal places) of the embedded point set; the RMS error $\left(= \sqrt{\sum_{i<j}(d_{ij} - \hat{d}_{ij})^2/\binom{n}{2}} \right)$ between $(d)_{n\times n}$ and $(\hat{d})_{n\times n}$ is 0.7654.

	p_1	p_2	p_3	p_4	p_5	p_6	p_7	p_8	p_9	p_{10}	p_{11}	p_{12}
p_1	0	41.77	28.92	30.89	38.19	31.69	23.49	12.27	10.71	23.37	35.63	13.35
p_2	41.77	0	24.01	24.42	9.33	10.85	18.57	30.39	38.81	22.94	11.08	30.30
p_3	28.92	24.01	0	34.07	27.74	21.04	13.72	17.40	20.75	26.31	13.11	25.00
p_4	30.89	24.42	34.07	0	15.54	15.72	20.43	25.68	34.76	8.88	28.00	17.87
p_5	38.19	9.33	27.74	15.54	0	7.45	17.54	28.38	37.67	16.28	16.87	25.39
p_6	31.69	10.85	21.04	15.72	7.45	0	10.09	21.21	30.36	12.24	12.37	19.61
p_7	23.49	18.57	13.72	20.43	17.54	10.09	0	11.82	20.51	12.65	12.76	14.02
p_8	12.27	30.39	17.40	25.68	28.38	21.21	11.82	0	9.48	16.85	23.43	9.72
p_9	10.71	38.81	20.75	34.76	37.67	30.36	20.51	9.48	0	26.06	30.41	17.48
p_{10}	23.37	22.94	26.31	8.88	16.28	12.24	12.65	16.85	26.06	0	22.86	10.02
p_{11}	35.63	11.08	13.11	28.00	16.87	12.37	12.76	23.43	30.41	22.86	0	26.71
p_{12}	13.35	30.30	25.00	17.87	25.39	19.61	14.02	9.72	17.48	10.02	26.71	0

	p_1	p_2	p_3	p_4	p_5	p_6	p_7	p_8	p_9	p_{10}	p_{11}	p_{12}
p_1	0	41.00	29.41	29.41	38.64	31.26	23.19	11.40	10.77	23.26	35.90	13.42
p_2	41.00	0	24.04	24.04	7.62	10.77	18.25	30.15	38.33	22.36	10.00	29.61
p_3	29.41	24.04	0	34.00	27.78	22.14	14.04	19.10	21.19	27.17	14.21	25.94
p_4	29.41	24.04	34.00	0	17.20	14.76	20.02	23.85	33.60	7.62	27.31	16.28
p_5	38.64	7.62	27.78	17.20	0	7.62	18.03	28.79	38.01	17.26	15.30	26.02
p_6	31.26	10.77	22.14	14.76	7.62	0	10.63	21.19	30.41	11.66	12.65	19.10
p_7	23.19	18.25	14.04	20.02	18.03	10.63	0	12.00	20.25	13.15	13.00	14.21
p_8	11.40	30.15	19.10	23.85	28.79	21.19	12.00	0	9.90	16.40	24.52	9.06
p_9	10.77	38.33	21.19	33.60	38.01	30.41	20.25	9.90	0	26.25	31.06	17.89
p_{10}	23.26	22.36	27.17	7.62	17.26	11.66	13.15	16.40	26.25	0	22.80	9.85
p_{11}	35.90	10.00	14.21	27.31	15.30	12.65	13.00	24.52	31.06	22.80	0	26.93
p_{12}	13.42	29.61	25.94	16.28	26.02	19.10	14.21	9.06	17.89	9.85	26.93	0

References

1. http://www.graphdrawing.org/. Accessed 27 Sept 2015
2. Blumenthal, L.M.: Theory and Application of Distance Geometry. Oxford University Press, Oxford (1953)
3. Brandes, U., Pich, C.: An experimental study on distance-based graph drawing. In: GD 2008, Revised Papers, pp. 218–229 (2008)
4. Bulusu, N., Estrin, D., Heidemann, J.: Scalable coordination for wireless sensor networks self-configuring localization systems. In: Proceedings of the ISCTA (2001)
5. Cheng, L., Wu, C., Zhang, Y., Wu, H., Li, M., Maple, C.: A survey of localization in wireless sensor network. Int. J. Distrib. Sens. Netw., 324–357 (2012)
6. Civril, A., Magdon-Ismail, M., Bocek-Rivele, E.: SDE: graph drawing using spectral distance embedding. In: Healy, P., Nikolov, N.S. (eds.) GD 2005. LNCS, vol. 3843, pp. 512–513. Springer, Heidelberg (2006)
7. Dong, Q., Wu, Z.: A linear-time algorithm for solving the molecular distance geometry problem with exact inter-atomic distance. J. Global Optim. **22**, 365–375 (2002)
8. Dong, Q., Wu, Z.: A geometric build-up algorithm for solving the molecular distance geometry problem with sparse distance data. J. Global Optim. **26**, 321–333 (2003)
9. Gibson, H., Faith, J., Vickers, P.: A survey of two-dimensional graph layout techniques for information visualization. Inf. Vis. **12**(3–4), 324–357 (2012)
10. Klette, R., Rosenfeld, A.: Digital Geometry: Geometric Methods for Digital Picture Analysis. Morgan Kaufmann, San Francisco (2004)
11. Lavor, C., Liberti, L., Maculan, N., Mucherino, A.: Recent advances on the discretizable molecular distance geometry problem. Computational Optimization and Applications (2012)
12. Lavor, C., Liberti, L., Mucherino, A.: The interval branch-and-prune algorithm for the discretizable molecular distance geometry problem with inexact distances. J. Global Optim. **56**, 855–871 (2013)
13. Liberti, L., Lavor, C., Maculan, N.: A branch-and-prune algorithm for the molecular distance geometry problem. Intl. Trans. Oper. Res. **15**, 1–17 (2008)
14. Liberti, L., Lavor, C., Mucherino, A.: Euclidean distance geometry and applications. SIAM Rev. **56**, 3–69 (2014)
15. Liberti, L., Lavor, C., Mucherino, A., Maculan, N.: Molecular distance geometry methods:from continuous to discrete. Intl. Trans. Oper. Res. **18**, 33–51 (2010)
16. Moré, J.J., Wu, Z.: Distance geometry optimization for protein structures. J. Global Optim. **15**, 219–234 (1999)
17. Mucherino, A., Lavor, C., Liberti, L., Maculan, N.: Distance Geometry: Theory, Methods, and Applications, 1st edn. Springer, New York (2013)

18. Savvides, A., Han, C.C., Strivastava, M.B.: Dynamic fine-grained localization in ad-hoc networks of sensors. In: Proceedings of the MobiCom 2001, pp. 166–179 (2001)
19. Saxe, J.B.: Embeddability of weighted graphs in k-space is strongly NP-hard. In: Proceedings of the 17th Allerton Conference in Communications, Control & Computing, pp. 480–489 (1979)
20. Sit, A.: Solving distance geometry problems for protein structure determination. P.h.D thesis, Iowa State University (2010)

Segmentation and Classification of Geoenvironmental Zones of Interest in Aerial Images Using the Bounded Irregular Pyramid

Mariletty Calderón, Rebeca Marfil, and Antonio Bandera$^{(\boxtimes)}$

Departamento de Tecnología Electrónica, Universidad de Málaga,
Campus Universitario de Teatinos, 29071 Málaga, Spain
ajbandera@uma.es

Abstract. The goal of this work is to automatically detect and classify a set of geoenvironmental zones of interest in panchromatic aerial images. Focused on a specific area, the zones to be detected are vegetation/mangrove, degradation/desertification, interface water-sediment and plain. These zones are very interesting from a geological point of view due to their spatial distribution and interrelation, which contribute to evaluate the natural anthropic impact level. The approach to unsupervisedly extract these zones from an input image has two steps. Firstly, the image is automatically segmented in homogeneous colored regions using the Bounded Irregular Pyramid (BIP). The BIP is a hierarchy of successively reduced graphs which produces accurate segmentation results with a low computational cost. Secondly, each obtained region is classified using texture features to determine if it belongs to one of the geoenvironmental zones of interest. As texture features, we have evaluated two variations of the Local Binary Pattern (LBP) descriptor: the Extended-LBP (ELBP) and the LBP variance (LBPV). Both methods include a local contrast measure. For classifying the obtained features, the Support Vector Machine (SVM) has been employed. At this stage, we have evaluated the use of linear and radial basis function (RBF) kernels. The whole framework was tested using images obtained from our specific area of interest: the location of Carenero, Miranda state (Venezuela), in years 1936 and 1992. They allow to study the variation of the geoenvironmental zones of interest of this location in this period of time. These images are low quality images and present significant variations in illumination. This makes difficult the texture classification of their zones. However, the obtained results show that the proposed approach provides good results in terms of identification of zones of geoenviromental interest in these images.

Keywords: Aerial image segmentation · Irregular pyramid · Texture classification

1 Introduction

The study of the anthropic impact produced in a given area is a key issue in geology. For this purpose, aerial and satellite images are being widely used to

© Springer International Publishing Switzerland 2016
A. Bac and J.-L. Mari (Eds.): CTIC 2016, LNCS 9667, pp. 290–301, 2016.
DOI: 10.1007/978-3-319-39441-1_26

obtain qualitative and quantitative geologic information [1,2] because they are non-invasive methods which allow to work in areas of difficult access. To do that, and depending on the geological features of the area to be studied, different geoenvironmental zones and their evolution over the time need to be studied. However, when the goal of an image processing algorithm is to divide the input image in a manner similar to human beings, the adopted strategy cannot simply be the grouping of image pixels into clusters (regions or boundaries) taking into account low-level photometric properties. Aerial and satellite images are generally composed of physically disjoint regions whose associated groups of image pixels may not be visually uniform. Hence, it is very difficult to formulate what should be recovered as a region or boundary or to segment complex regions from the image. With the aim of organizing low-level image features into higher level relational structures, the perceptual organization of the image content is usually thought as a process of grouping visual information into a hierarchy of levels of abstraction. Starting from the lower level of the hierarchy (i.e. the input image or an initial partition), each new layer groups the regions of the level below into a reduced set of regions. This grouping needs to define a region model (the features that describe each image region) and a dissimilarity measure (the metric on those features). This paper proposes a texture-based automatic system to identify a predefined set of geonvironmental zones in panchromatic aerial images. This system is divided in two steps: a pre-segmentation stage that accumulates local evidences from the original image to a single graph. This graph will encode a decomposition of the image into superpixels. This initial stage of the clustering process is guided by the principles described by Levinshtein et al. [11]. Thus, blobs represent connected sets of pixels without overlapping among them. They are compact and their boundaries coincide with the main image edges when the pre-segmentation stops. Then, a second stage categorizes the previously obtained blobs into a reduced set of perceptually significant classes. This stage characterizes every blob using a texture feature and then classifies them into one of a collection of predefined zones. The performance of the texture-based classification scheme has been evaluated using the Banja Luka dataset to measure its ability to deal with real images. After confirming the validity of this stage, the whole system has been applied to the detection and classification of vegetation/mangrove, degradation/desertification, interface water-sediment and plain zones in panchromatic aerial images captured in years 1936 and 1992 in the location of Carenero, Miranda state (Venezuela).

The rest of the paper is organized as follows: Sect. 2 provides an overview of the whole approach and describes how both stages are implemented. Experimental results showing the performance of the approach are presented at Sect. 3. Section 4 draws the conclusions and future work.

2   Proposed Method

Figure 1 provides an overview of the proposed method. In the pre-segmentation stage, the Bounded Irregular Pyramid [7] has been used for segmenting an equalized version of the input image. Instead of performing image segmentation based on a single representation of the input image, a pyramid segmentation algorithm describes the contents of the image using multiple representations with decreasing resolution. In this hierarchy, each representation or level is built by computing a set of local operations over the level below, being the original image the level 0 or base level of the hierarchy. Pyramid segmentation algorithms exhibit interesting properties with respect to segmentation algorithms based on a single representation. Thus, local operations can adapt the pyramidal hierarchy to the topology of the image, allowing the detection of global features of interest and representing them at low resolution levels. The pre-segmentation divides up the input image into a collection of non-overlapping blobs. These blobs are characterized using two variations of the Local Binary Pattern (LBP) descriptor: the Extended-LBP (ELBP) and the LBP variance (LBPV) [6]. For classifying the blobs as belonging to an specific environmental zone, the Support Vector Machine (SVM) has been employed using linear and radial basis function (RBF) kernels.

Fig. 1. Overview of the proposed method

2.1 Pre-segmentation Stage

After equalizing the input image, it is divided up into regions using a specific implementation of the Bounded Irregular pyramid (BIP). The BIP is an irregular, hierarchical procedure for image segmentation. For obtaining a new level from the level below, the BIP combines a regular and an irregular decimation procedures. The final result is the encoding of the image's content as a hierarchy of simple graphs [7]. Using this scheme, the BIP is able to obtain similar segmentation results to other irregular pyramids but in a faster way. Next, we describe the properties and metric employed for building the hierarchy and the decimation process.

Data Structure: Image Features and Metrics. Let $G_l = (N_l, E_l)$ be a hierarchy level where N_l stands for the set of regular and irregular nodes and E_l for the set of intra-level arcs. Let $\xi_{\mathbf{x}}$ be the neighborhood of the node $\mathbf{x}$ defined as $\{\mathbf{y} \in N_l : (\mathbf{x}, \mathbf{y}) \in E_l\}$. It can be noted that a given node $\mathbf{x}$ is not a member of its neighborhood, which can be composed by regular and irregular nodes. Each node $\mathbf{x}$ has associated a $\mathbf{v_x}$ value given by the averaged brightness of the image pixels linked to $\mathbf{x}$. Besides, each regular node has associated a boolean value $h_{\mathbf{x}}$: the homogeneity [7]. Only regular nodes which have $h_{\mathbf{x}}$ equal to 1 are considered to be part of the regular structure. Regular nodes with an homogeneity value equal to 0 are not considered for further processing. At the base level of the hierarchy G_0, all nodes are regular, and they have $h_\mathbf{x}$ equal to 1. In order to divide the image into a set of homogeneous blobs, the graph G_l is transformed in G_{l+1} using a pairwise comparison of neighboring nodes. At the first levels of the hierarchy, the pairwise comparison function, $g(\mathbf{v}_{\mathbf{x}_1}, \mathbf{v}_{\mathbf{x}_2})$, is true if the Euclidean distance between the HSV values $\mathbf{v}_{\mathbf{x}_1}$ and $\mathbf{v}_{\mathbf{x}_2}$ is under an user–defined threshold σ_{color}. When the hierarchy reaches level l_m and it is not possible to perform new mergings, the algorithm automatically changes the metric to add to the process the edge information. For this end, the roots of the blobs at level l_m constitute the first level of the new multiresolution output. Let $\mathcal{P}_{l_m}$ be the image partition at level l_m and $l > l_m \in \Re$ a level of the hierarchy, this second grouping process assigns a partition $\mathcal{Q}_l$ to the couple $(\mathcal{P}_{l_m}, l)$, satisfying that $\mathcal{Q}_{l_m}$ is equal to $\mathcal{P}_{l_m}$ and that

$$\exists l_n \in \Re^+ \quad : \quad \mathcal{Q}_l = \mathcal{Q}_{l_n}, \quad \forall l \geq l_n \tag{1}$$

That is, the grouping process is iterated until the number of nodes remains constant between two successive levels. In order to achieve the grouping process, a perceptual pairwise comparison function must be defined. In this case, the pairwise comparison function $g(\mathbf{v}_{\mathbf{y}_i}, \mathbf{v}_{\mathbf{y}_j})$ is implemented as a thresholding process, i.e. it is true if a distance measure between both nodes is under a given threshold σ_{percep}, and false otherwise. The defined distance integrates edge and region descriptors. Thus, it has two main components: the color contrast between image blobs and the edges of the original image computed using the Canny detector. In order to speed up the process, a global contrast measure is used instead of a local one. It allows to work with the nodes of the current working level, increasing the computational speed. This contrast measure is complemented with internal regions properties and with attributes of the boundary shared by both regions. The distance between two nodes $\mathbf{y}_i \in N_l$ and $\mathbf{y}_j \in N_l$, $\varphi^\alpha(\mathbf{y}_i, \mathbf{y}_j)$, is defined as

$$\varphi^\alpha(\mathbf{y}_i, \mathbf{y}_j) = \frac{d(\mathbf{y}_i, \mathbf{y}_j) \cdot min(b_{\mathbf{y}_i}, b_{\mathbf{y}_j})}{\alpha \cdot c_{\mathbf{y}_i \mathbf{y}_j} + \beta \cdot (b_{\mathbf{y}_i \mathbf{y}_j} - c_{\mathbf{y}_i \mathbf{y}_j})} \tag{2}$$

where $d(\mathbf{y}_i, \mathbf{y}_j)$ is the gray-level distance between $\mathbf{y}_i$ and $\mathbf{y}_j$. $b_{\mathbf{y}_i}$ is the perimeter of $\mathbf{y}_i$, $b_{\mathbf{y}_i \mathbf{y}_j}$ is the number of pixels in the common boundary between $\mathbf{y}_i$ and $\mathbf{y}_j$ and $c_{\mathbf{y}_i \mathbf{y}_j}$ is the set of pixels in the common boundary which corresponds to pixels of the edge detected by the Canny detector. α and β are two constant values used to control the influence of the Canny edges in the grouping process. They should be manually tuned depending on the application and environment.

The Decimation Process. The decimation algorithm runs two consecutive steps to obtain the set of nodes N_{l+1} from N_l. The first step generates the set of regular nodes of G_{l+1} from the regular nodes at G_l and the second one determines the set of irregular nodes at level $l+1$. This second process employs a union-find process which is simultaneously conducted over the set of regular and irregular nodes of G_l which do not present a parent in the upper level $l+1$. The decimation process consists of the following steps:

1. Regular decimation process. The $h_{\mathbf{x}}$ value of a regular node $\mathbf{x}$ at level $l+1$ is set to 1 if the four regular nodes immediately underneath $\{\mathbf{y}_i\}$ are similar and their $h_{\{\mathbf{y}_i\}}$ values are equal to 1. That is, $h_{\mathbf{x}}$ is set to 1 if

$$\{ \bigcap_{\forall \mathbf{y}_j, \mathbf{y}_k \in \{\mathbf{y}_i\}} g(\mathbf{v}_{\mathbf{y}_j}, \mathbf{v}_{\mathbf{y}_k})\} \cap \{ \bigcap_{\mathbf{y}_j \in \{\mathbf{y}_i\}} h_{\mathbf{y}_j}\} \qquad (3)$$

   Besides, at this step, inter-level arcs among regular nodes at levels l and $l+1$ are established. If $\mathbf{x}$ is an homogeneous regular node at level $l+1$ ($h_{\mathbf{x}}==1$), then the set of four nodes immediately underneath $\{\mathbf{y}_i\}$ are linked to $\mathbf{x}$.
2. Irregular decimation process. Each irregular or regular node $\mathbf{x} \in N_l$ without parent at level $l+1$ chooses the closest neighbor $\mathbf{y}$ according to the $\mathbf{v}_{\mathbf{x}}$ value. Besides, this node $\mathbf{y}$ must be similar to $\mathbf{x}$. That is, the node $\mathbf{y}$ must satisfy

$$\{\|\mathbf{v}_{\mathbf{x}} - \mathbf{v}_{\mathbf{y}}\| = \min(\|\mathbf{v}_{\mathbf{x}} - \mathbf{v}_{\mathbf{z}}\| : \mathbf{z} \in \xi_{\mathbf{x}})\} \cap \{g(\mathbf{v}_{\mathbf{x}}, \mathbf{v}_{\mathbf{y}})\} \qquad (4)$$

   If this condition is not satisfy by any node, then a new node $\mathbf{x}'$ is generated at level $l+1$. This node will be the parent node of $\mathbf{x}$. Besides, it will constitute a root node and the set of nodes linked to it at base level will be an homogeneous set of pixels according to the defined criteria. On the other hand, if $\mathbf{y}$ exists and it has a parent $\mathbf{z}$ at level $l+1$, then $\mathbf{x}$ is also linked to $\mathbf{z}$. If $\mathbf{y}$ exists but it does not have a parent at level $l+1$, a new irregular node $\mathbf{z}'$ is generated at level $l+1$. In this case, the nodes $\mathbf{x}$ and $\mathbf{y}$ are linked to $\mathbf{z}'$.

   This process is sequentially performed and, when it finishes, each node of G_l is linked to its parent node in G_{l+1}. That is, a partition of N_l is defined. It must be noted that this process constitutes an implementation of the union-find strategy [7].
3. Definition of intra-level arcs. The set of edges E_{l+1} is obtained by defining the neighborhood relationships between the nodes N_{l+1}. Two nodes at level $l+1$ are neighbors if their reduction windows, i.e. the sets of nodes linked to them at level l, are connected at level l.

2.2   Texture Classification

Texture Descriptors. The Local Binary Pattern (LBP) descriptor, originally proposed in [8] is a computational very simple algorithm which main advantage is its robustness against illumination variations. The original LBP operator forms labels for the image pixels by thresholding the 3×3 neighborhood of each pixel with the center value and considering the result as a binary number. This binary number is set to 1 if the neighbor is greater or equal than the central pixel and it is set to 0 in other case. The histogram of these $2^8 = 256$ different labels can then be used as a texture descriptor (Fig. 2). This descriptor was extended (ELBP) in [9] to use circular neighborhoods of different radius.

A formal description of the ELBP operator is shown in the following equation:

$$LBP_{P,R}(x_c, y_c) = \sum_{p=0}^{P-1} s(g_p - g_c)2^p \qquad (5)$$

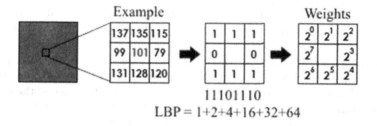

Fig. 2. Schematic description of the LBP approach

where (x_c, y_c) is the central pixel with an intensity value of g_c, g_p is the intensity of the neighbor pixel, P is the number of pixels, R the used radius and $s(x)$ is a function with the form:

$$s(x) = \begin{cases} 1 & \text{if } x \geq 0, \\ 0 & \text{otherwise.} \end{cases} \tag{6}$$

When a radius different of one is used, the neighbors are located in a circle with the center in the studied pixel. If a point of the circle does not correspond to an image pixel, it is interpolated.

In order to include local contrast information into the LBP descriptor and make it rotation invariant, Guo et al. [6] present the Local Binary Pattern Variance descriptor (LBPV).

$$LBPV_{P,R}(k) = \sum_{i=1}^{N}\sum_{j=1}^{M} w(LBP_{P,R}(i,j), k) \quad k \in [0, K] \tag{7}$$

$$w(LBP_{P,R}(i,j), k) = \begin{cases} VAR_{P,R}(i,j) & \text{if } LBP_{P,R}(i,j) = k, \\ 0 & \text{otherwise.} \end{cases} \tag{8}$$

$$VAR_{P,R} = \frac{1}{P}\sum_{p=0}^{P-1}(g_p - u)^2 \tag{9}$$

$$u = \frac{1}{P}\sum_{p=0}^{P-1} g_p \tag{10}$$

In this work, we have used the ELBP and the LBPV descriptors to characterize all nodes of the highest level of the hierarchy, $\mathbf{x} \in N_{l_h}$. The irregular shape is not a problem as the descriptor is locally computed for each pixel. Furthermore, contrary to what is done in other works, we divide up the receptive field of $\mathbf{x}$ into a grid of m regions. In each of these regions the descriptor is computed, having one histogram per region. These m histograms are concatenated, obtaining the final feature vector associated to the node. Both descriptors are evaluated at Sect. 3.

Classification. Once the texture of the nodes $\mathbf{x} \in N_{l_h}$ has been captured, the final step of the approach aims to categorize these nodes into a set of classes. To perform this step, the Support Vector Machine (SVM) classifier has been selected. Given a set of patterns where each pattern belongs to one of two possible classes, the goal of the SVM is to provide a model for classifying any new pattern. Basically, this model defines a separating hyperplane in the space of the patterns that maximizes the margin between the two classes. Training a SVM consists of finding the optimal hyperplane, that is, the one with the maximum distance from the nearest training patterns, called support vectors. However, it is not always possible to find a perfect separation. Otherwise the result of the model cannot be generalized for other incoming data. This problem is known as overfitting. In order to deal with it, the SVM uses an internal parameter, C, which controls the compensation between training errors and the rigid margins.

Whereas the easiest way to make the separation between classes is using a straight line, a plane or a n-dimensional hyperplane, it often happens that the sets to discriminate are not linearly separable in that space. To solve this issue, one solution is to map the original space into a higher-dimensional space, and to look for this hyperplane within this new space. This mapping is typically performed using a kernel function $k(x, y)$. Thus, the SVM classifier has the following form:

$$f(x) = sign(\sum_{i=1}^{1} \alpha_i y_i K(x, x_i) + b) \tag{11}$$

being $K(x, x_i)$ the kernel function.

In the proposed work, two different kernel functions have been used: the polynomial and the Gaussian radial basis function.

$$K(x_i, x_j) = (x_i * x_j)^n \qquad \text{Polynomial} \tag{12}$$

$$K(x_i, x_j) = \exp\left(-\gamma (x_i - x_j)^2\right) \qquad \text{Gaussian radial basis function} \tag{13}$$

3   Experimental Results

This section includes the verification of the texture description and classification stages and the validation of the whole system. For the first issue we have used a publicly available database of aerial images: the Banja Luka database[1]. The whole system has been tested using a collection of images from our specific area of interest, located at Carenero, Miranda state (Venezuela). These images were taken at 1936 and 1992. This will allow to analyze the anthropic impact produced in this area.

Banja Luka Database. The Banja Luka database contains 606 images of 128×128 pixels, which were manually classified into 6 classes: houses, cemetery, industry, field, river and trees. The distribution of images in these categories is highly uneven [3]. Figure 3 shows several images from the database.

---

[1] http://dsp.etfbl.net/aerial/.

Panchromatic versions of these images were used for testing. Table 1 shows the performance (mean classification accuracy and standard deviation) provided by different approaches. In these experiments, half of the images were used for training and the other half for testing. All approaches use SVM with radial basis function kernel. Gabor descriptors were computed at 8 scales and 8 orientations for all images, providing 128-dimensional vectors. Gabor (full) employs a Gabor descriptor composed by the means and standard deviations of all filter responses, while Gabor (mean) implies to use a descriptor obtained using only mean values. The Gist descriptor [4] was computed by first filtering the image by a filter bank of Gabor filters, and then averaging the responses of filters in each block on a 4 × 4 non-overlapping grid. The approach proposed by Lingua et al. [5] is used to provide the MSIFT results. The BoW descriptor [10] is obtained by computing SIFT descriptors on a regular grid and vector quantizing them using a codebook with 1000 codewords. Histogram of codeword occurrences is a 1000-dimensional BoW image descriptor. It can be noted that the proposed framework provides better results than these approaches.

Fig. 3. Examples of classes in the Banja Luka database

Carenero Database. This database is formed by panchromatic aerial images captured in the Carenero zone by the Geographical Institute of Venezuela (Simón Bolívar) in the years 1936 and 1992. Their scale is 1:25.000 and they have been geocited with dimensions of 675x471 pixels. The goal of this experiment

Table 1. Comparison of classification accuracy for in house dataset Banja Luka

Descriptor	Accuracy (%)	
Gist	88,75	± 2,07
MSIFT	84,33	± 1,70
Gabor(Full)	84,500	-
Gabor(Mean)	80,700	-
MBow	85,23	± 2,46
Proposed approach ($r=1$, $p=8$)	**97,5**	± 1,25

is to locate in these images the following zones: vegetation/mangrove, degradation/desertification, interface water-sediment and plain. For each class, 85 images of 40x40 pixels was manually obtained for training. Figure 4 shows one example of each category.

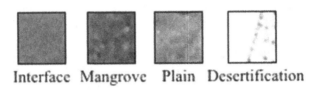

Interface   Mangrove   Plain   Desertification

Fig. 4. Categories of texture samples in the Carenero database

The pre-segmentation using the BIP algorithm depends on two threshold values: σ_{color} and σ_{percep}. The best results on the Caranero 1936 dataset were obtained using as threshold values $\sigma_{color}=80$ and $\sigma_{percep}=90$. But for Caranero 1992, the best values were $\sigma_{color}=30$ and $\sigma_{percep}=50$. There are a significant difference on their values, but it must be noted that these images were captured using two different sensors. Similar comments could be given respect to the

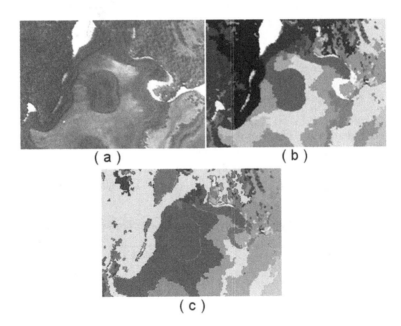

(a)   (b)

(c)

Fig. 5. (a) Original image from Carenero 1936, (b) image decomposition provided by the BIP approach, and (c) blobs associated to interface water-sediment (blue regions), plain (green regions), desertification (red regions) and the mangrove (yellow regions) (see text) (Color figure online).

parameters α and β used for weighting the impact between color contrast and edge information (typical values give more weight to color contrast).

Figure 5 shows the regions obtained by pre-segmenting one of the original images from Carenero 1936. The size of some of them are large and can be characterized by the ELBP-LBPV method. Other ones have got a small size and cannot be correctly characterized. They will remain as unclassified. Figure 5(c) shows the blobs that have been labeled by the approach as interface water-sediment (blue regions), plain (green regions), desertification (red regions) and the mangrove (yellow regions). The classifier uses the SVM model with a RBF kernel.

The global thematic accuracy, measured as a classification percentage, reached a value of 92,50 %. Besides, it was counted on the cross validation to identify the tuning parameters, for the radial gamma kernel (γ), in the linear kernel degree and C. Table 2 shows the performance (mean classification accuracy and standard deviation) for classes on Carenero 1936 and Carenero 1992 databases. Figure 6 shows the average accuracy for all texture classes on the Carenero 1992 dataset. It should be noted that the results are really good for several combinations of parameters. There are however problems for classifying the desertification areas. It can be also noted that there does not exist a pair of parameters that can provide the best results for all classes.

Table 2. Classification accuracy (mean and standard deviation) for Carenero Equalized 1936 and Carenero Equalized 1992

	Carenero 1936		Carenero 1992	
	$\bar{x}$	$\pm\sigma$	$\bar{x}$	$\pm\sigma$
ELBP + SVM (RBF)				
R=1, P=8	**92,500**	7,916	95,000	5,000
R=2, P=8	90,000	11,666	80,000	16,667
R=2, P=16	76,786	14,880	83,334	11,667
ELBP + SVM (LINEAR)				
R=1, P=8	88,095	11,905	93,333	6,667
R=2, P=8	91,694	8,305	83,333	13,333
R=2, P=16	78,125	13,542	83,333	11,667
LBPV + SVM (RBF)				
R=1, P=8	85,833	16,859	**99,999**	–
R=2, P=8	70,833	34,359	88,333	8,388
R=2, P=16	78,125	17,137	91,666	6,382
LBPV + SVM (LINEAR)				
R=1, P=8	71,875	29,141	**99,999**	–
R=2, P=8	61,458	28,336	88,333	8,389
R=2, P=16	51,041	42,542	90,000	6,667

The method is however considering that boundaries of the zones are abrupt, i.e. that there do not exist gradual transitions between zones. This problem is present on the images and we will need to work on how to deal with. Other approaches [13] use the (multiple) indicator kriging for take these gradual transitions into account. In our case, it will be necessary to include the uncertainty information within the categorization stage, providing for each region not only the category but also the probability associated to this process.

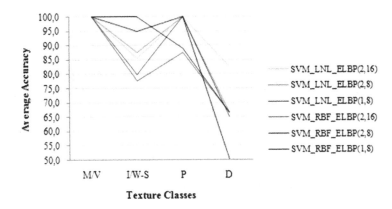

Fig. 6. Accuracy values for different geoenvironmental zones (Carenero 1992): vegetation/mangrove (V/M), degradation/desertification (D), interface water-sediment (I/W-S) and plain (P)

4   Conclusion

The main contribution of this paper is the evaluation of a whole framework for segmentation and classification of the regions composing an aerial image. The BIP approach is used for pre-segmenting the input image, providing a decomposition of the scene within uniform blobs. These blobs are arranged as a single graph, where simple adjacency relationships are encoded on the arcs of the graph. Then, these blobs are successfully categorized using texture. Regarding to this last stage, the proposed system is able to obtain better results than other popular approaches on the large Banja Luka dataset. Furthermore, the whole framework is able to deal with the automatic decomposition and labeling of complex, real images. Future work will focus on integrating within the framework the tools that can allow the user to easily redraw the segmentation and/or classification results (e.g. changing a provided label or choosing a lower level of decomposition for a specific region). We have also experience on introducing topological information within the hierarchical segmentation of natural images, changing the single graph by a combinatorial map [12]. This topological information will allow to consider inclusion and complex adjacency relationships, which could be useful to describe the geological evolution of the analyzed geoenvironmental area.

Acknowledgments. This paper has been partially supported by the Spanish Ministerio de Economía y Competitividad TIN2015-65686-C5 and FEDER funds.

References

1. Sulong, I., Mohd-Lokman, H., Mohd-Tarmizi, K., Ismail, A.: Mangrove mapping using landsat imagery and aerial photographs: Kemaman district, Terengganu, Malaysia. Environ. Dev. Sustain. **4**(2), 135–152 (2002)
2. Hirche, A., Salamani, M., Abdellaoui, A., Benhouhou, S., Valderrama, J.M.: Landscape changes of desertification in arid areas: the case of south-west Algeria. Environ. Monit. Assess. **179**(1–4), 403–420 (2011)
3. Risojevic, V., Momic, S., Babic, Z.: Gabor descriptors for aerial image classification. In: Adaptive and Natural Computing Algorithms, pp. 51–60 (2011)
4. Oliva, A., Torralba, A.: Modeling the shape of the scene: a holistic representation of the spatial envelope. Int. J. Comput. Vis. **42**(3), 145–175 (2001)
5. Lingua, A., Marenchino, D., Nex, F.: Performance analysis of the SIFT operator for automatic feature extraction and matching in photogrammetric applications. Sensors **9**(5), 3745–3766 (2009)
6. Guo, Z., Zhang, L., Zhang, D.: Rotation invariant texture classification using LBP Variance (LBPV) with global matching. Pattern Recogn. **43**, 706–719 (2010)
7. Marfil, R., Bandera, A.: Comparison of perceptual grouping Criteria within an integrated hierarchical framework. In: Torsello, A., Escolano, F., Brun, L. (eds.) GbRPR 2009. LNCS, vol. 5534, pp. 366–375. Springer, Heidelberg (2009)
8. Ojala, T., Pietikäinen, M., Harwood, D.: A comparative study of texture measures with classification based on feature distributions. Pattern Recogn. **19**(3), 51–59 (1996)
9. Ojala, T., Pietikäinen, M.: Multiresolution gray-scale and rotation invariant texture classification with local binary patterns. IEEE Trans. Pattern Anal. Mach. Intell. **24**(7), 971–987 (2002)
10. Yang, Y., Newsam, S.: Bag-of-visual-words and spatial extensions for land-use classification. In: Proceedings of the ACM SIGSPATIAL GIS, pp. 270–279 (2010)
11. Levinshtein, A., Stere, A., Kutulakos, K., Fleet, D., Dickinson, S., Siddiqi, K.: Turbopixels: fast superpixels using geometric flows. IEEE Trans. Pattern Anal. Mach. Intell. **31**(12), 2290–2297 (2009)
12. Antnez, E., Marfil, R., Bandera, J.P., Bandera, A.: Part-based object detection into a hierarchy of image segmentations combining color and topology. Pattern Recogn. Lett. **34**(7), 744–753 (2013)
13. Hengl, T., Toomanian, N., Reuter, H., Malakouti, M.: Methods to interpolate soil categorical variables from profile observations: Lessons from Iran. Geoderma **140**, 417–427 (2007)

Author Index

Printed in the United States
By Bookmasters